7·17·00
$109.-

MEDICINAL PLANTS

HOW TO ORDER THIS BOOK

BY PHONE: 800-233-9936 or 717-291-5609, 8AM–5PM Eastern Time

BY FAX: 717-295-4538

BY MAIL: Order Department
Technomic Publishing Company, Inc.
851 New Holland Avenue, Box 3535
Lancaster, PA 17604, U.S.A.

BY CREDIT CARD: American Express, VISA, MasterCard

BY WWW SITE: http://www.techpub.com

MEDICINAL PLANTS

Culture, Utilization and Phytopharmacology

Thomas S. C. Li, Ph.D.
Agriculture and Agri-Food Canada
Pacific Agri-Food Research Centre
Summerland, British Columbia

LANCASTER • BASEL

Medicinal Plants
a TECHNOMIC® publication

Technomic Publishing Company, Inc.
851 New Holland Avenue, Box 3535
Lancaster, Pennsylvania 17604 U.S.A.

Printed in the United States of America
10 9 8 7 6 5 4 3 2 1

Main entry under title:
Medicinal Plants: Culture, Utilization and Phytopharmacology

A Technomic Publishing Company book
Bibliography: p.
Includes index p. 475

Library of Congress Catalog Card No. 00-104016
ISBN No. 1-56676-903-5

TABLE OF CONTENTS

Foreword *vii*

Preface *ix*

Acknowledgments *xi*

Chapter 1. Major Constituents and Medicinal Values of Medicinal Plants. 1

Chapter 2. Toxicity of Medicinal Plants 67

Chapter 3. Essential Oils and Fractions from Medicinal Plants 111

Chapter 4. Value-Added Products and Possible Usage Derived from Medicinal Plants 167

Chapter 5. Cultivation and Harvesting. 225

Chapter 6. Major Diseases and Insects Found in Medicinal Plants . . . 299

Appendix 1: Major Constituents and Their Source *407*

Appendix 2: Essential Oils and Their Source *431*

Appendix 3: List of Common and Scientific Names *455*

Index *475*

FOREWORD

Medicinal plants are rapidly regaining the prominent position that they possessed in past centuries in Western medicine and that they have always held in most of the world. Explosive growth is occurring in the multibillion dollar business of medicinal plants, most evident in the increasing array of herbal offerings and supplements found in health food stores and supermarkets. The mass media too have become fascinated with the health-promoting and curative powers of medicinal plants and regularly feature the virtues of the latest species to catch the attention of the public. Those who grow, process, market, prescribe, or personally use medicinal plants or their components have a vital need to obtain accurate, contemporary, and correct information. Unfortunately, there are serious hazards and pitfalls associated with medicinal plants that can only be avoided when authoritative information is available.

There is certainly no shortage of information—there are hundreds of books that deal with the subject, and millions of presentations on the World Wide Web. However, most of the information provided in these sources is superficial, and an alarming amount is incorrect, misleading, and harmful. A very large proportion of important information on medicinal plants occurs in the foreign literature, unfamiliar to most people in the western world, or in the "gray literature," i.e., in obscure publications that are rarely retrievable using current bibliographic tools. Accordingly, there are relatively few books that are genuinely authoritative and comprehensive, such as the present volume by Dr. Thomas Li. This is a veritable treasure chest of information critical to all professionals who deal in one way or another with medicinal plants. Dr. Li, one of the world's most knowledgeable and well-known researchers on medicinal plants, has dedicated many years of effort to acquiring and condensing the information presented in this reference text. He is to be congratulated on this superb and invaluable synthesis.

ERNEST SMALL, PH.D.
Principal Research Scientist
Eastern Cereal and Oilseed Research Centre
Agriculture and Agri-Food Canada
Ottawa, Ontario, Canada

PREFACE

The use of medicinal plants for health reasons started thousands of years ago and is still part of medical practice in China, Egypt, India, and other developing countries. Over the centuries, the use of medicinal herbs has become an important part of daily life in the western world despite significant progress in modern medical and pharmaceutical research. Since World War II, the increasing availability of medicinal herbal products, a desire for nutraceuticals or functional foods and alternative medicines, and concerns about the possible side effects of some synthetic drugs have revived the use of medicinal herbs. Recently, there has been a tremendous surge of interest in medicinal plants or herbs, and their products have become a multibillion dollar industry in both North America and Europe.

Research on medicinal and cosmetic uses of herbs is contributing to the growth of the herbal industry. Increasing knowledge of metabolic processes and the effects of plants on human physiology have enlarged the range of application of medicinal plants. Some lesser known plants have been found to have significant medicinal value. According to the report by the World Bank in 1997 (Technical Paper No. 355), it is apparent that the significance of plant-based medicines has been increasing all over the world. Nearly 50 percent of medicines on the market are made of natural basic materials. Interestingly, the market demand for medicinal herbs is likely to remain high because many of the active ingredients in medicinal plants cannot yet be prepared synthetically.

In developed countries, the huge demand for medicinal plants or herbs largely reflects the growing interest in health enhancement, whereas in the developing world, because of the limited availability and high cost of modern medicines and medical and pharmaceutical services, medicinal plants or herbs continue to be used in medical practice based on a strong traditional belief in herbal medicine. One consequence of this high demand is the threatened existence of some of these natural resources. Some herbs are now listed as endangered species, and further harvesting is forbidden by law.

In many countries, medicinal plants or herbs are predominantly harvested from the wild in an unregulated manner. Yield and quality of herbs collected from the wild are unpredictable; both are significantly affected by the weather, pests, and other uncontrollable variables. Farming some of the popular herbs would help reduce problems of inconsistent supply and would thus regularize the trade. Furthermore, farmed products could be certified as to source, identity, and quality. However, cultivation of medicinal plants is presently constrained by a lack of suitable technology, which leads to low yield and products of poor quality.

This book is designed to provide manufacturers, researchers, and producers with easy access to information on medicinal herbs compiled from widely scattered sources in the literature. Each chapter presents data for more than 400 species in a table format arranged in alphabetical order by the scientific name followed by the common name. The individual chapters present current information on major constituents and medicinal values; toxicity or hazards; essential oils and their fractions; value-added products and their possible uses; cultivation and harvesting; and infectious diseases and insects. In addition to an index, three appendices cross-reference major active ingredients and their sources; essential oils and their derivations; and the common and scientific names of the medicinal plants cited in the tables.

The information in this book is primarily for reference and education. It is not intended to be a substitute for the advice of a physician. The uses of medicinal plants described in this book are not recommendations, and the author is not responsible for liability arising directly or indirectly from the use of information in this book.

ACKNOWLEDGMENTS

The author thanks Peggy Watson, librarian, for her constant and tireless effort in the literature search. Appreciation is also extended to M. Walker and T. Foreman for their help. I also thank my colleagues, Drs. Tom Beveridge, Benoit Girard, Dave Oomah, and Peter Sholberg, for their valuable editing assistance.

CHAPTER 1

Major Constituents and Medicinal Values of Medicinal Plants

Recent research on medicinal plants and herbs has generated a great deal of information about the biologically active chemical components that are responsible for the claimed medicinal effects. The level of active ingredients or chemical constituents has been used as a standard or marker for the quality of raw plant materials and value-added products.

Medicinal substances found in plants are the products of natural metabolic processes. However, each species has its own genetic structure that governs the presence of chemical components or bioactive molecules. In addition, the effects of environment and differences among varieties or cultivars within each species create variations in the quantity of compounds present. Thus, each plant species or variety produces chemical compounds differently, and some plants produce medicinally useful compounds, others do not or do so in very small quantities.

Until recently, almost all of the uses of medicinal plants for the treatment of human ailments were based on hearsay, folklore, or tradition, without scientific proofs, a practice that is potentially harmful and dangerous. The recent surge of interest in the use of medicinal plants has generated a great deal of research on major constituents and their effects on human health. The results of clinical studies have proven some of the claims. However, more research is needed to extend the search for potentially beneficial herbs from natural sources and determine their use in modern medicine.

The information provided in this chapter should not be used for the diagnosis, treatment, or prevention of diseases in humans or animals or replace the services of a physician. It is not intended to be used as a guide or a recommendation for medical treatment but, rather, as a source of information about medicinal plants, their potential, and their possibilities.

Table 1. Major constituents and medicinal values of medicinal plants.

Scientific name	Common name	Major constituents	Medicinal values*
Abies balsamea (L.) Mill.	Balsam fir	Oleo-resin, *l*-pinene[8, 30]	Antiseptic, stimulant for congestion, and chest infections.
Achillea millefolium L.	Yarrow	Achilleine, tannins, cineole, chamazulene, sesquiterpene lactones, menthol, camphor, sterols, triterpenes[8, 30, 40, 59, 120, 135]	Reduce fever, anti-inflammatory, treat common cold, diarrhea, dysentery, hypertension, and gastrointestinal complaints. Externally in poultices, lotions, and bath preparations.
Aconitum napellus L.	Monkshood	Aconitine, malonic acid, caffeic acids, hypaconitine, mesaconitine, neoline, napelline, benzol-aconitine[40, 120, 132]	Heart and nerve sedative, anticarcinogenic, and reduce fever.
Acorus calamus L. var. Americanus Wolff. *A. tatarinowii* L.	Calamus, sweet flag Shi Chang Pu	Acoric acid, asarone, linalool, palmitic acid, methylamine, saponin, mucilage[19, 30, 40, 135]	Used as a panacea. Antibacterial, antifungal, antiseptic, digestive upsets, fevers, and antiamebic, also used as a vermifuge, and antiprotozoal agent.
Actaea alba L. *A. rubra* (Ait.) Willd.	White baneberry Red baneberry	Resins in the root, trans-aconitic acid and protoanemonoid compounds in the seeds[120]	Used as rheumatic remedy, treat headache, insomnia, melancholy, and convulsions.
Aesculus hippocastanum L.	Horse chestnut	Aescin, citric acid, resin, saponin, tannin, uric acid, quercetin, kaempherol, flavonoids, coumarin derivatives[30, 40]	For antipyretic, antithrombin, and anti-exudative effects, treat lymphatic congestions, cerebral and pulmonary edema, crural ulcer, and hemorrhoidal complaints.

Scientific name	Common name	Major constituents	Medicinal values*
Agastache foeniculum L. *A. anethrodora* L.	Aniseed	Methylchavicol, anerhole, anisaldehyde[48, 120, 134]	Relieve abdominal distention, nausea or vomiting.
Agrimonia eupatoria L.	Agrimony	Tannins, coumarins, flavonoids luteolin, polysaccharides[30, 39]	Heal wounds and encourages clot formation, treat diarrhea, used as a tonic for digestion.
Agropyron repens (L.) Beauvois	Couch grass	Polysaccharides (triticin), mucilage, agropytene[30, 120]	Diuretic properties, treat urinary infections, enlarged prostate and cystitis, lower blood cholesterol level, goat, and rheumatism.
Alchemilla vulgaris L. *A. xanthochlora* Rothm.	Lady's mantle	Tannins, flavonoids, salicylic acid[30]	Treat mild diarrhea; as a wound healer, it reduces heavy menstrual bleeding, relieves menstrual cramps, and improves regular cycle.
Allium cepa L.	Onion	Thiamin, riboflavin, beta-carotene, ascorbic acid, sterols, alliin, allicin, quercetin, caffeic acid, linoleic acid [30, 134]	Relieve intestinal gas pains, reduce hypertension, inflammation, and cholesterol.
A. sativum L.	Garlic	Alliin, Iodine, diallyl trisulfide, 2-vinyl-4H-1,3-dithin, ajoene, linoleic acid, diallyl disulfide, scordinins, selenium[8, 30, 134]	Reduce serum cholesterol, lower blood pressure, and platelet aggregation. It is anticancer (intestinal tract), antimicrobial, and antithrombotic. With antioxidant value.
A. schoenoprasum L.	Chives	Alliin, sulphoxide, linoleic acid, sulphoevernan[61, 157]	Used as heart and blood circulation remedies.

Scientific name	Common name	Major constituents	Medicinal values*
Alnus crispus (Ait.) Pursh *A. incana* (L.) Moench. subsp. *tenufolia* *A. glutinosa* (L.) Gaertn.	Alder	Tannins, oils, resins, phlobaphenes, flavone glycoside, alnulin, taraxerol, protoalnulin, beta-sitosterol[40, 120]	As an astringent, with hemostatic function, reduce inflammation, and internal hemorrhage.
Aloe vera (L.) Burm. *A. barbadensis* L.	Aloe	Aloin, isobarbaloin, aloeresin A, B, aloesin glycone, aloesone, emodin, chrysophanic acid, 1,8-dihydroxy-anthracene derivatives, barbaloin, anthraquinone glycosides[27, 30, 40, 107]	Purgative, eupeptic, and cholagogue effect. With properties of laxative and cathartic. Juice from leaves used for cuts and possibly other skin problems.
Althaea officinalis L.	Marshmallow	Mucilage, asparagine, pectin, flavonoids, starch[30, 36]	Used as emollient, demulcent, antitussive, and expectorant. Treat bronchitis, asthma, and stomach disorder.
Amelanchier alnifolia Nutt.	Saskatoon	Anthocyanin, quinate, galacturonate, citrate, pyruvate, cis-aconitate, fumaric acid, oxalic acid[120, 122]	Treat diarrhea, watery stools, and prevent miscarriage.
Ananas comosus (L.) Merr.	Pineapple	Bromelain, vitamin A, C, citric acid, vanillin, methyl-n-propyl ketone, valerianic acid, isocaproic acid, acrylic acid, malic acid[40]	Unripe fruit improves digestion, increases appetite, and relieves dyspepsia. Used as uterine tonic. Ripe fruit reduces excessive gastric acid, juice as digestive tonic and diuretic.
Anaphalis margaritacea (L.) Bench and Hook	Everlastings	Monoterpenes, flavonoids, aglycones[120]	Used as pain reliever, treat lung congestion, sore throats and mumps, fevers, diarrhea, and irritable bowels.

Scientific name	Common name	Major constituents	Medicinal values*
Anethum graveolens L.	Dill	Carvone, limonene, flavonoids, coumarins, xanthones, triterpenes[30a, 40a, 134]	Used as infant colic, cough, cold and flu remedies. Relieve digestive disorders.
Angelica archangelica L.	Angelica	Angelicide, brefeldin A, ligustilide, n-butyldenephthalide, phyllandrene, tinnins, valeric acid, ferulic acid, limonene, coumarin, lactones[8, 30, 40, 134, 173]	Stimulate blood circulation, regulate menstruation, stimulate appetite, alleviating coughs and pain. It is carminative.
A. sinensis (Oliv.) Diels	Dong quai	Coumarins, vitamin B_{12}, beta-sitosterol[30]	Used as a tonic. It is antispasmodic, sedative, and promotes menstrual flow.
Antennaria magellanica Schultz	Everlastings	See *Anaphalis margaritacea*	
Anthemis nobilis L.	Chamomile	See *Chamaemelum nobile*	
Apium graveolens L.	Celery	Limonene, coumarins, apiin, oleic, linoleic, palmitic, paliloleic, petroselinic, petroselaidic, stearic myristic and myristoleic acids, bergapten[30, 33, 120]	It is carminative and antirheumatic.
Aralia racemosa L. *A. nudicaulis* L.	Spikenard	Volatile oil, tannins, diterpene acids, glucoside, arctiin[30, 120]	Treat rheumatism, asthma, and coughs. Externally for skin conditions, including eczema.

Scientific name	Common name	Major constituents	Medicinal values*
Arctium lappa L.	Burdock	Inulin, mueilage, tannins, resin, arctiin, arctic acid, arctiol, dehydrofukinone[30, 40]	For rheumatism, gout, and lung disease. It is laxative, diuretic, and perspiration inducer.
Arctostaphylos uva-ursi (L.) Spreng.	Bearberry, Uva-ursi	Arbutin, methylarbutin, tannins, monotropein, polyphenolic acid, (caffeic, chlorogenic) aspirin[40, 98, 135, 160]	Treat kidney stone, bladder, and urinary infections. It is antibacteria and diuretic.
Armoracia rusticana Gaertn, Mey and Scherb.	Horseradish	Asparagine, sinigrin, resin, vitamin C[30]	Internally for arthritis, gout, sciatica, and respiratory and urinary infections. Externally as a poultice for infected wounds, pleurisy, and arthritis.
Arnica latifolia Bong. *A. montana* L. *A. chamissonis* L. subsp. *foliosa* *A. condifolia* Hook *A. fulgens* Pursh *A. sororia* Greene	Arnica	Helenalin, sesquiterpene lactones, 11-α,13-dihydrohelenalin, arnidiol, carotenoides, arnicin, amidendiol, tannins, egin, anthoxanthine, inulin, phytosterol[27, 40, 77, 87, 135, 161]	Antineoplastic, antibacterial, antifungal, anti-inflammatory, antiseptic, reduce pain and swelling. Used as tonic, spasmodic agent, counterirritant. Applied locally to produce superficial inflammation to reduce pain.
Artemisia absinthium L.	Wormwood	Absinthol, tannins, thujyl alcohol, flavonoids, phenolic acid, lignans[30, 120]	Anthelmintic. External antiseptic.
A. annua L.	Qing Hao	Abrotamine, artemisinin, vitamin A[30]	Treat fever, headaches, dizziness, and tight-chested sensation.

Scientific name	Common name	Major constituents	Medicinal values*
A. dracunculus L.	Tarragon	Estragole, phelandrine, methyl chavicol, iodine, rutin, tannins, flavonoids, coumarins[21, 30, 40, 134]	Diuretic, used as an appetite stimulant.
A. tridentata Nutt.	Sagebrush	Volatile oil, furanoid, pentane[120]	Aromatic, bug repellent.
A. vulgaris L.	Mugwort	Cincole, thujone, ascorbic acid, thiamine, inulin, resin, tannin[40, 120]	Used as emmenagogue, appetite and uterine stimulant.
Asarum canadensis L.	Wild ginger	Pinenes, delta-linalool, borneol, terpineol, arislolochic acid[134]	Treat asthma, sore throats, stomach cramps, recurrence of herpes lesions.
Asparagus officinalis L. *A. racemosus* Willd.	Asparagus	Steroidal glycosides, asparagine, (asparagosides), flavonoids[30]	Diuretic, treat urinary problems, and rheumatic conditions. It has laxative or sedative effect.
Astragalus americana Bunge. *A. membranaceus* Bunge. *A. sinensis* L.	Astragalus Huang Qi	Asparagine, calcyosm, sterols, formononetin, kumatakenin[30]	Improve immune system, lowers blood pressure, as a diuretic, tonic, and endurance remedy.
Avena sativa L.	Oat	Proteins, vitamin B complex, saponin, carotenes[25, 120]	Antidepressant, heal skin disorders.
Baptisia tinctora (L.) R. Br. ex Ait.f.	Wild indigo	Isoflavones, flavonoids, alkaloids, coumarins, polysaccharides[30]	Antiseptic for cuts and wounds.
Bellis perennis L.	Daisy	Saponins, tannins, mucilage[40]	Used as an astringent and expectorant, relief gastritis, enteritis, and diarrhea.

Scientific name	Common name	Major constituents	Medicinal values*
Berberis aquifolium L.	Oregon grape	See *Mahonia aquifolium*	
B. vulgaris L.	Barberry	Berberine, tannins, resin, berbamine, berberubine[30, 40]	Improve liver function, antiseptic, and antidiarrhea.
Beta vulgaris L.	Beet root	Betaine, betanin[30]	Support gallbladder, influencing body metabolism and lower cholesterol level.
Betula lenta L.	Birch	Saponins, hyperoside, tannins, gallic acid, essential oil, methyl salicylate[120]	Treat headaches and rheumatic pain. Anti-inflammatory, used as blood cleanser with stimulating effect on kidney.
Borago officinalis L.	Borage	Tannins, saponins, source of Ca and K, mucilage, resin, silicic acid, arabinose, pentoses[40, 99, 120, 134]	Anti-inflammatory, demulcent, folk remedies for cancer, corns, sclerosis, tumor, bladder stone, and bronchitis.
Boswellia serrata Roxb. ex Colebr. *B. carteri* Birdwood	Frankincense	Gum resin (triterpenoids), beta-boswellic acid[40, 123]	Internally for bronchial and urinary infections, it is anti-inflammatory. Externally as an inhalant for mucus and a douche for vaginal infections.
Brassica nigra (L.) Kock.	Black mustard	Sinapine, mucilage, glucosinolates[16]	Treat lung congestion and bronchial problems.
Buxus sempervirens L.	Boxwood	Steroidal alkaloids, alpha-tocopherol[96, 151]	Used for recurrent fevers, rheumatism, and intestinal parasites.

Scientific name	Common name	Major constituents	Medicinal values*
Calamintha nepeta (L.) Savi *C. ascendens* L. *C. officinalis* L.	Basil thyme Calamint Mountain mint	Volatile oil (pulegone, menthone), flavonoids, isopinocamphone[30, 134]	Stimulate sweating, lower fevers, ease gas and indigestion, a cough and cold remedy and treat mild respiratory infections.
Calendula officinalis L.	Calendula, Marigold	Carotenoids, saponins, flavonoids, phytosterols, mucilage, triterpenes, resin[30, 40]	Anti-inflammatory, heal wounds, bed sores, ulcers, and skin rashes.
Camellia sinensis (L.) Kuntze	Green tea	Methylxanthines, caffeine, purine, polyphenols (-epigallocatechin-3-gallate (EGCg), ascorbic acid, beta-carotene, thiamine, niacin, theophylline[40, 76]	Antioxidant with stimulating effects.
Cananga odorata (Lam.) J. D. Hook and T. Thompson	Ylang-Ylang	Geraniol, linalool esters of acetic and benzoic acid, salicylic acid, essential oil[30, 40]	Antiseptic, sedative, relieve tension, lower blood pressure.
Cannabis sativa L.	Hemp	Tetrahydro-cannabinols, thiamine protein and carbohydrate. Seeds contain choline, inositol, xylose phytosterols, trigonelline[40, 120]	Induce euphoria and exhilaration, sedative, antispasmodic.
Capsella bursa-pastoris (L.) Medik.	Shepherd's purse	Amines choline, acetylcholine, bursine, histamine, flavonoids, polypeptides, tyramine[30, 120]	Control internal bleeding, profuse menstruation.

Scientific name	Common name	Major constituents	Medicinal values*
Capsicum annuum L. var. Annuum *C. annuum* L, var. Grossum *C. frutescens* L. *C. annuum* L. var. Acuminatum	Cayenne pepper Sweet pepper Chilli pepper	Capsaicin, solanidine, solanine, solasodine, scopoletin, chlorogenic acid, alanine, amyrin, caffeic acid, camphor, carvone, cinnamic, citric acid, linalool, linoleic acid, oleic, piperine, vitamin B_1, B_2,C, E[40, 134]	Improve the flow of blood to the skin and mucosa, treat rheumatism, sciatica, and pleurisy.
Carbenia benedicta L.	Blessed thistle	See *Cnicus benedictus*	
Carduus benedita L.	Blessed thistle	See *Cnicus benedictus*	
Carica papaya L.	Papaya	Proteolytic enzymes such as papain, malic acid, chymopapain, beta-carotene, ascorbic acid, vitamin E[40]	As a digestive agent, latex from the trunk apply externally for skin healing.
Carthamus tinctorius L.	Safflower	Carthamone, lignans, vitamin E, polysaccharide[52]	Reduce fever by inducing perspiration, it has laxative effect.
Carum carvi L.	Caraway	Carvone, limonene, flavonoids, polysaccharides[30, 51, 134]	Relieving gas pains, it is antispasmodic and carminative.
Cassia senna L. *C. angustifolia* Vahl.	Tinnevelly senna Alexandria senna	Authraquinone, β-sitosterol, rhein, dianthrone glucosides, sennosides A and B, napthalene glycosides, aloe-emodin, mucilage[30, 40]	Laxative, stimulant, cathartic, anticancer, cathartic.
Castanea sativa Mill	Chestnut	Tannins, plastoquinones, mucilage[30]	Treats whooping cough, bronchitis, and sore throats.

Scientific name	Common name	Major constituents	Medicinal values*
Catharanthus roseus (L.) Don	Periwinkle	See *Vinca minor*	
Caulophyllum thalietroides (L.) Michx. *C. giganteum* (F.) Loconte and W. H. Blackwell	Blue cohosh	Caulophylline, caulosaponin, methylcytisine, anagyrine, steroidal saponins, laburnine, magnoflorine[30, 135, 163]	Used during parturient, as an antispasmodic, diuretic, antirheumatic, promotes menstrual flow, and induces abortion. Externally treat dermatitis caused by poison ivy.
Cedrus libani A. Rich subsp. *atlantica*	Cedarwood	Essential oil, monoterpenes, sesquiterpenes, atlantol[26, 30]	Externally for skin diseases, ulcers, and dandruff. Internally for menstrual delay, rheumatism, arthritis.
Centaurea calcitrapa L.	Star thistle	Centapicrin, xanthone, gentianine[9]	Antipyretic and antidiabetic.
Centella asiatica (L.) Urb.	Gotu Kola	Triterpenoid, saponins, oleic acid, vellarin, hydrocotyline, sitosterol, asiatic, madecassic, madasiatic acids, asiaticoside[30, 40, 73, 136]	Treat skin disease, leprosy, antipyretic, detoxicant, diuretic, antirheumatic, mild diuretic, sedative, and peripheral vasodilator.
Chamaelirium luteum (L.) A. Gray	False unicorn root, fairy wand, devil's bit	Steroidal saponins, chamaelirin, helonin[30]	For menopausal complaints, infertility, pelvic inflammatory disease, pain reliever, diuretic, uterine disorders.
Chamaemelum nobile (L.) All.	Roman chamomile	Azulene, choline, chamezulene, flavonoid glycosides, angelic acid esters, nobilin[8, 40]	Carminative, antispasmodic, and sedative.
Chamaenerion angustifolium (L.) Scop.	Firewood	See *Epilobium angustifolium* L.	

Scientific name	Common name	Major constituents	Medicinal values*
Chenopodium album L.	Lamb's quarters	Tetain, ecdysteroids, triterpenoid saponins, flavonol glycoside, beta-sitosterol, oleanic acid, vitamin A and C[35, 120]	Constipation, iron supplement, tonic. Externally as poultices for inflammation, relieve headache.
C. ambrosioides L.	Wormseed	Ascaridole, saponins, myrcene, geraniol[30, 40, 134]	Anthelmintic and anthelmintic.
Chrysanthemum cinerarifolium (Trevir.) Vis.	Pyrethrum	Pyrethrins, cinerins, palmitic, linoleic acid, sesquiterpene lactones[21, 40, 134]	Externally used as contact insecticides.
C. parthenium (L.) Berhn.	Feverfew	Camphor, tannins, mucilage, sesquiterpene lactone[65]	Treat fevers, migraine, arthritis, colds, indigestion, diarrhea, and hysteria.
C. vulgare L.	Tansy	See *Tanacetum vulgare*	
Cichorium intybus L.	Chicory	Lactucin, intybin, anthocyanins, inulin, sesquiterpene[30, 134]	No medicinal value recorded, used as food.
Cimicifuga racemosa (L.) Nutt.	Black cohosh	Triterpene glycosides, isoflavones, isoferulic acid, resin, salicylates, sterols, methylcytisine, cimicifugin, actein[30, 40, 125, 135, 146]	Treat rheumatism, neuralgia, diarrhea, bronchitis, measles, whooping cough, tuberculosis, high blood pressure, migraine headaches, arthritis, relieve depression and suppress hot flashes.

Scientific name	Common name	Major constituents	Medicinal values*
Cinnamomum verum J. Presl. *C. cassia* (Nees) Nees & Eberm *C. camphora* (L.) Nees & Eberm.	Cinnamon Cinnamon Camphor	Tannins, coumarins, mucilage Camphor, camphene, dipentene, limonene, phyllandrene, pinene, cinnamaldehyde d-Camphor[30, 40]	A warming stimulant, it is carminative, antispasmodic, antiseptic, and antiviral.
Citrus limon (L.) Burm. *C. bergamia* Risso & Poit *C. reticulata* Blanco *C. aurantifolia* Swingle *C. aurantium* L. *C. paradisi* Macfad	Lemon Bergamot orange Mandarin Lime Bitter orange Grapefruit	Coumarins, bioflavonoids, vitamins A, B, C, mucilage, volatile oils Limonene, *l*-linalool, d-limonene, linalyl acetate, flavonoids, benzoic acid, cinnamic acid, coumarins, carotenoids[30, 97]	Antiseptic, anti-rheumatic, antibacterial, antioxidant, carminative, laxative, liniment, and vermifuge. Juice is used to treat tumors and warts.
Cnicus benedictus L.	Blessed thistle	Enicin, tannins, mucilage, minerals, silymarin[8, 30]	A contraceptive, treat cancer, heart and liver ailments.
Cochlearia amorucia L.	Horseradish	See *Armoracia rusticana*	
Codonopsis pilosula (Franch.) Nannfeldt. *C. tangshen* Oliver	Codonopsis, Dang Shen, bellflower	Triterpenoid saponins, sterins, perlolyrin, alkenyl, polysaccharides, alkenyl glycoside, tangshenoside[30, 80]	As an adaptogen, stimulant, and tonic.
Coffea arabica L.	Coffee	Caffeine, theobromine, theophylline, tannins[30]	A stimulant for central nervous system.
Colchicum autumnale L.	Crocus Meadow saffron	Colchicine, flavonoids[30, 39]	Anti-inflammatory.

Scientific name	Common name	Major constituents	Medicinal values*
Collinsonia canadensis L.	Pilewort	Anemonin, tannins, saponins, volitile oil[30]	Diuretic and anti-inflammatory, a tonic for digestive system, kidney, and urinary stones.
Commiphora molmol Engl. ex Tschirch *C. myrrha* (Nees) Engl.	Myrrh	Gum, acidic polysaccharides, resin[30]	A stimulant, antiseptic, astringent, and expectorant, it is anti-inflammatory, antispasmodic, and carminative.
Convallaria majalis L.	Lily of the valley	Cardiac glycosides such as convallatoxin and convallatoxol saponins[30]	Treat heart conditions.
Convolvulus arvensis L. *C. sepium* L.	Bindweed	Convolvuline, resinous glycoside, tannins[120]	Jaundice, dropsy, digestive illness.
Coriandrum sativum L.	fruit: Coriander leaf: Cilantro	Linalool, proteins, vitamin C, alpha-pinene, terpinene[30, 134]	A digestive tonic, it is carminative and sedative.
Cornus canadensis L.	Bunchberry	Cornine, cornic acid, quercitin, tannins[120]	Decreases inflammation, pain, and fever. Crushed berries for burns.
C. florida L.	Dogwood	Verbenalin, saponins, tannins, resin, gallic acid, malic, tartaric, gallic and tannic acids[40, 70]	A tonic, astringent, and hemostatic.
Corylus cornuta Marsh. *C. rostrata* Marsh. *C. avellana* L.	Hazelnut	Tannins, essential oil, ferric oxide, beta-sitosterol[120]	Hazelnut milk for coughs and colds, leaves are diuretic, prostaglandin inhibition, and anti-inflammatory.

Scientific name	Common name	Major constituents	Medicinal values*
Crataegus monogyna Jacq. *C. pinnatifida* Bge. *C. oxyacantha* L.	English hawthorn Chinese hawthorn Hawthorn	Flavonoid glycosides, procyanidins, catechins, triterpenoid acid, Vitamin C, B complex, pectins, amygdalin, proanthocyanidins, emulsin, tartaric acid, tannins, crataegus acid, rutin, coumarins, quercitin, amines[3, 30, 40, 128, 143]	Therapeutic treatment of heart insufficiency, hypotensive, coronary blood supply, and arrhythmia.
Crocus sativus L.	Saffron	Crocine glycosides, beta-carotene, phytoene, phytofluene[40, 134]	Saffron stomachic, antispasmodic, it has emmenagogue properties.
Cryptotaenia japonica Hassk.	Japanese parsley, Milsuba	See *Petroselinum crispum*	
Cucurbita pepo L.	Pumpkin	Fatty acid, proteins, cucurbitin, resin[30]	Anthelmintic, treat intestinal worms.
Cuminum cyminum L.	Cumin	Pinene, alpha-terpineol, cimin-aldehyde[30, 134]	Relieves flatulence and bloating, stimulates digestive process.
Curcuma domestica L. *C. xanthorrhiza* L. *C. longa* L.	Curcuma, Jiang Huang	Curcuminoids, essential oil, resin[30]	Biliary disorder, anti-inflammatory, lipid reducing sedative.
Cuscuta epithymum Murr. *C. chinensis* Lam.	Dodder	Flavonoids, hydroxycinnamic acid, bergegin[120]	Remedy for kidney disorder and liver disease. It is laxative.

Scientific name	Common name	Major constituents	Medicinal values*
Cymbopogon citratus (DC. ex Nees) Stapf. *C. nardus* L. *C. winterianus* Jowitt *C. martinii* (Roxb) Wats	Lemon grass Citronella Palmarosa, ginger grass	Volatile oil (citral), citronellal[34, 40, 134, 145]	Treat digestive problems, relieve cramping pains.
Cynara scolymus L. *C. cardunculus* L.	Artichoke Wild artichoke	Luteolin, caffeic acid, cynarin, phenolics, flavanoids, glavanoid glycosides, sesquiterpenoids, polyacetylenes, chlorogenic acid, cynaroside, saponins[24, 34]	Inhibition of cholesterol synthesis, lower lipid levels, antioxidant effects on the liver.
Cypripedium calceolus L.	Lady's slipper	Cypripedin (a resinoid substance) tannic, gallic acid[40]	Treat headache, nervousness, and anodyne, it is antispasmodic and sedative. A tonic.
Cytisus scoparius (L.) Link	Broom	Sparteine, scoparoside, flavone[40, 151]	Diuretic and cathartic.
Daucus carota L.	Carrot	Thiamine, nicotinic acid, phytin, lipids, carotenes, vitamin B complex and C[40]	It has anthelminthic and diuretic properties.
Digitalis purpurea L.	Foxglove, Digitalis	Purpurea-glycosides A and B, digoxin, digitoxin, caffeic acid, lanatoside, choline, saponins, chlorogenic acid[40]	Improving blood circulation to the kidneys, it has cardiologic effect.

Scientific name	Common name	Major constituents	Medicinal values*
Dioscorea oppositae Thunb.	Chinese yam (Shen yao)	Steroidal saponins, albuminoides, diosgenin, progestoron, sapogenin[50]	Hormonal effect, treat vaginal discharge. It has diuretic, and anti-inflammatory properties.
D. villosa L.	Wild or Mexican yam	Steroidal saponins, beta-sitosterol, tannins, alkaloids, diosgenin[30, 50]	Anti-inflammatory, antispasmodic, antirheumatic, diuretic, increase sweating.
Echinacea angustifolia DC. Cornq. *E. pallida* (Nutt.) Nutt. *E. pururea* (L.) Moench.	Narrow leaf echinacea Pale-flower echinacea Purple coneflower	Caffeic acid derivatives, alkylamides, polyacetylenes, glycoproteins, polysaccharides, echinacosides, isobutylamides[11, 12, 13, 30, 91, 93, 144]	Immune modulator activity, treat common cold and flu, and antimicrobial.
Echinopanax horridus (Sm.) Decne. and Planch. ex H. A. T. Harms	Devil's club	See *Oplopanax horridus*	
Elaeagnus angustifolia L.	Russian olive	Caffeic acid, isorhamnetin, fatty acids[55, 120]	Antimicrobial.
Eleutherococcus senticosus (Rupr. ex Maxim) Maxim	Siberian ginseng	Eleutherosides, lignans, coumarins, phenylroparnoids, isofraxin, pectin, saponins, resins[30, 40, 60]	An adaptogen, tonic, stimulant, protects the immune system.
Elymus repens L.	Couch grass	See *Agropyron repens*	

Scientific name	Common name	Major constituents	Medicinal values*
Ephedra distachya L.	Ephedra	Alkaloids, ephedrine, l-ephedrine, d-pseudoephedrine[39, 111]	Treat fevers, relieve kidney pain, for hypertensive aid, treat asthma, nose and lung congestions, hay fever.
E. nevadensis Wats.	Mormon tea	Pseudoephedrine, l-ephedrine, d-pseudoephedrine[40]	A decongestant and asthma remedy, for hypertensive aid, hay fever.
E. sinica Stapf.	Ma Huang	Protoalkaloids, tannins, saponin, flavone, volatile oil, ephedrine, l-ephedrine, d-pseudoephedrine[30, 39]	Increase sweating, dilates bronchioles, a stimulant, diuretic, and raises blood pressure.
Epilobium angustifolium L.	Fireweed	3-O-β-D-glucuronide, mucilage, tannins[34, 135]	Treat skin irritation and burns, infused flower for gargle sore throat and laryngitis. It is anti-irritant and used as a mild sun screen, and inhibiting microbial growth.
Epilobium parviflorum Schreb.	Small-flowered willow herb	Flavonoids, esters of sitosterol, gallic acid derivatives[16]	Antiphlogistic.
Equisetum arvense L. *E. telmateia* Ehrh.	Horsetail	Silicic acid, trace of nicotine, equisitine, silicates[30, 40, 120]	Treat bleeding wounds, antibiotic, mouth wash for oral infection, antidiaphoretic.
Erythroxylum coca Lam.	Coca	Cocaine, nicotine, benzoylecgonine, cinnamylcoaine, ecgonine, methyl salicylate[40]	Anesthetic, aphrodisiac, stimulant.
Eschscholtzia california Cham.	California poppy	Protopine, cryptopine, chelidonine, flavone glycoside[30, 40]	As a pain reliever, it is sedative for headache and insomnia.

Scientific name	Common name	Major constituents	Medicinal values*
Eucalyptus citriodora Hool. *E. globulus* Labill.	Gum tree Eucalyptus	Essential oil (cineole or eucalyptol), caffeic, coumaric, gallic, gentisic, hydroxybenzoic, syringic, vanillic[30, 40]	Externally for athlete's foot, dandruff, herpes, and an inhalation for fevers and asthma.
Eugenia caryophyllata L.	Clove	See *Syzygium aromaticum*	
Eupatorium perfoliatum L.	Boneset	Sesquiterpene lactones, eupatorin, polysaccharides, flavonoids, tannic acid, gallic acid[40]	Antipyretic, diaphoretic.
Euphrasia officinalis L.	Eyebright	Tannins, ash, vitamin C, carotene, gallotanins, auculin, aucubigenin[40]	Astringent, collyrium, laxative, ophthalmic and tonic.
Fagopyrum tuturicum L.	Buckwheat	Rutin, vitamin B_1, B_2, choline, bioflavonoids[30]	Reduce cholesterol, prevent high blood pressure, keep capillaries and arteries strong and flexible.
Fagus grandifolia Ehrh. *F. sylvatica* L.	Beech European beech	Creosote, suberin, polyphenols, phospholipids[118]	Internally for chronic bronchitis and upper respiratory tract infection, externally for skin diseases.
Filipendula ulmaria (L.) Maxim.	Meadow sweet	Salicylates, flavonol glycosides, heliotropin, vanillin, tannins[21, 30, 40]	It is a laxative, treat headache.
Foeniculum vulgare Mill.	Fennel	Anethole, fenchone[40, 134]	Antispasmodic.
Forsythia suspensa (Thunb.) Vahl.	Forsythia Lian Qiao	Forsythin, vitamin P[75]	Antiseptic, remedy for colds, flu, sore throats and tonsilitis.

Scientific name	Common name	Major constituents	Medicinal values*
Fragaria vesca L.	Strawberry	Tannins, vitamin C, pectin, citric and malic acids[120]	Stimulate appetite, antidyspeptic.
Fraxinus americana L. *F. excelsior* L.	White ash	Coumarins, flavonoids, tinnins, sugar and volatile oil[30]	A tonic, astringent, it is laxative, a diuretic and treat fevers.
Fucus vesiculosus L.	Kelp	Phenols, polysaccharides, iodine[81, 103, 116, 135, 150, 158]	Treat iodine deficiency, antibiotic, promote hormone production.
Galium aparine L.	Cleavers	Iridoids, valepotriates, polyphenolic acids, anthraquinones, tannins[30, 134]	Remedy of vitamin C deficiency.
Gaultheria procumbens L.	Wintergreen, teaberry	Methyl salicylate[40]	Antiseptic, carminative, and diuretic.
Geranium macrorrhizum L.	Geranium, American cranesbill	Tannins[30]	Treat stomach disorder, aphrodisiac, and colitis, peptic ulcer, as a diarrhea tonic.
Ginkgo biloba L.	Ginkgo	Ginkgocide A,B,C,J,M, flavonoids, bilobalide, sciadopitysin, ginkgetin, isoginkgetin, bilobetin, carotenoids, 4'-0-methylpyridoxine[2, 4, 30, 165]	Treat dementia and cerebral insufficiency, relieve asthma, treat cerebral disorders.
Glechoma hederacea L.	Ground ivy	Glechomine, tannins, flavonoids, resins, saponins, sesquiterpene[30]	For mucous problems (respiratory), glue ear, lung congestions, and urine retention.

Scientific name	Common name	Major constituents	Medicinal values*
Glycine max (L.) Merrill	Soybean	Protein, fatty acid, vitamins, carbohydrates, lecithin, globulin, glycine, mineral, daidzin astrogen, caffeic acid, coumestrol, fiber, choline, tocopherol, saponins, phytic acid, idoflavones [25, 43]	Prevent arteriosclerosis and coronary heart disease, an astringent, treat hypercholesterol, a starting source of stigma sterol.
Glycyrrhiza glabra L.	Licorice	Glycorrhizin, mucilage, flavonoids, glycyrrhetinic acid, saponin, glabridin, tannic acid, 2-β-glucuronosyl, glucuronic acid[30, 40, 64, 67, 134]	Antiulcerative activity, stomach, duodenal ulcers, anti-inflammatory, antiallergic, antihepatitis. A cough remedy, an expectorant, demulcent, it is laxative.
G. uralensis Fisch ex DC.	Chinese licorice	Triterpene saponins, chalcones flavonoids, isoflavonoids[30]	Sweat-tasting tonic, treat sore throats, wheezing, coughs, canker sores, peptic ulcer, and gastritis.
Gossypium hirsutum L.	Cotton	Root bark contains gossypol and flavonoids seed oil contains gossypol, flavonoids[30]	Oil causes infertility in men, treat heavy menstrual bleeding, root bark for stimulating uterine contractions, encourages the blood to clot and the secretion of breast milk.
Grindelia robusta Nutt. *G. squarrosa* (Pursh) Donal	Gumweed	Grindelic acid, acacetin, quercitin, kumatakenin, tannins[120]	Catarrh of the upper respiratory tract, circulatory stimulant, relaxes heart muscle.

Scientific name	Common name	Major constituents	Medicinal values*
Hamamelis virginiana L.	Witch hazel	Catechins, proanthocyanidins, tannins (hamamelitannin), gallic acid, caffeic acid, volatile oil, quercitin, kaempferol[40, 44, 82, 135, 171]	An astringent, anti-inflammatory, treat external and internal bleeding, antihemorrhagic, treat inflammation of the gums and mucous membranes of the mouth.
Harpagophytum procumbens DC. ex Meisn	Devil's claw	Harpagoside, phytosterols, flavonoide, harpagoqside, stachyose[30, 40]	Anti-inflammatory, analgesic, digestive stimulant.
Hedeoma pulegiodes (L.) Pers.	Pennyroyal	Pulegone, pinene, limonene, formic, and salicylic acid[40]	Treat pain and headache. Decoction used for flatulence, diaphoretic in catarrh, and cold.
Hedera helix L.	English ivy	Tannins, hederin, aglycone, iodine, beta-elemone, elixen, hederacoside B and C, germacrene B[40]	Applied to cuts, sores, and skin eruptions. An expectorant with antispasmodic and cardiac actions.
Helianthus annuus L.	Sunflower	Linolenic and oleic, palmitic and arachic[24]	Internally relieves constipation, externally on cuts and bruises.
Helichrysum angustifolium (Lam.) DC.	Curry plant	Acylphloroglucinal derivatives, terpenoids, cinnamic acid, azulene, lignans[49, 71, 72]	No medicinal values recorded.
Helonias dioica L.	False unicorn root	See *Chamaelirium luteum*	

Scientific name	Common name	Major constituents	Medicinal values*
Heracleum maximum Bart. *H. lanatum* Michx *H. sphondylium* L.	Cow parsley Cow parsnip Cow parsnip	Sphondin, psoralen, heraclein, glutamine, essential oils[120]	For headaches, poor memory, melancholy and agitation, indigestion, and asthma. Externally for healing and rheumatic pain and palpitations.
Hibiscus sabdariffa L. *H. rosa-sinensis* L.	Hibiscus	Mucilage, citric, malic, tartaric acids, hibiscus acid, thiamine, gossypetin, anthocyanin, myristic acid, palmitic acid[40, 134]	Soothing effect on mucous membranes that line the respiratory and digestive tracts. Seeds used for cramps, flower as an astringent.
Hierochloe odorata (L) Beauv.	Sweet grass	Coumarin, massoilactone, lactone[135]	Treat coughs and sore throats, venereal infections, bleeding after childbirth, chapped or wind-burned skin and eye irritations.
Hippophae rhamnoides L	Sea buckthorn	Carotenoid, flavonoid, volatile oil, essential oil, saturated and unsaturated fatty acids, tannins, quercitin, provitamin A, vitamin C, B complex, and E[92, 93, 115, 120]	Improve resistance to infection, skin irritation and eruptions.
Humulus lupulus L.	Hop	Humulone, lupulone, humulene, alpha, beta-acids, polyphenols, steroids, resins, tannins[61, 135, 141]	Sedative effect, hypnotic, stomachic, diuretic. Against gram-positive organisms and tuberculosis.
Hydrangea arborescens L.	Hydrangea	Flavonoids, cyanogenic glycoside, saponins, hydrangein, tannin[30, 40]	Treat kidney and bladder stones.

Scientific name	Common name	Major constituents	Medicinal values*
Hydrastis canadensis L.	Goldenseal	Hydrastine, berberine, berberastine, canadine, isoquinoline alkaloids, resin, volatile oil[15, 30, 40, 135]	Affect circulation, uterine functions, central nervous system, and constricts peripheral vessels, decrease blood pressure and inhibit synthesis of DNA and proteins and oxidation of glucose. It has antiseptic, hemostatic, and anti-inflammatory effects.
Hypericum perforatum L.	St. John's wort	Hypericin, hyperoside, rutin, quercitin, chlorogenic acid, pseudohypericin, flavonoids[30, 40, 120]	Antidepressant, anti-inflammatory, diuretic, antiseptic and astringent properties.
Hyssopus officinalis L.	Hyssop	Pinene, limonene, pinecamphene, hesperidin, tannins. terpenes[30, 134]	Treat respiratory problem, coughs, sore throat, hoarseness, asthma, and bronchitis.
Ilex aquifolium L.	English holly	Triterpenoids, salicylic acid, caffeine, isophthalic acid[16]	Relieve menstrual cramps, calm nervous stomachs.
Inula helenium L.	Elecampane	Inulin, resin and mucilage, helenalin, dammaranedienol[30]	For asthma, chest cold, stomach ulcers, antitussive, diuretic, antiseptic.
Iris versicolor L.	Blue flag	Triterpenoids, salicylic acid, isophthalic acid, alpha-phytosterol, myricylalchol[40]	Relieve menstrual cramps, calm nervous stomachs.
Jasminum grandiflorum L. *J. auriculatum* Vahl.	Royal jasmine	Essential oil, isoquercitrin, ursolic acid, 2-3,4-dihydroxyphenyl-ethanol[22, 74]	Treat high fever, sunstroke, cancer (Hodgkin's disease)

Scientific name	Common name	Major constituents	Medicinal values*
Juglans regia L. *J. nigra* L.	Walnut Black walnut	Tannin, juglandin, juglone, hydrojuglone[16]	Astringent, hemostatic, anti-inflammatory, antispasmodic, antiphlogistic and mild sedative.
Juniperus communis L.	Juniper	Resin, pinene, borneol, inositol, juniperin, limonene, cymene, terpinene[40, 120, 134]	For dropsy, bladder and kidney disorders, rheumatic pain.
Laminaria digitata (Hudss.) Lank *L. saccharina* (L.) Lamk.	Kelp	See *Fucus vesiculosus*	
Lamium album L.	Nettle	Saponins, flavones, mucilage, tannins[30]	An astringent and demulcent. Used as a uterine tonic, intermenstrual bleeding, reduce excessive menstrual flow.
Larrea tridentata (Sesse. and Moc. ex DC.) Coville.	Chapatral	Resin, nordihydroguaiaretic acid[40]	Treat rheumatic disease, venereal and urinary infection, leukemia.
Laurus nobilis L.	Bay laurel	Cineole, linalool, alpha-pinene, alpha-terpineol, acetate, mucilage, tannin, resin, eucalyptol[8, 40]	Internally for indigestion and appetite. Externally for dandruff, rheumatism, sprains, and bruises.
Lavandula angustifolia Mill. *L. spica* L. *L. officinalis* L.	Lavender	Linalyl acetate, linalool, camphor, borneol, tannins, coumarins, flavonoids, ursolic acid[30, 40, 134]	A mild sedative, carminative, rubefacient, antispasmodic, relieve intestinal gas. Essential oil used as a stimulant with activating and antirheumatic effects.

Scientific name	Common name	Major constituents	Medicinal values*
Ledum latifolium Jacq.	Labrador tea	Arbutin, essential oil[120]	Treat bronchial congestion and diarrhea. Tinctures used against mosquitoes, lice and fleas.
Lemna minor L.	Duckweed	Arginine, lysine, iron manganese[120]	For fever, skin disease, and rash, water retention.
Leonurus cardiaca L.	Motherwort	Leonurin, leonuride, pyrogallol, catechins, choline, saponins[40]	Emmenagogues, cardiologic, antispasmodic, hypotensive, and astringent properties.
Levisticum officinale W. Koch. *L. levisticum* L.	Lovage	Terpineol, butyl phthalidine, coumarins, beta-sitosterol, resins, gums, alkyl-phthalides[16, 30]	Carminative, diuretic, relieve gas pains and breath deodorizer.
Ligustium lucidium W.T. Aiton. *L. vulgare* L.	Ligustrum fruit Privet	Essential oil, phthalides, terpenoides[40]	Prevent bone marrow loss, treat acquired immune deficiency syndrome, respiratory tract infections, hypertension, Parkinson's disease, and hepatitis.
Linum usitatissimum L.	Flax	Linseed oil, linoleic acid, linolenic acid, stearic acid, oleic acid, mucilage, linamarin[30]	Externally as a poultice for boils, burns. Relieve constipation, demulcent, laxative.
Liquidambar styraciflua L.	Gum tree, Storax	Levant storax: cinnamic acid, cinnamyl cinnamate, phenylpropyl cinnamate, triterpene acid[30]	Internally for sore throats, coughs, colds, asthma, bronchitis and vaginal discharge. Externally for sores, hemorrhoids, ringworm.

Scientific name	Common name	Major constituents	Medicinal values*
L. orientalis L.	Oriental sweet gum	Levant storax:cinnamic acid, cinnamyl cinnamate, phenylpropyl cinnamate, triterpene acid[30]	Internally for strokes, infantile convulsions, coma, heart disease, and pruritus.
Lobelia inflata L.	Lobelia	Piperidine alkaloids (lobeline, lobelidiol, lobelanidine), carboxylic acid[30, 40, 105]	Respiratory stimulant, antispasmodic, expectorant, induces vomiting.
L. siphilitica L.	Blue lobelia	Alkaloids[24]	Cure syphilis.
L. pulmonaria L.	Lungwort	d-Usnic acid, thamnolic, polysaccharides, anthraquinones[24]	Stimulate immune system, antitumor, cancer.
Lomatium dissectum (Nutt.) Math. & Const.	Lomatium	Flavonoid, glycoside, 7-O-β-D-glucosyl luteolin[120, 155]	Treat tuberculosis.
Lonicera caerulea L. *L. caprifolium* L.	Mountain fly honeysuckle, Dutch honeysuckle	Various sugars, sorbitol, inositol, limonic and malic acids, citric acid, tannins, salicylic acid[120]	Hypotensive, sedative and antipyretic, a tonic.
Lycium barbarum L. *L. chinense* Mill. *L. pallidum* L.	Lycium Chinese wolfberry Wolfberry	Betaine, beta-sitosterol. In berry: physalin, carotene vitamins B_1, B_{12}, and C. In root: cinnamic acid, psyllic acid[30]	Berry: internally for high blood pressure, a tonic protect liver, menopausal complaint. Root: treat chronic fevers, lower blood pressure, internal hemorrhage, tuberculosis.

Scientific name	Common name	Major constituents	Medicinal values*
Lycopodium clavatum L.	Clubmoss	Lycopodine, dihydrolycopodine, myristic, resins, ployphenols, flavonoids, triterpenes[30]	A diuretic for kidney and bladder complaints.
Lysimachia vulgaris L.	Loosestrife	Saponins, flavonoids, tannins, benzoquinene[30]	An astringent, treat gastrointestinal conditions such as diarrhea and dysentery, stop bleeding.
Lythrum salicaria L.	Purple loosestrife	Tannin, triacylglycerols, salicarin, vitexin[30]	Lower serum cholesterol, glucose, and triglyceride levels, and antiatherosclerotic action. Relieve diarrhea, gargle for sore throat, clean wounds.
Macropiper excelsum G. Forst	Kava-Kava	See *Piper methysticum*	
Mahonia aquifolium (Pursh) Nutt.	Oregon grape	Berberine, protoberberine alkaloids, oxyberberine, magnoflorine and columbamine[40, 100, 104]	Eczema, gall bladder disorder, chronic hepatitis B, gastritis, diarrhea, antipsoriasis.
Marrubium vulgare L.	Horsehound	Stachydrine, tannins, saponin, diterpenes, marrubiin, marrubenol betonicial[30]	An expectorant, used as an appetite stimulant, antispasmodic and with emmenagogic properties.
Matricaria chamomilla L. *M. recutita* L.	Chamomile German chamomile	Flavonoid, glycosides, tannins, n-coumaria acid, herniarin, luteolin, umbelliferone, cynaroside, alpha-bisabolol), azulene, anthemidin, luteolin, coumarins[30, 40, 134]	Antispasmodic for relieving cramps Nervous digestive upsets, insomnia. Also used as an antiallergenic agent and as in hear conditioner.

Scientific name	Common name	Major constituents	Medicinal values*
Matteuccia struthiopteris (L.) Tod.	Ostrich fern	Palmitic acid, astragalin, caffeic acid, chlorogenic, p-coumaric, p-hydroxybenzoic, ferulic, vanillic, protocatechuic, beta-sitosterol, campesterol, stigmasterol, oleoresins[147]	Expel parasites, treat inflammation of lymphatic glands.
Medicago sativa L.	Alfalfa	Isoflavones, coumarins, alkaloids, vitamins, porphyrins, stachydrine, 1-homostarchydrine[40, 120]	For menstruation and menopause.
Melaleuca alternifolia (Maid. & Bet.) Cheel *M. cojuputi* Auct.	Melaleuca, medicinal tea tree	Cineole, beta-pinene, alpha-terpineol[43]	An antiseptic, externally for thrush, vaginal infection, acne, athlete's foot, verruca, wart, insect bites, cold sore, and nits.
Melilotus officinalis (L.) Lamk. *M. arvensis* L.	Sweet clover Melilot	Flavonoids, coumarins, resin, tannins, volatile oil, dicoumarol[120]	Help varicose veins and hemorrhoids, reduce the rash of phlebitis and thrombosis.
Melissa officinalis L.	Lemon balm Balm	Citral, linalool, geraniol, citronellal, terpenic acid, tannins, polyphenolds, flavonoids, triterpenes[30, 134]	Relieve feverish colds and headaches, it has carminative, antispasmodic, stomachic, diaphoretic and sedative properties.
Mentha x piperita L. *M. spicata* L.	Peppermint Spearmint	Menthol, menthone, isomenthone, pinene, myrcene, limonene, cineole, cymene, terpinene, carvone, luteolin[30, 40, 43, 134]	Carminative, stomachic, mild antispasmodic, expectorant, antiseptic, and local anesthetic properties.

Scientific name	Common name	Major constituents	Medicinal values*
M. pulegium L.	Pennyroyal	Pulegone, isopulegone, menthol, terpenoids[30]	Digestive tonic, relieves flatulence and colic.
Monarda odoratissima Benth. *M. punctata* T. J. Howell	Horsebalm Horsemint	Oxytocics, albitocin, thymol, carvacrol[8, 120]	Emmenagogues/uterine stimulant.
Myosotis scorpioides L.	Lily of the valley	See *Convallaria majalis*	
Myrica penxylvanica Lois. *M. cerifera* L.	Bayberry	Triterpenes, flavonoids, tannins, phenols, resins[30]	Increase circulation, stimulate perspiration.
Myristica fragrans Houtt.	Nutmeg	Safrole, myristicin, lauric acid, oleic acid stearic acid, hexadecenoic acid, linoleic acid, d-camphene[8, 43]	Internally for diarrhea, dysentery, vomiting, abdominal distention, indigestion, and colic.
Myrtus communis L.	Myrtle	Tannins, flavonoids, d-pinene, eucalyptol[30, 134]	Leaves are astringent, tonic, and antiseptic.
Narcissus pseudonarcissus L.	Daffodil	Acetylated alkaloids, lectins[10, 20, 84]	Treat coughs and colds, as an emetic and purgative, inhibits HIV-1 and -2 infection of cells in vitro.
Nasturtium officinale L.	Watercress	Glucosinolate, Fe, Mn, Ca, provitamin A, vitamins B complex, C, D and E, iodine[30, 134]	As an appetizer, for digestive and gall bladder disorders, antiscorbutic, for coughs and asthma.

Scientific name	Common name	Major constituents	Medicinal values*
Nepeta cataria L.	Catnip, catmint	Carvacrol, thymol, tannins, iridoides[30, 134]	Relieve intestinal cramps, gas pains, toothache, carminative, antidiarrheal, diuretic, antipyretic.
Nymphaea alba L.	White water lily	Tinnins, nupharine, nymphaeine, resin[30]	Astringent, cardiologic and antispasmodic properties, a proprietary medicine to reduce sexual drive.
Ocimum basilicum L.	Basil	Linalool, tannins, glycosides and saponin, methyl chavicol, uronic acid, methyl eugenol[8, 40, 134]	Relieve nausea, gas, pains, and dysentery. Carminative, expectorant, antispasmodic, a mild sedative.
Ocotea bullata (Birth) E. May	Stinkwood	Safrole, iso-ocobullenone, a neolignan ketone[37, 38]	Internally for gastrointestinal complaints, colic, menstrual pain, headaches, inflammatory diseases.
Oenothera biennis L.	Evening primrose	Linoleic, gamma-linolenic acid, linoleic acid, phenolics, flavanoids, tannins[21, 102, 135,]	Treat asthmatic, arteriosclerosis, cardiovascular diseases, atopic eczema, schizophrenia, diabetic neuropathy, multiple sclerosis, antitumor.
Olea europaea L.	Olive	Quercitrin, luteolin, luteolin-7-glucoside, chlorogenic acid, oleoropine, oleasterol, leine[30]	Antiseptic, lower fever and blood pressure, laxative, improve cardiovascular functions.

Scientific name	Common name	Major constituents	Medicinal values*
Oplopanax horridus (Sm.) Miq.	Devil's club	Polyynes, sesquiterpene[135, 152]	Hypoglycemia effects, reducing serious implications caused by diabetes such as kidney and heart diseases. Treat arthritis, rheumatism, stomach and digestive problems.
Opuntia fragilis (Nutt.) Haw.	Prickly pear	Calcium exalate, tannins, mucilage, saponins[120]	Lessen pain and inflammation from gum infection and mouth sores, ease urination pain, hypoglycemia, prostatic hypertrophy.
Origanum majorana L.	Sweet marjoram	Terpins, terpineol, carotenes, caffeic acid, vitamin C, resmarinic acid, flavonoids[30, 134]	Remedy for stomach disorder, diuretic, carminative, and antispasmodic.
Origanum vulgare L. subsp. *hirtum* (Link) Ietswaart.	Marjoram, Oregano	Tannins, resin, sterols, flavonoids[30]	As an astringent and expectorant. It has antispasmodic, antiseptic, mild tonic, stomachic, and carminative effects.
Paeonia lactiflora Pall.	White peony, Bia shao	Monoterpenoid glycosides, benzoic acid, albiflorin, paeonol, astragalin, gallotannin, pentagallotannin, beta-sitosterol, benzoic acid, palmatic acid, myoinositol, pentagalloyl glucoside[30, 166]	Antispasmodic, tonic, astringent, analgesic, sedative, anti-inflammatory, prophylactic effect on stress ulcer and hypotension.
P. officinalis L.	European peony	Glycosides, tannins, anthocyanidin, peregrinine, paeonine[16, 30]	Antispasmodic, diuretic, vase constrictive, sedative properties.

Scientific name	Common name	Major constituents	Medicinal values*
P. suffruiticosa Andr.	Tree peony, Mu Dan Pi	Monoterpenoid glycosides, benzoic acid[24, 30]	Antispasmodic, tonic, astringent, analgesic.
Panax ginseng C. A. Meyer. *P. notoginseng* (Burk.) F. H. Chen *P. quinquefolium* L.	Asian ginseng Tian Qi American ginseng	Ginsenosides, acetylenic compounds, polysaccharides, panaxosides[90, 93, 115, 135, 142, 156]	A stimulant, tonic, adaptogen, diuretic, stomachic agent, carminative, aphrodisiac, healing properties, provides energy, retards the aging process. American ginseng may lower the blood pressure.
Papaver bracteatum L. *P. rhoens* L.	Iranian poppy Poppy, corn poppy	Thebaine, oripavine (source of papaverine), morphine, codeine, thebaine[30, 40]	Mild sedative to induce sleep in babies, ease cough, relieve pain, narcotic analgesic, antitussive.
Passiflora incarnata L.	Passion flower	Passiflorine, flavonoids, maltol, cyanogenic glycosides, indole alkaloids[30]	Sedative, relieve pain.
Pelargonium capitatum (L.) L'Her *P. graveolens* L'Her. ex. Aiton. *P. crispum* (L.) L'Her ex. Aiton	Wild rose geranium Lemon geranium Rose geranium	Geraniol esters, essential oil[8, 29, 134]	Internally for minor digestive ailments, kidney and bladder disorders. Externally for rashes and calloused and cracked skin.
Peltigera canina L.	Ground liverwort, Dog lichen	d-Usnic acid, thamnolic, nostoclide I and II[170]	Laxative effect, a liver tonic, treat throat congestion.

Scientific name	Common name	Major constituents	Medicinal values*
Petroselinum crispum (Mill.) Nym.	Parsley	Apiole, myristicin, pinene, apiin, havonoids, phthalides, coumarins[30, 40, 134]	Diuretic, stomachic, carminative, irritant, and emmenagogue property.
Phataris canariensis L.	Canary creeper	See *Tropaeolum majus*	
Phellodendron chinensis Schneid.	Chinese corktree (Huang Bai)	Isoquinoline alkaloids, plant steroels, sesquiterpene lactones[58]	Treat diarrhea, dysentery, jaundice, vaginal infection.
Phoradendron leucarpum (Raf) M. C. Johnston *P. serotinum* (Raf) M. C. Johnston *P. flavescens* (Push.) Nutt.	Mistletoe American mistletoe	Phoratoxin, viscotoxins, beta-phenylethylamine, tyramine[40]	Antihypertensive, sedative, increased uterine and intestinal contractions.
Phyllanthus emblica L.	Myrobalan, emblic	Vitamin C, emblicanins A and B, pedunculagin[67, 129]	Anti-inflammatory and antipyretic.
Physalis alkekengi L.	Chinese lantern	Physalin, vitamin C, alkaloids, flavoroids, sterols[30]	Diuretic, treat kidney and urinary disorder.
Phytolacca americana L.	Pokeweed	Caryophyllene, isobetanine, isoprebetanine[40]	Treat catarrh, dyspepsia, granular conjunctivitis, and rheumatism.
Picea mariana (Mill) Black *P. glauca* (Moench) Voss. *P. ables* (L.) Karst	Black spruce White spruce Norway spruce	Tannins, procyanidins, thiamine, protein, monoterpene hydrocarbons, flavonols[120]	Treat bronchial and sinus congestion, vaginal discharge, prolapsed uterus, hemorrhoids.

Scientific name	Common name	Major constituents	Medicinal values*
Pimpinella anisum L.	Anise	Anethole, fatty oil, protein, creosol, coumarin, acetylinic, flavonoids[40]	Antispasmodic, expectorant, carminative, diuretic, relieve gas pains.
Pinus mugo Turra var. Pumilio *P. palustris* Mill. *P. strobus* L. *P. albicaulis* Engelm. *P. contorta* Dougl. ex. Loud.	Dwarf mountain pine Southern pitch pine White pine Lodgepole pine	Bishomophinolenic acid, resins, mallol, borneol acetate, tannins, vit. A and C, glucose, galactose, alpha- and beta-pinenes, anthocyanin[8, 120]	Relieve fever, bronchial and nasal congestion, and improves blood flow locally, in cough syrups. Anthocyanin extraction from bark has antioxidant activity, inhibit the enzymes that cause inflammation.
Piper methysticum G. Forst	Kava-Kava	Kava lactones, piperidine alkaloids, kawine, yangonin, dihydrokawain, dihydromethysticin, methysticin[8, 30, 40]	Sedative, analgesic, anticonvulsing, muscle-relaxant, as local anesthetics, improved cognitive performance and stabilized emotions (side effect, such as rash, on long-term use and large doses).
Piper nigrum L.	English pepper	Piperine, volatile oil, protein, *l*-phyllandrene, caryophyllene[40]	Stimulant effect on digestive and circulatory system.
Plantago asiatica L. *P. psyllium* L.	Psyllium	Mucilage, linoleic, oleic, palmitic acids, fiber[24, 40]	Demulcent, laxative, antidiarrheal.
Plantago lanceolata L. *P. major* L.	Plantain	Aucubin, mucilage, carotenes, carotenes, tannin, chlorogenic acid[40, 120]	Expectorant, emollient, demulcent, vulneraria and astringent, soothing effect.

Scientific name	Common name	Major constituents	Medicinal values*
Podophyllum peltatum L.	Mayapple	Lignans, flavonoids, podophyllum resin, alpha-peltatin, beta-peltatin, podophyllotoxin (starting compound for etoposide and teniposide) Pharmaceutical drug Vepeside is extracted from this plant[49, 135, 138]	External as a poultice, lotion. Internally, used for the treatment of lung, testicular, and skin cancer and leukemia. Expel parasitic worms, typhoid fever, cholera, dysentery, hepatitis, rheumatism, kidney, bladder, and prostate problems.
Polygala senega L.	Seneca snakeroot	Triterpenoid saponins, phenolic acids, polygalitol, methyl salicylate, sterols[30, 101, 135, 149]	Treat rattlesnake bite, cough, bronchitis, and asthma.
Polygonum bistorta L. *P. multiflorum* Thunb.	Bistort root Flowery knotweed (Fu-ti)	Chrysophanic acid, anthraquinones, lecithin[30, 120]	Mildly sedative, nourishes the blood, tonic.
Populus balsamifera L. *P. candicans* L.	Poplar, Balm of gilead	Flavonoids, phenolic glycoside[120]	Antiseptic, sore throats, dry irritable coughs.
P. tremuloides Michx.	American aspen	Phenolic glycosides including salicin and populin. Tannins[30]	Fever reducing, pain relieving and anti-inflammatory.
Primula vulgaris Huds. *P. veris* L.	Primrose Cowslip	Triterpenoid sponins, flavonoids, phenols, tannins, volatile oil[30]	Internally for bronchitis, respiratory tract of infections, insomnia, anxiety, rheumatic disorder.

Scientific name	Common name	Major constituents	Medicinal values*
Prunella vulgaris L.	Heal all, selfheal	Tannins, saponins, aucubin, vitamin B, C, and K, caffeic acid, urosolic, betulinic, and deanolic acid[30, 120]	Astringent, anti-inflammatory, hemostatic, gargle for sore throat, clean wounds.
Prunus affricana L.	African prune	Liposterolic, beta-sitosterol, 3-0-glucoside, beta-sitostenone, triterpenic acids, oleanolic[6]	Antiestrogenic, anti-inflammatory, antiendermatic and immune-stimulating. actions, treat benign prostatic hyperplasia, genital urinary conditions.
P. dulcis (Mill.) D. A. Webb	Almond	Fatty oil, proteins, enzymes, vitamins, cyanogenic glycoside amygdalin[24]	Treat cough, nausea, vomiting and retching.
P. mume Siebold & Zucc.	Japanese apricot	Laetrile, cyanide, beta-carotene, thiamine, ascorbic acid, malic and citric acid, oligopeptides, polysaccharide[31, 45]	Internally for chronic coughs, externally for fungal skin infections, warts, improving blood fluidity, it has immunochemical characterization.
P. serotina J. F. Ehrb	Black cherry, wild cherry	Prunasin (a cyanogenic glycoside), benzaldehyde, coumarins, tannins, eudesmic acid[30]	Bark counters chronic dry and irritable coughs, eases indigestion and irritable bowel syndrome.
P. virginiana L.	Chokecherry	Prunin, malic acid, cyanogenic glycoside, scopoletin, coumarins[120]	Cough remedies, sedative, pectoral.
Pueraria lobata (Willd) Ohwi	Kudzu	Daidzin, daidzein, isoflavonoids, puerarin, sterol[30]	Internally for colds, influenza, feverish illness, thirst in diabetes, externally for snake bite.

Scientific name	Common name	Major constituents	Medicinal values*
Pygeum africanum L.	Pygeum	Beta-sitosterol, estrogens, ferulic acids, n-tetracosanol triterpenes[30]	Reduce benign prostate hyperplasia (BPH).
Pyrethrum cinerarifolium L.	Pyrethrum	See *Chrysanthemum cinerarifolium*	
Quercus robur L.	White oak	Tannins, cutins, suberins[30]	Sore throats, tonsillitis, astringent.
Ranunculus occidentalis Nutt.	Buttercup	Anemonin[120]	A stimulant, externally relieve chronic sciatica.
R. ficaria L.	Pilewort	See *Collinsonia canadensis*	
Raphanus sativus L.	Radish	Glucosinolates, arginine, histidine, vitamin A, B and C[120]	Leaf is diuretic, laxative, root for hemorrhoids.
Rhamnus catharticus L. *R. frangula* L. *R. purshianus* L.	Buckthorn Alder buckthorn Cascara sagrada	Anthraquinone glycosides, phenolic flavonols, pectin, vitamin C, glucofrangulin A and B, frangulin A and B, emodin, chrysophanol, physcion[40, 135, 154]	Laxative, diuretic, constipating, astringent, antibacterial, purgative, digestive complaints.
Rheum palmatum L. *R. officinal* Baill. *R. tanguticum* L.	Chinese rhubarb Rhubarb Rhubarb	Cinnamic acid, gallic acid, emodin, rhein, rhein anthrones, catechin, anthraquinone compounds, tannin, calcium oxalate[30, 40, 134]	Treat diarrhea, stimulate appetite, chronic constipation, laxative, cathartic.
Rhodiola rosea (L.) Scop.	Roseroot	See *Sedum rosea*	

Scientific name	Common name	Major constituents	Medicinal values*
Rhus radicans L.	Poison ivy	Toxicodendrol, urushiol, 3-n-pentadecylcatechol[120]	Sympathetic stimulant, restore nerve function, facial neuritis, ulcerated sores on the lips, mouth, and nasal membranes.
R. toxicodendron L.	Roseroot	See *Sedum rosea* subsp. *integrifolium*	
Ribes nigrum L. *R. lacustre* (Pers.) Poir.	Black currant Black gooseberry	Anthocyanosides, antiprotease, tannins vitamin B_1, B_2, C, and P, citric acid, pectin[120]	Diuretic and diaphoretic properties, urinary infection, rheumatism and diarrhea, infusion of gooseberry leaf to lessen the pain associated with female menstrual cycle.
Robertium macrorrhizum Pic.	Geranium	See *Geranium macrorrhizum*	
Rosa canina L. *R. damascena* Mill.	Dog rose Damask rose	Vitamin C, B complex carotenes, malic and citric acids, pectin, geraniol, *l*-citronellol[120, 131, 134]	Tonic, astringent, mild diuretic and laxative effect. Excellent source of vitamin C when it's fresh.
Rosmarinus officinalis L.	Rosemary	Cineole, camphor and ursolic acid borneol, picrosalvin, rosmarinic acid, tannins, apigernin, diosmin[30, 40, 134]	Carminative, antispasmodic, antirheumatic liniments and ointments.
Rubus chamaemorus L.	Cloudberry	Tocopherol, benzoic acid, salicylic acid, ascorbic acid, vitamin C[96, 120]	Laxative, tonic, treat cough and fever.
R. fruiticosus L.	Blackberry	Tannins, organic acids, vitamin C[24]	Mild astringent, antiseptic, antifungal, diuretic and tonic properties.

Scientific name	Common name	Major constituents	Medicinal values*
R. idaeus L.	Raspberry	Tannins, pectin, vitamin C. Fruit: anthocyanins, pectin, flavonoids, gallic acid[30]	Treat diarrhea, antispasmodic for astringent, and diuretic. As a tonic.
Rúmex acetosella L. *R. obtusifolia* L.	Sorrel Dock	Oxaltes, anthraquinones, phanol, physcion, tannic acid[21, 40, 120]	Antiseptic, laxative, rheumatic pains.
Ruscus aculeatus L.	Butcher's broom	Saponin glycosides (ruscogenin, neoruscogenin)[30]	Diuretic and mildly laxative.
Ruta graveolens L.	Rue, herb of grace	Rutoside, coumarins, rutaverine, arborinine, methyl ketones (toxic), rutin[117, 134]	Antispasmodic, menstruation, sedative, cholagogic, diaphoretic, anthelmintic.
Salix alba L. *S. discolour* Muhlenb. *S. caprea* L.	White willow Pussy willow Pussy willow	Salicin, tannins, phenolic, flavonoid, glycosides, salicortin, triandrin[30, 120]	Antipyretic, diaphoretic, antirheumatic, and analgesic.
Salvia officinalis L. *S. sclarea* L.	Sage Clary sage	Thujone, borneol, cineole, camphor, salvin, tannin, fumaric, malic, and oxalic acid[40, 95, 134]	Carminative, lower fever, antiseptic, antifungal, astringent, diuretic, carminative, antidiarrheal, antispasmodic.
Sambucus nigra L. *S. canadensis* L.	Elderberry	Flower: flavonoids, phenolic, triterpenes, sterols; leaf: cyanogenic glycosides; fruit: flavonoids, anthocyanins, vitamin A and C[30]	Increases sweating, diuretic, anti-inflammatory.

Scientific name	Common name	Major constituents	Medicinal values*
S. racemosa L.	Elder	Rutin, tannins, cyanogenic, glucans, glycosides, baldrianic acid[120]	Antiulcer, antimutagens, anticoagulant.
Sanguinaria canadensis L.	Bloodroot	Benzophenanthridine alkaloids, sanguinarine, isoquinoline alkaloids, pharmaceutical drug Viadent is extracted from this plant[134, 135, 156]	Treat rheumatism, asthma, bronchitis, laryngitis and fevers. Antimicrobial and antitumor, with bactericidal and bacteriostatic properties.
Sanicula marilandica L.	Black snakeroot	Saponins, allantoin, volatile oil, tannins, chlorogenic and rosmarinic acid[24]	Healing wounds, treat internal bleeding.
Santalum album L.	Sandalwood	Dihydro-beta-agarofuran, curcumin, sequiterpene hydrocarbons, dendrolasin, santalols[8, 40]	Internally for genitourinary disorder, fever, sunstroke, externally for skin disorder.
Saponaria officinalis L.	Soapwort	Saponins, resin, sapogenin, sterol, trace of volatile oil[63]	As an expectorant, bronchitis, coughs, asthma, rheumatic and arthritic pain.
Sarothamnus scoparius L.	Broom	See *Cytisus scoparius*	
Sassafras albidum (Nutt.) Nees	Sassafras	Safrole, camphor, methyleugenol, boldine, cinnamolaurine, isoboldine, norboldine[40, 62]	Carcinogenic and mutagenic activities, internally for gastrointestinal complaints, colic, menstrual pain, externally for sore eye, lice, and insect bites.
Satureja hortensis L. *S. montana* L.	Savory	Carvacrol, cymene, linalool, thymol, ursolic acid[30, 40, 134]	Astringent, carminative, antiseptic.

Scientific name	Common name	Major constituents	Medicinal values*
Schisandra chinensis (Turcz.) Ball	Schizandra (wu wei zi)	Lignans, phytosterols, vitamin C and E[30]	Tonic, adaptogenic, protects liver.
Scutellaria baicalensis Georgi. *S. lateriflora* L. *S. galericulata* L.	Baical skullcap Virginia skullcap Skullcap	Scutellarin, flavonoids such as baicalin, baicalein, wogonin, benzoic acid, catalpol, tannins, sterols such as beta-sitosterol, camphesterol, stigmasterol[30, 120, 167]	Sedative and antispasmodic, prevent epileptic seizures nerving tonic, antispasmodic, antiallergic.
Sedum acre L.	Stonecrop, smallhouseleek	Sedacrine, n-methyl anabasine, sedinine, sedacryptine, flavanol glycosides[120, 140]	Insomnia, depressant, hemorrhoidal pain, for excessive menstrual flow especially around menopause.
S. rosea (L.) Scop. subsp. *interifolium*	Roseroot	Rhodioloside, flavanol glycosides[135, 170]	As an adaptogen. Improve learning and memory and reduce stress, anticancer properties, stimulate the central nervous system and protect the liver.
Sempervivum tectorum L.	Houseleek, hens and chicks	Tannins, mucilage, malic and formic acid, flavonoids, lycophilized extract, polymeric poly-phenols, oligomeric polyphenols[1, 17, 30, 140]	Antioxidant, leaf and juice are used for cooling and astringent effect, leaf chewed to relieve toothache. Internal use is not advised.
Senecio vulgris L.	Groundsel	Volatile oil, seneciphyline, jacoline, pyrrolizidine, senecionine, tinnins, resin[23]	As a poultice ointment or lotion to relieve pain and inflammation.
Senna alexandrina L	Alexandrina	See *Cassia angustifolia*	

Scientific name	Common name	Major constituents	Medicinal values*
Serenoa repens (Bartr.) Small.	Saw palmetto	Sitosterol, tannins, saponins, liposterolic, beta-sitosterol and its glucoside, stigmasterol, campesterol[7, 40]	Estrogenic properties, antiandrogenic action, reduce benign prostate hyperplasia (BPH), treat irritations of the bladder and urethra.
Sesamum indicum L. *S. orientale* L.	Sesame	Phenol, lignan, 55% seed oil including oleic and linoleic acids, protein, vitamins B, E, folic acid[32, 113]	An antioxidant, antitumor, antimitotic, antiviral, prevent breast cancer. Internally for premature hair loss and graying, strengthens bones and teeth.
Shepherdia canadensis (L.) Nutt.	Buffalo berry	Tannins (shephagenins A and B)[169]	Inhibitory activity against HIV-1 reverse transcriptase.
Silybum marianum (L.) Gaertn.	Milk thistle	Flavonolignans, silibinin, silymarin[114, 137, 148]	Treat hepatitis, cirrhosis, regeneration of diseased liver, liver poisoning, digestion.
Simmondsia chinensis (Link) C. Schneid	Jojoba	Protein, fatty acids (lysine, histidine, arginine, glutamic acid, oil (liquid wax), flavonoid[40]	Externally for dry skin and hair, psoriasis, acne and sunburn, treat poison ivy.
Solidago virgaurea L. *S. canadensis* L.	Goldenrod	Tannins, saponins, polygalic acid, cariaester, inulin, salicylic acid[40]	Alleviate intestinal gas, relieve fever.
Sorbus aucuparia L.	Mountain ash	Organic acids, tannins, vitamin C, pectin[120]	Source of vitamin C, astringent for hemorrhoids and diarrhea.
Spirea ulmaria L.	Meadowsweet	See *Filipendula ulmaria*	

Scientific name	Common name	Major constituents	Medicinal values*
Stachys officinalis (L.) Trev.	Betony, woundwort	Tannins, stachydrine, betonicine, betaine, choline[30, 40, 134]	Stop bleeding from open wound, a tonic, stomachic, and antiseptic.
Stellaria media (L.) Vill.	Chickweed	Triterpenoid saponins, vitamin C, coumarins, flavonoiids, linolenic acid, octadecatetraenic acid[30, 40]	Treat internal and external inflammations, irritated skin.
Stevia rebaudiana (Bertoni) Bertoni	Stevia Sweet herb of Paraguay	Stevioside, diterpenoid glycosides, steviobioside, rebaudiosides A and B, caffeic and chlorogenic acid, polysaccharides[43]	Treatment for diabetes and hypoglycemia.
Symphytum officinale L.	Comfrey	Tannins, mucilage, allantoin, cadaverine, putrescine, spermidine, pyrrolizidine alkaloids, rosmarinic acid, asparagine[30, 66, 134]	Accelerates healing of surface wounds and sores and broken bones.
Syringa vulgaris L.	Lilac	Lilacin, ligustrin, lignans, hydroxyphenylethanol glycosides[28, 120]	Tonic, neurotrophic, adaptogenic, immune stimulating, antimicrobial from leaf.
Syzygium aromaticum (L.) Merr. & Perry	Clove	Sesquiterpenes, volatile oil (eugenol), tannins, gum[43, 172]	Internally for gastroenteritis and intestinal parasites, externally for toothache and insect bites. In China, used for nausea, vomiting, hiccups, stomach chills, impotence, therapeutic antiherpes simplex virus.

Scientific name	Common name	Major constituents	Medicinal values*
Tagetes lucida Cav.	Mexican mint marigold	Coumarin derivatives, resin, gallic acid, tannins, glucose, pectin, gum[134]	Diarrhea, indigestion, nausea, externally for smooth muscle, scorpion bites, and to remove ticks.
Tanacetum parthenium (L.) Schultz.	Feverfew	See *Chrysanthemum parthenium*	
T. cinerarifolium L.	Pyrethum	See *Chrysanthemum cinerarifolium*	
T. vulgare L.	Tansy	Thujone, borneol, camphor[8, 120]	Insect repellent, antispasmodic, vermifuge, emmenagogues.
Taraxacum officinale G. H. Weber ex Wigg.	Dandelion	Taraxacin, taraxerol, taraxasterol, inulin, gluten, gum, choline, levulin, pulin, tannins, provitamin A, vitamin B and C[85, 86, 135]	Tonic, diuretic, diuretic, stimulate appetite, digestion, treat fever, insomnia, jaundice eczema, rheumatism, and arthritis.
Taxus x media Rehd. *T. brevifolia* Nutt.	Yew	Taxol (paclitaxel), resin[135, 159, 160]	Treat cancer, gout and rheumatism, arthritis.
Thymus citriodorus (Pers.) Schreb.	Lemon thyme	Thymol, carvacrol, saponins[40]	Sedative, antiseptic, diuretic, expectorant.
Thymus vulgaris L	Thyme	Thymol, tannins, carvacrol, saponins, apigenin, luteolin[30, 40, 134]	Antispasmodic, antitussive, relieve coughing.

Scientific name	Common name	Major constituents	Medicinal values*
Tillia cordata Mill. *T. europœa* L.	Linden, small leaved lime	Mucilage, tannins, flavonoid, caffeic acid, taraxerol, tiliadine, vanillin, phytosterols, mucilage, glycosides[30, 40]	Diaphoretic, antispasmodic, diuretic, mild sedative.
Tribulus terrestris L.	Puncture vine	Sitosterol, tannins, saponins, tribulusamide A and B, N-trans-feruloyltyramine, terrestriamide, N-trans-coumaroyltyramine[89]	Estrogenic properties, antiandrogenic action, reduce benign prostate hyperplasia (BPH).
Trifolium incarnatum L.	Clover (crimson)	Flavonoids, salicylic acid[24]	Treat skin conditions, spasmodic coughs.
T. pratense L.	Clover (red)	Tannins, phenolic glycosides, p-coumaric acid, silicic acid, caffeic acid, salicylic acid[40, 120]	Remedy for sore throat, colds, coughs, bronchitis, diarrhea, chronic skin.
Trigonella foenum-gracecum L.	Fenugreek	Protein, linoleic, oleic linolenic, and palmitic acids, trigonelline, choline, coumarin, nicotinic acid[40, 78, 134]	Reduce total cholesterol and triglycerides without affecting the HDL, reduce blood sugar.
Tropaeolum majus L.	Canary creeper Nasturtium	Glucocyanates, myrosin, oxalic acid, vitamin C[14, 30]	Antibiotic, clear nasal and bronchial congestion, juice treat scrofula.
Tussilago farfara L.	Colisfoot	Mucilage, sterols, pigments, inulin, gallic, malic, tartaric acids, tannins, pyrrolizidine alkaloids[30, 40]	Expectorant, demulcent, astringent, anti-inflammatory.

Scientific name	Common name	Major constituents	Medicinal values*
Ulmus rubra Muhl. *U. procera* L.	Slippery elm, sweet elm English elm	Tannins, mucilage, cholesterol, campesterol, beta-sitosterol, pentoses[30, 40]	Astringent, demulcent, anti-inflammatory.
Umbelluslaria california Nutt.	California laurel Bay laurel	1,8-cineole (eucalyptol), thujene, umbellulone, cis-sabinene hydrate, trans-sabinene hydrate, flavonoids[56, 57, 109, 119]	Internally for headache, neuralgia, intestinal cramps, and gastroenteritis, externally for headache and fainting.
Uncaria tomentose (Willd.) DC. *U. guianensis* (Aubl.) Gmel.	Cat's claw	Oxindole alkaloids such as pentacyclic oxindole alkaloids, tetracyclic oxindole alkaloids[79, 110, 164]	Stimulating immunologic system, antiviral, anti-inflammatory and antitumor. An antioxidant.
Urtica dioica L. *U. urens* L.	Stinging nettle	Phytosterols (stigmast-4-3-one), stigmasterol, beta-sitosterol, polysaccharides (*Urtica dioica* agglutinin), aretylcholine, serotonin, quercitin, histamine, choline, glucoquinone[7, 30, 120]	Treat benign prostatic hyperplasia, hair tonic and growth stimulant, use in anti-dandruff shampoo.
Vaccinium macrocarpon Ait. *V. vitis-idaea* L.	Cranberry, Mountain cranberry Cowberry, foxberry	Anthocyanosides, hippuric acid, vitamin C, vitamin A[36, 108, 112,126, 135]	Treat urinary complaint including infection and stones, as a preventive for urinary infections. Juice may have antioxidant effect on clogged heart arteries, thus reducing cardiovascular disease.

Scientific name	Common name	Major constituents	Medicinal values*
V. myrtilloides Michx. *V. myrtillus* L. *V. oreophilum* Rydb.	Blueberry Bilberry	Tannins, arbutin, iridoids, insulins, anthocyanosides, myrtocyan[47, 112, 135]	Strengthening cardiovascular system, improve vision, treat diabetes, digestive disorder, urinary disorder. It has anti-oxidant properties.
Valeriana officinalis L.	Valerian	Essential oil, valtrate, valepotriates, bornyl esters, alkaloids, isovaltrate[30, 40]	Sedative for nervous disorders, antispasmodic.
Verbascum thapsus L.	Mullein	Saponins, mucilage, pigment, tannins[120]	Demulcent, reduces fluid accumulation treat for respiratory ailments, colds, asthma, bronchitis.
Veronica officinalis L.	Speedwell	Tannins, essential oil, aucuboside, vitamin C, flavonoids, acetopenone glucoside[30]	Diuretic, expectorant.
Vetiveria zizanioides L. (Nash.)	Vetiver	Vetivene, vetivenols, vetivenyl, vetivenates, vetivenic acid, palmitic acid, benzoic acid[8, 43]	Internally for nervous and circulatory problems, externally for lice.
Viburnum opulus L.	Highbush cranberry	Hydroquinones, coumarins, tannins, resin[30, 120]	Antispasmodic, sedative, an astringent.

Scientific name	Common name	Major constituents	Medicinal values*
Vinca minor L.	Periwinkle	Alkaloids, tannins, saponins, pectin (source of vincristine) phenolic resin, oleoresin, aldehydes, dimeric indole alkaloids, vinblastine, sesquiterpenes[40, 161]	Treat diabetes, leukemia, reduce blood pressure. Anticancer, Hodgkin's disease, hypotensive, sedative and tranquillizing.
Viola tricolor L.	Pansy	Saponins, mucilage, violin, salicylic compounds, tannins[30, 134]	Diuretic, diaphoretic, tonic, anti-inflammatory and blood purifying properties, skin ailments, expectorant for loosening phlegm.
Viscum album L.	European mistletoe	Galactoside-specific lectin, lignans, viscotoxin, choline, alkaloids, resin, acetylcholine, protein, flavonoids, caffeic acid, viscin, carotenoids[30, 40, 53]	Lower blood pressure, stimulate heart action, and treat arteriosclerosis.
Vitex agnus-castus L.	Chaste tree	Flavonoids, iridoids, agnuside, aucubin, cineol, casticin, viticine[30, 162]	Treatment of mastopathy, premenstrual, syndrome and luteae insufficiency. Regulates hormones, progesterogenic, increases breast milk production.
V. negundo L.	Chinese chaste tree	Iridoids, flavonoids, volatile oil[30]	For malaria, poisonous bites, arthritis and breast cancer.

Scientific name	Common name	Major constituents	Medicinal values*
Vitis labrusca L. *V. vinifera* L.	Grape	Fatty acids from seed (70% linoleic, 18% oleic, 8% palmitic, and 5% stearic acid), flavonoids, malic acid, anthocyanins, tartaric, tannins, monoterpene glycosides[30, 68]	Antioxidant, internally for varicose veins, excessive menstruation, menopausal syndrome, hemorrhage, and hypertension.
Withania somnifera Dunal.	Ashwagandha, Indian ginseng	Alkaloids, steroidal lactones (with anolides), anahygrine, beta-citosterol[30, 133]	Adaptogenic, tonic, sedative.
Yucca aloifolia L.	Yucca	Saponins, glycoside	Treat arthritis.
Yucca glauca L.	Soapweed	Astringent saponin	Ease arthritic pain, used in cosmetics, soap.
Zanthoxylum americanum Mill	Prickly ash, toothache tree	Chelerythrine, herclavin, asarinin, neoherculin, tannins, resins[30]	Circulatory stimulant, increases sweating.
Zea mays L.	Corn	Saponins, fatty acids, tannins, resin, maysin, essential oil, thiamin, mucilage[30]	Cystitis, urethritis, prostatitis, urinary stones.
Zingiber officinale Roscoe	Ginger	Volatile oil, gingerol, shogaols, *l*-zingiberene[18, 30, 124, 134]	Antisemitic, carminative, circulatory stimulant, anti-inflammatory, antiseptic.

*This information should not be used for the diagnosis, treatment, or prevention of diseases in humans. The information contained herein is in no way intended to be a guide to medical practice or a recommendation that herbs be used for medicinal purposes. The information is presented here mainly for educational purposes and should not be used to promote the sale of any product or replace the service of a physician.

References

1. Abram, V. and M. Donko. 1999. Tentative identification of polyphenols in *Sempervivum tectorum* and assessment of the antimicrobial activity of *Sempervivum* L. J. Agric. Food Chem. 47: 485-489.

2. Ahlemeyer, B. and J. Krieglstein. 1998. Neuroprotective effects of *Ginkgo biloba* extract. In: Lawson, L. D. and R. Bauer (eds.). Phytomedicines of Europe, Chemistry and Biological Activity. Am. Chem. Soc., Washington, DC. p. 210-220.

3. Ahn, K., M. Hahm, E. Park and H. Lee. 1998. Corosolic acid isolated from the fruit of *Crataegus pinnatifida* var. psilosa is a protein kinase C inhibitor as well as a cytotoxic agent. Planta Medica 64: 468-470.

4. Arenz, A., M. Klein, K. Fiehe, J. Gross, C. Drewke, T. Hemscheidt and E. Leistner. 1996. Occurrence of neurotoxic 4′-O-methylpyridoxine in *Ginkgo biloba* leaves, Ginkgo medications and Japanese Ginkgo food. Planta Medica 62: 548-551.

5. Arkhangel'skaya, A. D. 1935. The Carmine-Producing Coccids (*Margarodes*) of Middle Asia and Species of an Allied Genus *Neomargarodes*. Tashkent, Izd. Komit. Nauk USSR. 1935. 36 p.

6. Awang, D. 1997. Saw palmetto, african prune and stinging nettle for benign prostatic hyperplasic (BPH), Can. Pharmaceutical J. Nov., 1997. p. 37-40, 43-44, 60.

7. Baguena, C. L. 1942. Notes on some longicorns harmful to cultivated trees, including Cacao and Coffee, in the Spanish territories in the gulf of Guinea. Publ. Direcc. Agric. Territ. Exp. Golfi. Guinea 6: 39-91.

8. Balandrin, M. F. and J. A. Klocke. 1988. Medicinal, aromatic, and industrial materials from plants. In: Bajaj, Y. P. S. (ed.). Biotechnology in Agriculture and Forestry 4. Medicinal and Aromatic Plants I. Springer-Verlag Co., New York. p. 3-36.

9. Baresova, H. 1988. *Centaurium erythraea* Rafn: micropropagation and the production of secoiridoid glucosides. In: Bajaj, Y. P. S. (ed.). Biotechnology in Agriculture and Forestry 4. Medicinal and Aromatic Plants I. Springer-Verlag Co., New York. p. 350-366.

10. Bastida, J., S. Bergonon, F. Viladomat and C. Codina. 1994. Alkaloids from *Narcissus primigenius.* Planta Medica 60: 95-96.

11. Bauer, R. 1998. *Echinacea*: biological effects and active principles. In: Lawson, L. D. and R. Bauer (eds.). Phytomedicines of Europe. Chemistry and Biological Activity. Am. Chem.

Soc. Washington, DC. 324 p.

12. Bauer, R. and H. Wagner. *Echinacea* species as potential immunostimulatory drugs. In: H. Wagner and N. R. Farnsworth (eds.). Economical and Medicinal Plant Research. Vol. 5. Academic Press, London. p. 253-321.

13. Bauer, R., P. Remiger and E. Alstat. Alkamides and caffeic acid derivatives from the roots of *Echinacea tennesseensis*. Planta Medica 56: 533-534.

14. Bello, A. 1967. Data on the geographical distribution of *Halenchus dumnonicus* (Nematoda). Boletin-de-la-Real-Cociedad-Espanola-de-Historia-Natural-Biologica 65:1-2, 81-82.

15. Bergner, P. 1997. The Healing Power of Echinacea, Goldenseal and Other Immune System Herbs. Prima, Rocklin, CA. 322 p.

16. Bisset, N. G. 1994. Herbal Drugs and Phytopharmaceuticals. CRC Press, London. 566 p.

17. Blazovics, A., L. Pronal, J. Feher, A. Kery and G. Petri. 1993. A natural antioxidant extract from *Sempervivum tectorum*. Phytotherapy Res. 7: 95-97.

18. Bordia, A., S. K. Verma and K. C. Srivastava. 1997. Effect of ginger (*Zingiber officinale* Rosc.) and fenugreek (*Trigonella foenumgraecum* L.) on blood lipids, blood sugar and platelet aggregation in patients with coronary artery diseses. Prostaglandins Leukotrienes and Essential Fatty Acids. 56: 379-384.

19. Bown, D. 1987. *Acorus calamus* L: a species with a history. Aroideana (International Aroid Society, South Miami Fla) 10: 11-14.

20. Bracco, V. 1973. Determination of polycyclic aromatic hydrocarbons: technique and application to coffee oil. Rivista Italiana delle Sostanze Grasse 50: 166-176.

21. Bremness L. 1994. The Complete Book of Herbs. Dorling Kindersley Ltd., London. 304 p.

22. Brinda, S., U. W. Smitt, V. George and P. Pushpangadan. 1998. Angiotensin converting enzyme (ACE) inhibitors from *Jasminum azoricum* and *Jasminum grandiflorum*. Planta Medica 64: 246-250.

23. Brown, M. S. and R. J. Molyneux. 1996. Effects of water and mineral nutrient deficiencies on pyrrolizidine alkaloid content of *Senecio vulgaris* flowers. J. Sci. Food Agric. 70: 209-211.

24. Bunney, S. 1992. The illustrated encyclopedia of herbs, their medicinal and culinary uses. Chancellor Press. London. 320 p.

25. Caballero, R., M. Haj-Ayed, J. F. Galvez, P. J. Hernaiz and M. H. Ayed. 1995. Yield components and chemical composition of some annual legumes and oat under continental Mediterranean conditions. Agricoltura Mediterranea 125: 222-230.

26. Canard, D., O. Perru, V. Tauzin, C. Devillard and J. P. Bonhoure. 1997. Terpene composition variations in diverse provenances of *Cedrus liboni* (A.) Rich. and *Cedrus atlantica* Manet. Trees Structure and Function 11: 504-510.

27. Cavallini, A., L. Natali and I. Castorena Sanchez. 1991. *Aloe barbadensis* Mill. (=*A. vera* L.). In: Bajaj, Y. P. S. (ed.). Biotechnology in Agriculture and Forestry 15. Medicinal and Aromatic Plants III. Springer-Verlag Co., New York. p. 95-106.

28. Chapple, C. C. S. and B. E. Ellis. 1991. *Syringa vulgaris* L. (Common lilac): in vitro culture and the occurrence and biosynthesis of phenylpropanoid glycosides. In: Bajaj, Y. P. S. (ed.). Biotechnology in Agriculture and Forestry 15. Medicinal and Aromatic Plants III. Springer-Verlag Co., New York. p. 478-497.

29. Charlwood, B. V. and K. A. Charlwood. 1991. *Pelargonium* spp. (Geranium): in vitro culture and the production of aromatic compounds. In: Bajaj, Y. P. S. (ed.). Biotechnology in Agriculture and Forestry 15. Medicinal and Aromatic Plants III. Springer-Verlag Co., New York. p. 339-352.

30. Chevallier, A. 1996. The Encyclopedia of Medicinal Plants. Dorling Kindersley Ltd. London. 336 p.

31. Chuda, Y., H. Ono, M. Ohnishi-Kameyama, K. Matsumoto, T. Nagata and Y. Kikuchi. 1999. Mumefural, citric acid derivative improving blood fluidity from fruit-juice concentrate of Japanese apricot (*Prunus mume* Sieb. Et Zucc.). J. Agric. Food Chem. 47: 828-831.

32. Chung, C. H., Y. J. Yee, D. H. Kim, H. K. Kim and D. S. Chung. 1995. Changes of lipid, protein, RNA and fatty acid composition in developing seasame (*Sesamum indicum* L.) seeds. Plant Sci. Limerick 109: 237-243.

33. Collin, H. A. and S. Isaac. 1991. *Apium graveolens* L. (Celery): in vitro culture and the production of flavors. In: Bajaj, Y. P. S. (ed.). Biotechnology in Agriculture and Forestry 15. Medicinal and Aromatic Plants III. Springer-Verlag Co., New York. p. 73-94.

34. Cordeiro, M. C., M. S. Pats and P. E. Brodelius. 1998. *Cynara cardunculus* subsp. *flavescens* (cardoon): in vitro culture, and the production of cyprosins—milk-clotting enzymes. In: Bajaj, Y. P. S. (ed.). Biotechnology in Agriculture and Forestry 41. Medicinal and Aromatic Plants X. Springer-Verlag Co., New York. p. 132-153.

35. Corto-Cositet, M. F., L. Chapuis and J. P. Delbecque. 1998. *Chenopodium album* L. (fat hen): in vitro cell culture, and production of secondary metabolites (phytosterols and ecdysteroids). In: Bajaj, Y. P. S. (ed.). Biotechnology in Agriculture and Forestry 41. Medicinal and Aromatic Plants X. Springer-Verlag Co., New York. p. 97-112.

36. Cunio, L. 1994. *Vaccinium myrtillus*. Australian J. Medical Herbalism 5: 81-85.

37. Drewes, S. E., M. M. Horn, B. M. Sehlapelo, N. Ramesar, J. S. Field, R. S. Shaw and P. Sandor. 1995. Iso-ocobullenone and a neolignan ketone from *Ocotea bullata* bark. Phytochemistry 38: 1505-1508.

38. Drewes, S. E., M. M. Horn, and S. Mavi. 1997. *Cryptocarya liebertiana* and *Ocotea bullata*—their phytochemical relationship. Phytochemistry 44: 437-440.

39. Ducrey, B., J. L. Wolfender, A. Marston and K. Hostenmann. 1955. Analysis of flavonol glycosides of thirteen *Epilobium* species (Onagraceae) by LC-UV and thermospray LC-MS. Phytochemistry 38: 129-137.

40. Duke, J. A. 1985. CRC Handbook of Medicinal Herbs. CRC Press, Inc., Boca Raton, FL. 677 p.

41. Duke, J. A. 1992. Handbook of Biologically Active Phytochemicals and Their Activities. CRC Press, Boca Raton, FL.

42. Duke, J. A. 1992. Handbook of Phytochemical Constituents in GRAS Herbs, Plant Foods and Medicinal Plants. CRC Press, Boca Raton, FL.

43. Duke, J. A. and J. L. duCellier. 1993. CRC Handbook of Alternative Cash Crops. CRC Press, London. 536 p.

44. Erdelmeler, C. A., J. Clinatl, H. Rabenau, H. W. Doer, A. Biber and E. Koch. 1996. Antiviral and antiphlogistic activities of *Hamamelis virginiana* bark. Planta Medica 62: 241-245.

45. Fang, T. T., J. H. Huang, T. T. Fang, J. H. Huang. 1999. Extraction, fractionation and identification of bitter oligopeptides and amino acids in mei fruit (*Prunus mume* Sieb. et Zucc.). J. Beijing Forestry Univ. 21: 61-71

46. Farnsworth, N. R. 1973. Importance of secondary plant constituents as drugs. In: L. P. Miller (ed.). Phytochemistry vol. 3. Inorganic Elements and Special Groups of Chemicals. Van Nostrand, New York. p. 351-380.

47. Fraisse, D., A. Carnat and J. L. Lamaison. 1996. Polyphenolic composition of the leaf of bilberry. Am. Pharm. Fr. 54: 280-283.

48. Fuentes-Granados R. G., M. P. Widrlechner and L. A. Wilson. 1998. An overview of *Agastache* research. J. Herbs, Spices & Medicinal Plants 6: 69-97.

49. Fujii, Y. 1991. *Podophyllum* spp.: in vitro regeneration and the production of podophyllotoxins. In Bajaj, Y. P. S. (ed.). Biotechnology in Agriculture and Forestry 15. Medicinal and Aromatic Plants III. Springer-Verlag Co., New York. p. 362-375.

50. Furmanowa, M. and J. Guzewska. 1989. *Dioscorea*: in vitro culture and the micropropagation of diosgenin-containing species. In: Bajaj, Y. P. S. (ed.). Biotechnology in Agriculture and Forestry 7. Medicinal and Aromatic Plants II. Springer-Verlag Co., New York. p. 162-184.

51. Furmanowa, M., D. Sowinska and A. Pietrosiuk. 1991. *Carum carvi* L. (Caraway): in vitro culture, embryogenesis, and the production of aromatic compounds. In: Bajaj, Y. P. S. (ed.). Biotechnology in Agriculture and Forestry 15. Medicinal and Aromatic Plants III. Springer-Verlag Co., New York. p. 177-192.

52. Furuya, T. and T. Yoshikawa. 1991. *Carthamus tinctorius* L. (Safflower): production of vitamin E in cell culture. In: Bajaj, Y. P. S. (ed.). Biotechnology in Agriculture and Forestry 15. Medicinal and Aromatic Plants III. Springer-Verlag Co., New York. p. 142-155.

53. Gabius, H. J. and S. Gabius. 1998. Phytotherapeutic immunomodulation as a treatment modality in oncology: lessons from research with mistletoe. In: Lawson , L. D. and R. Bauer (eds.). Phytomedicines of Europe, Chemistry and Biological Activity. Am. Chem. Soc., Washington, DC. p. 278-286.

54. Genest, K. and W. Hughes. 1969. Natural products in Canadian pharmaceuticals IV. Can. J. Pharm. Sci. 4: 41-45.

55. Goncharva, N. P. and A. I. Glushenkova. 1990. Lipids of Elaegnus fruit. Hort. Abst. 61: 3566.

56. Goralka, R. J. L. and J. H. Langenheim. 1995. Analysis of foliar monoterpenoid content in the California bay tree, *Umbellularia california*, among populations across the distribution of the species. Biochem. Systematics and Ecology 23: 439-448.

57. Goralka, R. J. L., M. A. Schumaker and J. H. Langenheim. 1996. Variation in chemical and physical properties during leaf development in California bay tree (*Umbellularia californica*): predictions regarding palatability for deer. Biochem Systematica and Ecology 24: 93-103

58. Gray, A. I., P. Bhandari and P. G. Waterman. 1988. New protolimonoids from the fruits of *Phelloderdron chinense*. Phytochemistry 27: 1805-1808.

59. Guedon, D., P. Abbe and J. L. Lamaison. 1993. Leaf and flower head flavonoids of *Achillea millefolium* L. subspecies. Biochem. Syst. Ecol. 21: 607-611.

60. Halstead, B. W. and L. L. Hood. 1942. *Eleutherococcus senticosus*, siberian ginseng: an introduction to the concept of adaptogenic medicine. Oriental Healing Arts Inst. Long Beach, CA. 94 p.

61. Heale, J. B., T. Legg and S. Connell. 1989. *Humulus lupulus* L.: in vitro culture: attempted production of bittering components and novel disease resistance. In: Bajaj, Y. P. S. (ed.). Biotechnology in Agriculture and Forestry 7. Medicinal and Aromatic Plants II. Springer-Verlag Co., New York. p. 264-285.

62. Heikes, D. L. 1994. SFE with GC and MS determination of safrole and related allybenzenes in sassafras teas. J. Chromatographic Sci. 32: 253-258.

63. Henry, M. 1989. *Saponaria officinalis* L.: in vitro culture and the production of triterpenoidal saponins. In: Bajaj, Y. P. S. (ed.). Biotechnology in Agriculture and Forestry 7. Medicinal and Aromatic Plants II. Springer-Verlag Co., New York. p. 431-442.

64. Henry, M., A. M. Edy, P. Desmarest and J. Du Manuir. 1991. *Glycyrrhiza glabra* L. (Licorice): cell culture, regeneration, and the production of glycyrrhizin. In: Bajaj, Y. P. S. (ed.). Biotechnology in Agriculture and Forestry 15. Medicinal and Aromatic Plants III. Springer-Verlag Co., New York. p. 270-282.

65. Heptinstall, S. and D. V. C. Awang. 1998. Feverfew: a review of its histroy, its biological and medicinal properties, and the status of commercial preparations of the herb. In: Lawson, L. D. and R. Bauer (eds.). Phytomedicines of Europe, Chemistry and Biological Activity. Am. Chem. Soc., Washington, DC. p. 158-175.

66. Huizing, H. J. and J. H. Sietsma. 1991. *Symphytum officinale* (Comfrey): in vitro culture, regeneration, and biogenesis of pyrrolizidine alkaloids. In: Bajaj, Y. P. S. (ed.). Biotechnology in Agriculture and Forestry 15. Medicinal and Aromatic Plants III. Springer-Verlag Co., New York. p. 464-477.

67. Ihantola, V. A., J. Summanen, and H. Kankaanranta 1997. Anti-inflammatory activity of extracts from leaves of *Phyllanthus emblica*. Planta Medica 63: 518-524.

68. Ikan, R., V. Weinstein, Y. Milner, B. Bravdo, O. Shoseyov, D. Segal, A. Altman, I. Chet, D. Palevitch and E. Putievsky. 1993. Natural glycosides as potential odorants and flavorants. Internat. Symp. Medicinal Aromatic Plants, Tiberias on the Sea of Galilee, Israel,, 22-25 Mar. 1993. Acta Hort. 344: 17-28.

69. Ishikura, N. 1989. *Crytomeria japonica* Don (Japanese cedar): in vitro production of volatile

oils. In: Bajaj, Y. P. S. (ed.). Biotechnology in Agriculture and Forestry 7. Medicinal and Aromatic Plants II. Springer-Verlag Co., New York. p. 129-134.

70. Ishimaru, K., N. Tanaka, T. Kamiya, T. Sato and K. Shimomura. 1998. *Cornus kousa* (Dogwood): in vitro culture, and the production of tannins and other phenolic compounds. In: Bajaj, Y. P. S. (ed.). Biotechnology in Agriculture and Forestry 41. Medicinal and Aromatic Plants X. Springer-Verlag Co., New York. p. 113-131.

71. Jakupovic, J., V. P. Pathak, F. Bohlmann, R. M. King, and H. Robinson. 1987. Obliquin derivatives and other constituents from Australian *Helichrysum* species. Phytochemistry 26: 803-807.

72. Jakupovic, J., C. Zdero, M. Grenz, and F. Tsichritzis. 1989. Twenty-one acylphloroglucinol derivatives and further constituents from South African *Helichrysum* species. Phytochemistry 28: 1119-1131.

73. Jayatilake, G. S. and A. J. MacLeod. 1987. Volatile constituents of *Centella asiatica*. In: M. Martens, G. A. Dalen and H. Russwurm Jr. (eds.). Flavour Science & Technology, John Wiley & Sons Ltd. p. 79-82.

74. Jonard, R. 1989. *Jasminum* spp. (Jasmine): micropropagation and the production of essential oils. In: Bajaj, Y. P. S. (ed.). Biotechnology in Agriculture and Forestry 7. Medicinal and Aromatic Plants II. Springer-Verlag Co., New York. p. 315-331.

75. Kameoka, H., M. Miyazawa and K. Haze. 1975. 3-Ethyl-7-hydroxyphthalide from *Forsythia japonica*. Phytochemistry 14: 1676-1677.

76. Kato, M. 1989. *Camellia sinensis* L. (Tea): in vitro regneration. In Bajaj, Y. P. S. (ed.). Biotechnology in Agriculture and Forestry 7. Medicinal and Aromatic Plants II. Springer-Verlag Co., New York. p. 82-98.

77. Kaziro, G. S. 1990. Metronidazole (Flagyl) and *Arnica montana* in the prevention of post-surgical complications, a comparative placebo controlled clinical trial. Br. J. Clin. Pract. 44: 619-621.

78. Kaushalya, G., K. K. Thakral, S. K. Arora, M. L. Chowdhary and K. Gupta. 1996. Structural carbohydrate and mineral contents of fenugreek seeds. Indian Cocoa, Arecanut and Spices J. 20: 120-124.

79. Keplinger, K., G. Laus, M. Wurm, M. P. Dierich and H. Teppner. 1999. *Uncaria tomentosa* (Willd.) DC.: ethnomedicinal use and new pharmacological, toxicological and botanical results. J. Ethnopharmacol. 64: 23-34.

80. Khodzhimatov, K., S. F. Fakhrutdinov, G. S. Aprasidi, N. P. Kuchni and K. Karimov. 1987. *Codonopsis clematidea* Schreuk—a valuable medicinal plant. Referativnyi Zhurnal 10: 55, 790.

81. Kirchin, A. 1998. Focus on seaweed. Ingredient, health & nutrition, formulation, markets & technologies. Summer 1998.

82. Korting, H. C., M. Schaefer-Korting, L. Hart, P. Laux and M. Schmid. 1993. Anti-inflammatory activity of *Hamamelis distillate* applied topically to the skin: influence of vehicle and dose. Rur. J. Clin. Pharmacol. 44: 315-318.

83. Kosaku, T., B. Jeffrey, B. Harborne and R. Self. 1986. Identification and distribution of malonated anthocyanins in plants of the compositae. Phytochemistry 25: 1337-1342.

84. Kreh, M., R. Matusch and L. Witte. 1995. Acetylated alkaloids from *Narcissus pseudonarcissus*. Phytochemistry 40: 1303-1306.

85. Kussi, T., K. Hardh and H. Kanon. 1984. Experiments on the cultivation of dandelion for salads use. II. The nutritive value and intrinsic quality of dandelion leaves. J. Agric. Sci. Finland 56: 23-31.

86. Kuusi, T., H. Pyysalo and K. Autio. 1985. The bitterness properties of dandelion II. Chemical investigations. Lebensm. Wiss. Technol. (Zurich) 18: 349-359.

87. Levin, W. and G. Willuhn. 1987. Sesquiterpene lactones from *Arnica chamissionis* Less. VI. Identification and quantitative determination by high performance liquid and gas chromatography. J. Chromatogr. 41: 329-342.

88. Lewis, W. H. 1992. Plants used medically by indigenous peoples. In Nigg, H. N. and D. Seigler (eds.). Phytochemical Resources for Medicine and Agriculture. Plenum Press, New York. 445 p.

89. Li, J. X., Q. Shi, Q. B. Xiong, J. K. Prasain, Y. Texuka. 1998. Tribulusamide A and B, new hepatoprotective lignanamides from the fruits of *Tribulus terrestris*: indications of cytoprotective activity in murine hepatocyte culture. Plant Medica 64: 628-631.

90. Li, T. S. C. 1995. Asian and American ginseng, a review. HortTechnology 5: 27-34.

91. Li, T. S. C. 1998. Echinacea: cultivation and medicinal value. HortTechnology 8: 122-129.

92. Li, T. S. C. and W. R. Schroeder. 1996. Sea buckthorn (*Hippophae rhamnoides* L.): A multipurpose plant. HortTechnology 6: 370-380.

93. Li, T. S.C. and L. C. H. Wang. 1998. Physiological components and health effects of ginseng, echinacea and sea buckhtorn. In: G. Mazza (ed.). Functional Foods, Biochemical and Processing Aspects. Technomic Publishing Co., Inc. Lancaster, PA. 460 p.

94. Longo, R. and U. Fumagalli. 1965. Method of chemical analysis of the anthron components of *Cascara sagrada*. Boll. Chim. Farm. 104: 824-827.

95. McGimsey J. 1993. Sage, *Salvia officianalis.* WWW. Crop.cri.nz/broadshe/sage.htm.

96. Mallet, J. F., C. Cerrati, E. Ucciani, J. Gamisans and M. Gruber. 1994. Antioxidant activity of plant leaves in relation to their alpha-tocopherol content. Food Chem. 49: 61-65.

97. Mansell, R. L. and C. A. McIntosh. 1991. *Citrus* spp.: in vitro culture and the production of naringin and limonin. In: Bajaj, Y. P. S. (ed.). Biotechnology in Agriculture and Forestry 15. Medicinal and Aromatic Plants III. Springer-Verlag Co., New York. p. 193-210.

98. Matsuda, H., S. Nakamura, H. Shiomoto, T. Tanaka and M. Kubo. 1992. Pharmacological studies on leaf of *Arctostaphylos uva-ursi* (L.) Spreng. IV. Effect of 50 percent methanolic extract from *Arctostaphylos uva-ursa* (L.) Spreng (bearberry lear) on melanin synthesis. Yakugaku Zasshi 112: 276-282.

99. Mierendorff, H. J. 1995. Determination of pyrrolizidine alkaloids by thin layer chromatography in the oil of seeds of *Borage officinalis* L. Fett-Wissenschaft-Technologie 97: 33-37.

100. Misik, V., L. Bezakova, L. Malekova and D. Kostalova. 1995. Lipoxygenase inhibition and antioxidant properties of protoberberine and aporphine alkaloids isolated from *Mahonia aquifolium*. Planta Medica 61: 372-373.

101. Moes, A. 1966. A parallel study of the chemical composition of *Polygala senega* and of "*Securidaca longepedunculata*" Fres. Var. parvifolia, a Congolese polygalacea. J. Phar, Belg. 21: 347-362.

102. Mulherjee, K. D. and I. Kiewitt. 1987. Formation of gamma linolenic acid in the higher plant evening primrose (*Oenothera biennis* L.). J. Agric. Food Chem. 35: 1009-1012.

103. Muller, D., A. Carnat and J. L. Lamaison. 1991. *Fucus*: comparative study of *Fucus vesiculosus* L. *Fucus serratus* L. and *Ascophyllum nodosum* Le Jolis. Plantes Medicinales et Phytotherapie 25: 194-201.

104. Muller, K. and K. Ziereis. 1994. The antipsoriatic *Mahonia aquifolium* and its active consitutents. I. Pro-and antioxidant properties and inhibition of 5-lipoxygenase. Planta Medica 60: 421-424.

105. Murray, M. T. 1995. The healing power of herbs. Prima Publishing, Rocklin, CA. 410 p.

106. Nandi, P., G. Talukder and A. Sharma. 1997. Dietary chemoprevention of clastogenic effects of 3,4-benzo(a) pyrene by *Emblica officinalis* Gaertn. Fruit extract. British J. Cancer. 76: 1279-1283.

107. Natali, A. C. and I. C. Sanchez. 1991. *Aloe barbadensis* Mill. (= *A. vera* L.) In: Bajaj, Y. P. S. (ed.). Biotechnology in Agriculture and Forestry 15. Medicinal and Aromatic Plants III. Springer-Verlag Co., New York. p. 95-106.

108. Nazarko, L. 1995. Infection control. The therapeutic uses of cranberry juice. Nurs. Stand. 9: 33-35.

109. Neville, H. A., B. A. Bohm. 1994. Flavonoids of *Umbellularia california*. Phytochemistry 36: 1229-1231.

110. Obregon Vilches, L. E. 1995. Cat's claw. 3 rd ed. Institute de Fitoterapia Americano, Lima, Peru. 169 p.

111. O'Dowd, N. A., P. G. McCauley, G. Wilson, J. A. N. Parnell, T. A. K. Kavanagh and D. J. McConnell. 1998. *Ephedra* species: in vitro culture, micropropagation, and the production ephedrine and other alkaloids. In: Bajaj, Y. P. S. (ed.). Biotechnology in Agriculture and Forestry 41. Medicinal and Aromatic Plants X. Springer-Verlag Co., New York. p. 154-193.

112. Ofek, I., J. Goldhar and N. Sharon. 1996. Anti-*Escherichia coli* adhesin activity of cranberry and blueberry juices. Adv. Exp. Med. Biol. 408: 179-183.

113. Ogasawara, T., K. Chiba and M. Tada. 1998. *Sesamum indicum* L. (sesame): in vitro culture, and the production of naphthoquinone and other secondary metabolites. In: Bajaj, Y. P. S. (ed.). Biotechnology in Agriculture and Forestry 41. Medicinal and Aromatic Plants X. Springer-Verlag Co., New York. p. 366-393.

114. Omer, E. A., A. M. Refaat, S. S. Ahmed, A. Kamel and F. M. Hammouda. 1993. Effect of spacing and fertilization on the yield and active constituents of milk thistle, *Silybum marianum*. J. Herbs Spices and Medicinal Plants 1: 17-23.

115. Oomah, D., L. Stephanie and D. V. Godfrey. 1999. Properties of sea buckthorn (*Hippophae rhamnoides* L.) and ginseng (*Panax quinquefolium* L.) seed oils. Proc. Canadian Inst. Food Sci. Technology Annual Conf. Kelowna, BC. p. 55.

116. Ortega-Calvo, J. J., C. Mazuelos, B. Hermosin and C. Saiz-jimenez. 1993. Chemical composition of Spirulina and eukaryotic algae food products marketed in Spain. J. Applied Phycology 5: 425-435.

117. Petit-Paly, G., K. G. Ramawat, J. C. Chenieux and M. Rideau. 1989. *Ruta graveolens*: in vitro production of alkaloids and medicinal compounds. In: Bajaj, Y. P. S. (ed.). Biotechnology in Agriculture and Forestry 7. Medicinal and Aromatic Plants II. Springer-Verlag Co., New York. p. 488-505.

118. Polle, A. and B. Morawe. 1995. Seasonal changes of the antioxidative systems in foliar buds and leaves of field-grown beech trees (*Fagus xylvatica* L.) in a stressful climate. Botanica Acta 108: 314-320.

119. Reynolds, T., J. V. Dring and C. Hughes. 1991. Lauric acid containing triglycerides in seeds of *Umbellularia californica* Nutt. (Lauraceae) J. Am. Oil Chemists Society 68: 976-977.

120. Rogers, R. D. 1997. Sundew, Moonwort, Medicinal Plants of the Prairies. Vol. 1 & 2. Edmonton, Alberta. 282 p.

121. Rogers, R. and L. Szott-Rogers. 1997. Prairie Deva, Flower Essences. 40 p.

122. Rogiers, S. Y. and N. R. Knowles. 1997. Physical and chemical changes during growth, maturation, and ripening of saskatoon. Can. J. Bot. 75: 1215-1225.

123. Safayhi, H., E. R. Sailer and H. P. T. Ammon 1996. 5-Lipoxygenase inhibition by acetyl-11-keto-beta-boswellic acid (AKBA) by a novel mechanism. Symp. Salai guggal—*Boswellis serrata*, Halle (Saale), Germany 3-7 Sept. 1995. Phytomedicine 3: 71-72.

124. Sakamura, F. and T. Suga. 1989. *Zingiber officinale* Roscoe (Ginger): in vitro propagation and the production of volatile constituents. In: Bajaj, Y. P. S. (ed.). Biotechnology in Agriculture and Forestry 7. Medicinal and Aromatic Plants II. Springer-Verlag Co., New York. p. 524-538.

125. Sakurai, N. and M. Nagai. 1996. Chemical constitutents of original plants of *Cimicifuga rhizoma* in Chinese medicine. Yakugaku Zasshi 116: 850-865.

126. Schmidt, D. R. and A. E. Sobota. 1988. An examination of the anti-adherence activity of cranberry juice on urinary and nonurinary bacterial isolates. Microbios. 55: 173-181.

127. Schumacher, H. M. 1988. Biotechnology in the production and conservation of medicinal plants. In: Akerele, O, V. Heywood and H. Synge (eds.). The Conservation of Medicinal Plants. Cambridge Univ. Press. New York. 362 p.

128. Shahat, A. A., S. I. Ismail, F. N. Hammouda and S. A. Azzam. 1998. Anti-HIV activity of flavonoids and proanthocyanidins from *Crataegus sinalica*. Phytomedicine 5: 133-136.

129. Shishoo, C. J., S. A. Shah, I. S. Rathod and S. G. Patel. 1997. Determination of vitamin c

content of *Phyllanthus emblica* and Chyavanprash. Indian J. Pharmaceutical Sci. 59: 268-270.

130. Shoji, J. and Y. Tsukitani. 1972. On the structure of senegin, 3 of *Senegae radix*. Chem. Pharm. Bull. (Tokyo) 20: 424-426.

131. Short, K. C. and A. V. Roberts. 1991. *Rosa* spp. (roses): in vitro culture, micropropagation, and the production of secondary products. In: Bajaj, Y. P. S. (ed.). Biotechnology in Agriculture and Forestry 15. Medicinal and Aromatic Plants III. Springer-Verlag Co., New York. p. 376-397.

132. Shoyama, Y., I. Nishioka and K. Hatano. 1991. *Aconitum* spp. (Monkshood): somatic embryogenesis, plant regeneration, and the production of aconitine and other alkaloids. In Bajaj, Y. P. S. (ed.). Biotechnology in Agriculture and Forestry 15. Medicinal and Aromatic plants III. Springer-Verlag Co., New York. p. 58-72.

133. Singh, S., S. Kumar, S. Singh and S. Kumar. 1998. *Withania somnifera*: the Indian gineng ashwagandha. Central Inst. Medicinal and Aromatic Plants, India. 293 p.

134. Small, E.(ed.). 1997. Culinary Herbs. NRC Research Press. Ottawa. 710 p.

135. Small, E. and P. M. Catling (eds.) 1999. Canadian Medicinal Crops. NRC Research Press. 240 p.

136. Solet, J. M., A. Simon-Ramiasa, L. Cosson and J. L. Guignard. 1998. *Centella asiatica* (L.) urban. (Pennywort): cell culture, production of terpenoids, and biotransformation capacity. In: Bajaj, Y. P. S. (ed.). Biotechnology in Agriculture and Forestry 41. Medicinal and Aromatic Plants X. Springer-Verlag Co., New York. p. 81-96.

137. Sonnenbichler, J., I. Sonnenbichler and F. Scalera. 1998. Influence of the flavonolignan silibinin of milk thistle on hepatocytes and kidney cells. In: Lawson , L. D. and R. Bauer (eds.) Phytomedicines of Europe, Chemistry and Biological Activity. Am. Chem. Soc., Washington, DC. p. 263-277.

138. Stahein, H. F. and A. Warburg. 1991. The chemical and biological route from podophyllotoxin glucoside to etoposide: ninth Cain memorial award lecture. Cancer Res. 51: 5-51.

139. Standard, S.A. , P. Vaux and C.M. Bray. 1985. High-performance liquid chromatography of nucleotides and nucleotide sugars extracted from wheat embryo and vegetable seed. J. Chromatogr. 318: 433-439.

140. Stevens, J. F., H. Hart, E. T. Elema, A. Bolck and H. Hart. 1996. Flavonoid variation in

Eurasian *Sedum* and *Sempervivum*. Phytochemistry 41: 503-512

141. Stevens, R. 1967. The chemistry of hop constituents. Chem. Rev. 67: 19-71.

142. Sticher, O. 1998. Biochemical, pharmaceutical, and medical perspectives of ginseng. In: Lawson, L. D. and R. Bauer (eds.). Phytomedicines of Europe, Chemistry and Biological Activity. Am. Chem. Soc., Washington, DC. p. 221-240.

143. Sticher, O. and B. Meier. 1998. Hawthorn (*Crataegus*): biological activity and new strategies for quality control. In: Lawson, L. D. and R. Bauer (eds.). Phytomedicines of Europe, Chemistry and Biological Activity. Am. Chem. Soc., Washington, DC. p. 241-262.

144. Stoll, A., J. Renz and A. Brack. 1950. Isolierung und konstitution des echinacosids cines glykosids aus den wurzeln von *Echinacea angustifolia* D. C. Helv. Chim Acta 33: 1877-1893.

145. Strenath, H. L. and K. S. Jagadishchandra. 1991. *Cymbopogon* Spreng. (Aromatic grasses): in vitro culture, regeneration, and the production of essential oils. In: Bajaj, Y. P. S. (ed.). Biotechnology in Agriculture and Forestry 15. Medicinal and Aromatic Plants III. Springer-Verlag Co., New York. p. 211-236.

146. Struck, D., M. Tegtmeier and G. Harnischfeger. 1997. Flavones in extract of *Cimicifaga racemosa.* Planta Med. 63: 289.

147. Syrchina, A. I., N. N. Pechurina, A. L. Vereshchagin, A. G. Gorshkov, I. E. Tsapalova and A. A. Semenov. 1993. A chemical investigation of *Matteuccia struthiopteris*. Chem. Natural Compounds 29: 535-536.

148. Szentimihalyi, K., M. Then, V. Illes, S. Perneczky, Z. Sandor, B. Lakatos, and P. Vinkler. 1998. Phytochemical examination of oils obtained from the fruit of milk thistle (*Silybum marianum* L. Gaertner) by suercritical fluid extraction. Zeitschrift-fur-Naturforschung Sec. C-biosciences 53: 9-10, 779-784.

149. Takeda, O., S. Azuma, H. Mizukami, T. Ikenaga and H. Ohashi. 1986. Cultivation of *Polygala senega* var. latifolia: II. Effect of soil moisture content on the growth and senegin content. Shoyakugaku Zasshi. 40: 434-437.

150. Teas, J. 1973. The dietary intake of *Laminaria*, a brown seaweed, and breast cancer prevention. Nutr. Cancer 4: 217-222.

151. Tominaga, T. and D. Dubourdieu. 1997. Identification of 4-mercapto-4-methylpentan-2-one from the box tree (*Buxus sempervirens* L.) and broom [*Sarothamnus scoparius* (L.) Koch.]. Flavour and Fragrance J. 12: 373-376.

152. Turner, N. J. 1982. Traditional use of devil's-club (*Oplopanax horridus*; Araliaceae) by native peoples in western North America. J. Ethnobiol. 2: 17-38.

153. Tyler, V. E. 1986. Plant drugs in the twenty-first century. Economic Bot. 40: 279-280.

154. Van den Berg, A. J. J. and R. P. Labadie. 1988. *Rhamnus* spp.: in vitro production of anthraquinones, anthrones, and dianthrones. In: Bajaj, Y. P. S. (ed.). Biotechnology in Agriculture and Forestry 4. Medicinal and Aromatic Plants I. Springer-Verlag Co., New York. p. 513-528.

155. Vanwagenen, B. C. and J. H. H. Cardellina. 1986. Native American food and medicinal plants 7. Antimicrobial tetronic acids from *Lomatium dissectum*. Tetrahedron 42: 1117-1122.

156. Walker, C. 1990. Effects of sanguinarine and *Sanguinaria* extract on the microbiota assoicated with the oral cavity. J. Can. Dent. Assoc. 56 (7 Suppl.): 13-30.

157. Weiler, B. E., H. C. Kreuter, and R. Voth. 1990. Sulphoevernan, a polyanionic polysaccharide, and the narcissus lectin potently inhibit human immunodeficiency virus infection by binding to viral envelop protein. J. Gen. Virol. 71: 1957-1963.

158. Walkiw, O. and D. E. Douglas. 1975. Health food supplements prepared from kelp—a source of elevated urinary arsenic. Clin. Toxicol. 8: 325-331.

159. Whiterup, K. M. , S. A. Look, M. W. Stasko, T. J. Ghiorzi, G. M. Muschik and G. M. Cragg. 1990. *Taxus* spp. needles contain amounts of taxol comparable to the bark of *Taxus brevifolia:* analysis and isolation. J. Nat. Prod. 53: 1249-1255.

160. Wickremesinhe, E. R. M. and R. N. Arteca. 1998. *Taxus* species (yew): in vitro culture, and the production of taxol and other secondary metabolites. In: Bajaj, Y. P. S. (ed.). Biotechnology in Agriculture and Forestry 41. Medicinal and Aromatic Plants X. Springer-Verlag Co., New York. p. 415-442.

161. Willuhn, G. 1998. *Arnica* flowers: pharmacology, toxicology and anlysis of the sesuiterpene lactonesn, their main active substance. In: Lawson, L. D. and R. Bauer (eds.). Phytomedicines of Europe, Chemistry and Biological Activity. Am. Chem. Soc., Washington, DC. p. 118-132.

162. Winterhoff, H. 1998. *Vitex agnus-castus* (Chastle tree): pharmacological and clinical data. In: Lawson , L. D. and R. Bauer (eds.). Phytomedicines of Europe, Chemistry and Biological Activity. Am. Chem. Soc., Washington, DC. p. 299-307.

163. Woldemariam, T. Z., J. M. Betz and P. J. Houghton. 1997. Analysis of aporphine and quinolizidine alkaloids from *Caulophyllum thalictroides* by densitometry and HPLC. J. Pharm. Biomed. Anal. 15: 839-843.

164. Wurm, M., L. Kacani, G. Laus, K. Keplinger and M. P. Dierich. 1998. Pentacyclic oxindole alkaloids from *Uncaria tomentosa* induce human endothelial cells to release a lymphocyte proliferation regulating factor. Planta Medica 64: 701-704.

165. Xing, S., G. Huangpu, Y. Zhang, J. Hou, X. Sun, F. Han, and J. Yang. 1997. Analysis of the nutritional components of the seeds of promising ginkgo cultivars. J. Fruit Sci. 14: 39-41.

166. Yamamoto, H. 1988. *Paeonia* spp.: in vitro culture and the production of paeoniflorin. In Bajaj, Y. P. S. (ed.). Biotechnology in Agriculture and Forestry 4. Medicinal and Aromatic Plants I. Springer-Verlag Co., New York. p. 464-483.

167. Yamamoto, H. 1991. *Scutellaria baicalensis* Georgi: in vitro culture and the production of flavonoids. In Bajaj, Y. P. S. (ed.). Biotechnology in Agriculture and Forestry 15. Medicinal and Aromatic Plants III. Springer-Verlag Co., New York. p. 398-418.

168. Yang, M., X. Pan, X. Zhao, J. Wang, M. Yang and X. Pan. 1996. Study on the chemical constituents of the essential oil of berries of *Physalis pubescene*. J. NorthEast Forestry Univ. 24: 94-98.

169. Yang, X., Y. Shimizu, J. R. Steiner and J. Clardy 1993. Nostoclide I and II, extracellular metabolites from a symbiotic cyanobacterium, *Nostoc* sp., from the lichen *Peltigera canina*.Tetrahedron Letters 34: 761-764.

170. Zapesochnaya, G. G. and V. A. Kurkin. 1983. The flavonoids of the rhizomes of *Rhodiola rosea* II. A flavonolignan and glycosides of herbacetin. Chem. Nat. Comp. 19: 21-29.

171. Zeyistra, H. 1998. *Hamamelis virginiana*. Br. J. Phytotherapy 5: 23-28.

172. Zheng, G. Q., P. M. Kenney and L. K. T. Lam. 1992. Sesquiterpenes from clove (*Eugenia caryophyllata*) as potential anticarcinogenic agents. J. Natural Products 55: 999-1003.

173. Zhang, S. and K. Cheng. 1989. *Angelica sinensis* (Oliv.) Diels: in vitro culture, regeneration, and the production of medicinal compounds. In: Bajaj, Y. P. S. (ed.). Biotechnology in Agriculture and Forestry 7. Medicinal and Aromatic Plants II. Springer-Verlag Co., New York. p. 1-22.

CHAPTER 2

Toxicity of Medicinal Plants

In the past, the medicinal values of herbs were based on tradition and accidental discovery. There are many herbal preparations that are safe and may help ease minor ailments. A few are potent and dangerous to use. Some medicinal plants that are widely available should not be taken internally because the safety of their prolonged use is in question. Others are very poisonous, and great care should be taken to prevent children and livestock from eating them. Toxicity does not only refer to lethal effects but also to minor body reactions such as allergy, irritation, and sensitivity.

There is limited information in the literature regarding the proper usage of medicinal herbs, such as dosage, frequency and usage period, physical condition and sensitivity of the user, and possible interaction with prescribed drugs. More and more herbal products on the market are mixtures of two or more herbs. This is a serious concern, because there is limited research on the effect of combinations of herbs on humans. To avoid possible toxicity, thorough research and the guidance of a physician or naturopathic physician are strongly recommended.

Table 2. Toxicity of medicinal plants.

Scientific name	Common name	Toxicity *
Abies balsamea (L.) Mill.	Balsam fir	Herbs that can be safely consumed when used appropriately.[13]
Achillea millefolium L.	Yarrow	This herb is not considered toxic. Following use restrictions apply, unless otherwise directed by an expert qualified in the use of the described substance: not to be used during pregnancy. May cause itching and inflammatory changes in the skin, development of photosensitivity, avoid large doses and prolonged use.[4, 6, 13, 17]
Aconitum napellus L.	Monkshood	Dry roots are highly toxic if ingested, active ingredients, aconitine, hypaconitine, and mesaconitine are toxic at 2-3 mg in human.[6, 15]
Acorus calamus L. var. Americanus Wolff. *A. tatarinowii* L.	Calamus, sweet flag Shi Chang Pu	In 1968, US Food & Drug Administration declared this species "UNSAFE." It is poisonous, producing disturbed digestion, constipation, gastroenteritis, and bloody diarrhea. Note: Canadian plants appear free of carcinogenic beta-asarone.[1, 5, 8, 17]
Actaea alba L. *A. rubra* (Ait.) Wild.	White baneberry Red baneberry	Berry, root, and sap are irritant if ingested, sometimes externally vesicant. This herb is on the poisonous plants list.[1, 5, 6]
Aesculus hippocastanum L.	Horse chestnut	Symptoms of heat in the mouth and throat, redness of the lining membrane of the gullet. Children have been poisoned in Europe after ingesting large quantities of nuts. Classified by FDA as an unsafe herb.[6]
Agastache foeniculum L. *A. anethrodora* L.	Aniseed	Herbs that can be safely consumed when used appropriately.[13]

Scientific name	Common name	Toxicity *
Agrimonia cuparoria L.	Agrimony	Herbs that can be safely consumed when used appropriately.[13]
Agropyron repens (L.) Beauvois	Couch grass	No information based on the databases searched is available.
Alchemilla vulgaris L. *A. xanthochlora* Rothm.	Lady's mantle	Herbs that can be safely consumed when used appropriately.[13]
Allium cepa L.	Onion	Following use restrictions apply, unless otherwise directed by an expert qualified in the use of the described substance: not to be used while nursing. A child, after ingesting nine stalks of onion, experienced a severe rash on face and body.[13]
A. sativum L.	Garlic	Following use restrictions apply, unless otherwise directed by an expert qualified in the use of the described substance: not to be used while nursing.[13]
A. schoenoprasum L.	Chives	Following use restrictions apply, unless otherwise directed by an expert qualified in the use of the described substance: not to be used while nursing.[13]
Alnus crispus (Ait.) Pursh *A. incana* (L.) Moench. subsp. *tenufolia* *A. glutinosa* (L.) Gaertn.	Alder	Herbs for which insufficient data are available for classification.[13]

Scientific name	Common name	Toxicity *
Aloe vera (L.) Burm. *A. barbadensis* L.	Aloe	Following use restrictions apply, unless otherwise directed by an expert qualified in the use of the described substance: not to be used during pregnancy. Externally, may delay wound healing following laparotomy or caesarean delivery. The latex is poisonous if ingested, may cause kidney irritation.[6]
Althaea officinalis L.	Marshmallow	Herbs that can be safely consumed when used appropriately.[13]
Amelanchier alnifolia Nutt.	Saskatoon	Herbs for which insufficient data are available for classification.[13]
Ananas comosus (L.) Merr.	Pineapple	Unripe pineapple is poisonous, causing violent purgation.[5, 6, 8]
Anaphalis margaritacea (L.) Bench & Hook	Everlastings	Herbs that can be safely consumed when used appropriately. The FDA classified as an herb of undefined safety.[6, 13]
Anethum graveolens L.	Dill	Do not take the essential oil internally from this species except under professional supervision, may cause dermatitis.[6]
Angelica archangelica L.	Angelica	Following use restrictions apply, unless otherwise directed by an expert qualified in the use of the described substance: not to be used during pregnancy. Avoid prolonged exposure to sunlight, may evoke photodermatitis.[6, 13]
A. sinensis (Oliv.) Diels	Dong quai	Following use restrictions apply, unless otherwise directed by an expert qualified in the use of the described substance: not to be used during pregnancy.[13]

Scientific name	Common name	Toxicity *
Antennaria magellanica Schultz	Everlastings	See *Anaphalis margaritacea*
Anthemis nobilis L.	Chamomile	See *Chamaemelum nobile*
Apium graveolens L.	Celery	Following use restrictions apply, unless otherwise directed by an expert qualified in the use of the described substance: not to be used during pregnancy. Individuals with renal disorders should use with caution.[13] Wild celery considered to be poisonous.[6]
Aralia racemosa L. *A. nudicaulis* L.	Spikenard	Following use restrictions apply, unless otherwise directed by an expert qualified in the use of the described substance: not to be used during pregnancy.[13]
Arctium lappa L.	Burdock	Herbs that can be safely consumed when used appropriately, may irritate skin and cornea.[6, 13]
Arctostaphylos uva-ursi (L.) Spreng	Bearberry, Uva-ursi	Following use restrictions apply, unless otherwise directed by an expert qualified in the use of the described substance: not to be used during pregnancy. Health Canada listed as an herb that is unacceptable as a nonprescription drug product for oral use. May cause nausea, stomach distress, and vomiting. The recommended dose varies from 13 g, 3-6 times daily to about 10 g daily. Avoid long- term use especially in children.[1, 6, 13, 17]
Armoracia rusticana Gaertn, Mey & Scherb.	Horseradish	Contraindicated with inflammation of the gastric mucosa and kidney disorder, not to be used by children under 4 years old.[4]

Scientific name	Common name	Toxicity *
Arnica latifolia Bong. *A. montana* L. *A. chamissonis* subsp. *foliosa* *A. condifolia* Hook *A. fulgens* Pursh *A. sororia* Greene	Arnica	Externally, do not use on open wounds or broken skin, and it may cause contact dermatitis. Health Canada listed arnica as unacceptable as a nonprescription drug product for oral use. The FDA classified it as an unsafe herb.[1, 5, 6, 17]
Artemisia absinthium L.	Wormwood	Following use restrictions apply, unless otherwise directed by an expert qualified in the use of the described substance: not to be used during pregnancy. It is toxic and should not be taken internally.[4, 13]
A. annua L.	Chinese wormwood (Qing Hao)	Following use restrictions apply, unless otherwise directed by an expert qualified in the use of the described substance: not to be used during pregnancy.[13]
A. dracunculus L.	Tarragon	Herbs that can be safely consumed when used appropriately. The main constituent of essential oil, estragole is reported to produce tumor in mice.[6, 13]
A. tridentata Nutt.	Sagebrush	Toxic if in large doses.[6]
A. vulgaris L.	Mugwort	Toxic if in large doses, cause epileptic spasms.[6]
Asarum canadensis L.	Wild ginger	Following use restrictions apply, unless otherwise directed by an expert qualified in the use of the described substance: not to be used during pregnancy. A few cases are documented of dermatitis after handling leaves. Not for long-term use and do not exceed recommended dose.[8, 13]

Scientific name	Common name	Toxicity *
Asparagus officinalis L. *A. racemosus* Willd.	Asparagus	No information based on the databases searched is available.[13]
Astragalus americana Bunge. *A. membranaceus* Bunge *A. sinensis* L.	Astragalus Huang Qi	Herbs that can be safely consumed when used appropriately This herb is on the poisonous plant list. No information on humans; however, poisoning and death in chickens, horses, cattle, and sheep are reported.[1, 5, 13]
Avena sativa L.	Oat	Herbs that can be safely consumed when used appropriately.[13]
Baptisia tinctora (L.) R.Br. ex. Ait. f.	Wild indigo	Herbs that can be safely consumed when used appropriately; however, it was reported that the entire plant is toxic.[13]
Bellis perennis L.	Daisy	Herbs for which insufficient data are available for classification.[13]
Berberis aquifloium L.	Oregon grape	See *Mahonia aquifolium*
B. vulgaris L.	Barberry	Following use restrictions apply, unless otherwise directed by an expert qualified in the use of the described substance: not to be used during pregnancy. Those who handle the wood may develop colic and diarrhea.[6, 13]
Beta vulgaris L.	Beet root	No information based on the databases searched is available.
Betula lenta L.	Birch	Herbs that can be safely consumed when used appropriately.[13]
Borago officinalis L.	Borage	Following use restrictions apply, unless otherwise directed by an expert qualified in the use of the described substance: for external use only, not to be used during pregnancy, not to be used while nursing, and long-term use is not recommended.[13]

Scientific name	Common name	Toxicity *
Boswellia sacra Roxb. ex Colebr. *B. carteri* Birdwood	Frankincense	Herbs that can be safely consumed when used appropriately.[13]
Brassica nigra (L.) Kock.	Black mustard	Internally, this herb can be safely consumed when used appropriately. Externally, duration of use should not exceed 2 weeks, not for children under 6 years of age.[13]
Buxus sempervirens L.	Boxwood	It is highly toxic, causes vomiting, convulsions, and death. To be used only under the supervision of an expert qualified in the appropriate use of this substance. Labeling must include proper use information: dosage, contraindications, potential adverse effects and drug interactions, and any other relevant information related to the safe use of this substance.[13]
Calamintha nepeta (L.) Savi *C. ascendens* L. *C. officinalis* L.	Basil thyme Calamint Mountain mint	No information based on the databases searched is available.
Calendula officinalis L.	Marigold	Herbs that can be safely consumed when used appropriately. [13]
Camellia sinensis (L.) Kunize	Green tea	Fermented black teas are not recommended for excessive or long-term use. There is evidence that the condensed catechin tannin of tea is linked to high rate of esophageal cancer in some areas where tea is heavily consumed.[6]
Cananga odorata (Lam) J. D. Hook & T. Thompson	Ylang-Ylang	Herbs for which insufficient data are available for classification. Oil can produce dermatitis.[6, 13]

Scientific name	Common name	Toxicity *
Cannabis sativa L.	Hemp	Pollen cause allergic rhinitis, cannabis may cause impotence, chromosomal damage, irreversible brain damage, bronchitis. This plant is on the poisonous plants list.[1, 5, 6]
Capsella bursa-pastoris (L.) Medik.	Shepherd's purse	Following use restrictions apply, unless otherwise directed by an expert qualified in the use of the described substance: not to be used during pregnancy. Class 2b, individuals with a history of kidney stones should use cautiously.[13]
Capsicum annuum L. var. Annuum *C. annum* L. var. Grossum *C. frutescens* L. *C. annum* L. var. Acuminatum	Cayenne pepper Sweet pepper Chilli pepper	Internally, this herb can be safely consumed when used appropriately. Externally, contraindicated on injured skin or near eyes. May be carcinogenic.[6]
Carbenia benedicta L.	Blessed thistle	See *Cnicus benedictus*
Carduus benedita L.	Blessed thistle	See *Cnicus benedictus*
Carica papaya L.	Papaya	Herbs that can be safely consumed when used appropriately. Externally the latex is irritant.[6, 13]
Carthamus tinctorius L.	Safflower	Following use restrictions apply, unless otherwise directed by an expert qualified in the use of the described substance: not to be used during pregnancy. Contraindicated in patients with hemorrhagic diseases or peptic ulcers.[13]
Carum carvi L.	Caraway	Herbs that can be safely consumed when used appropriately.[13]

Scientific name	Common name	Toxicity *
Cassia senns L. *C. angustifolia* Vahl.	Tinnevelly senna Alexandria senna	Nursing mothers should avoid this herb, causes dermatitis on contact of leaves.[6]
Castanea sative Mill	Chestnut	Herbs that can be safely consumed when used appropriately.[13]
Catharanthus roseus (L.) Don	Periwinkle	See *Vinca minor*
Caulophyllum thalietroides (L.) Michx. *C. giganteum* (F.) Loconte & W. H. Blackwell	Blue cohosh	Following use restrictions apply, unless otherwise directed by an expert qualified in the use of the described substance: not to be used during pregnancy. Berries and roots are cytotoxic. The dust of the powdered root is strongly irritating to mucous membranes.[13, 17]
Cedrus libani Riich subsp. *atlantica*	Cedarwood	Herbs for which insufficient data are available for classification.[13]
Centaurea calcitrapa L.	Star thistle	Herbs for which insufficient data are available for classification.[13]
Centella asiatica (L.) Urb.	Gotu Kola	Herbs that can be safely consumed when used appropriately.[13]
Chamaelirium luteum (L.) A. Gray	False unicorn root, fairywand	Canadian regulations do not allow false unicorn root as a non-medicinal ingredient for oral use products.[1]
Chamaemelum nobile (L.) All.	Roman chamomile	Following use restrictions apply, unless otherwise directed by an expert qualified in the use of the described substance: not to be used during pregnancy. May cause allergy, asthma, and urticaria.[4, 6, 13]
Chamaenerion angustifolium (L.) Scop.	Firewood	See *Epilobium angustifolium* L.

Scientific name	Common name	Toxicity *
Chenopodium album L.	Lamb's quarters	Herbs for which insufficient data are available for classification. This plant is on the poisonous plants list. It was reported that poisoning occurred in Europe when lamb's quarters was eaten in large quantities.[1, 2, 13]
C. ambrosioides L.	Wormseed	Herbs for which insufficient data are available for classification. Oil is poisonous.[6, 13]
Chrysanthemum cinerarifolium (Trevir.) Vis.	Pyrethrum	Causes urticaria, asthma, rhinitis. Overdose causes headache, tinnitus, facial pallor, nausea, syncope. May cause allergy.[4, 6]
C. parthenium (L.) Burm.	Feverfew	Following use restrictions apply, unless otherwise directed by an expert qualified in the use of the described substance: not to be used during pregnancy. Contact with the plant may cause skin irritations.[6, 13]
C. vulgare L.	Tansy	See *Tanacetum vulgare*
Cichorium intybus L.	Chicory	Herbs that can be safely consumed when used appropriately.[13]
Cimicifuga racemosa (L.) Nutt.	Black cohosh	Not to be used during pregnancy. Overdoses can cause headaches, nausea, vomiting, slow pulse rate, dizziness, and visual disturbances. Others include abnormal blood clotting, liver problems, and promotion of breast tumors. Health Canada listed as an herb that is unacceptable as a nonprescription drug product for oral use.[1, 5, 13, 319]

Scientific name	Common name	Toxicity *
Cinnamomum verum J. Presl.	Cinnamon	Following use restrictions apply, unless otherwise directed by an expert qualified in the use of the described substance: not to be used during pregnancy, not for long-term use, do not exceed recommended dose.[13]
C. cassia (Nees) Nees & Eberm.	Cassia	Following use restrictions apply, unless otherwise directed by an expert qualified in the use of the described substance: not to be used during pregnancy, not for long-term use, do not exceed recommended dose.[13]
C. camphora (L.) Nees & Eberm.	Camphor	Following use restrictions apply, unless otherwise directed by an expert qualified in the use of the described substance: not to be used during pregnancy. Avoid facial regions of infants and small children.[6, 13]
Citrus limon (L.) Burm.	Lemon	Herbs that can be safely consumed when used appropriately.[13]
C. bergamia Risso & Poit	Bergamot orange	Following use restrictions apply, unless otherwise directed by an expert qualified in the use of the described substance: not to be used during pregnancy.[13]
C. reticulata Blanco	Mandarin	Herbs that can be safely consumed when used appropriately.[13]
C. aurantifolia Swingle	Lime	Herbs that can be safely consumed when used appropriately.[13]
C. aurantium L.	Bitter orange	Herbs that can be safely consumed when used appropriately.[13]

Scientific name	Common name	Toxicity *
C. paradisi Macfad	Grapefruit	Herbs that can be safely consumed when used appropriately.[13]
Cnicus benedictus L.	Blessed thistle	Following use restrictions apply, unless otherwise directed by an expert qualified in the use of the described substance: not to be used during pregnancy.[13]
Cochlearia amorucia L.	Horseradish	See *Armoracia rusticana*
Codonopsis pilosula (Franch.) Nannfeldt. *C. tangshen* Oliver	Codonopsis, Dang Shen, bellflower	Herbs that can be safely consumed when used appropriately.[13]
Coffea arabica L.	Coffee	Following use restrictions apply, unless otherwise directed by an expert qualified in the use of the described substance: not to be used during pregnancy, not recommended for excessive or long-term use, contraindicated for people with gastric ulcer, glaucoma.[13]
Colchicum autumrnale L.	Crocus Meadow saffron	All parts are highly toxic if eaten, handling of corms may cause skin allergy. Burning sensation in mouth and throat is reported.[8]
Collinsonia canadensis L.	Pilewort	Herbs that can be safely consumed when used appropriately.[13]
Commiphora molmol Engl. ex Tschirch *C. myrrha* (Nees) Engl.	Myrrh	Following use restrictions apply, unless otherwise directed by an expert qualified in the use of the described substance: not to be used during pregnancy. Doses over 2.0-4.0 g may cause irritation of the kidneys and diarrhea.[6, 13]

Scientific name	Common name	Toxicity *
Convallaria majalis L.	Lily of the valley	Highly toxic, this herb for which significant data exist to recommend the following labeling: "To be used only under the supervision of an expert qualified in the appropriate use of this substance." Labeling must include proper use information: dosage, contraindications, potential adverse affects and drug interactions and any other relevant information related to the safe use of this substance.[13]
Convolvulus arvensis L. *C. sepium* L.	Bindweed	Herbs for which insufficient data are available for classification.[13]
Coriandrum sativum L.	Coriander (fruit) Cilantro (leaf)	Herbs for which insufficient data are available for classification.[13]
Cornus canadensis L. *C. florida* L.	Bunchberry Dogwood	Herbs for which insufficient data are available for classification.[13]
Corylus cornuta Marsh. *C. rostrata* Marsh. *C. avellana* L.	Hazelnut	Herbs that can be safely consumed when used appropriately.[13]
Crataegus monogyna Jacq. *C. pinnatifida* Bge.	English hawthorn Chinese hawthorn	Herbs that can be safely consumed when used appropriately.[13]
Crocus sativus L.	Saffron	Following use restrictions apply, unless otherwise directed by an expert qualified in the use of the described substance: not to be used during pregnancy.[13]
Cryptotaenia japonica Hassk.	Japanese parsley, Mitsuba	Herbs for which insufficient data are available for classification.[13]

Scientific name	Common name	Toxicity *
Cucurbita pepo L.	Pumpkin	Herbs for which insufficient data are available for classification.[13]
Cuminum cyminum L.	Cumin	Herbs that can be safely consumed when used appropriately.[13]
Curcuma domestica L. *C. xanthorrhiza* L. *C. longa* L.	Curcuma, Jiang Huang	Following use restrictions apply, unless otherwise directed by an expert qualified in the use of the described substance: not to be used during pregnancy. Therapeutic quantities should not be taken by people with bile duct obstruction or gallstones.[13]
Cuscuta epithymum Murr. *C. chinensis* Lam.	Dodder	Herbs that can be safely consumed when used appropriately.[13]
Cymbopogon citratus (DC. ex Nees) Stapf.	Lemon grass	Following use restrictions apply, unless otherwise directed by an expert qualified in the use of the described substance: not to be used during pregnancy.[13]
C. martinii (Roxb) Wats	Palmorosa, gingergrass	Herbs for which insufficient data are available for classification.[13]
C. nardus L. *C. winterianus* Jowitt	Citronella	Herbs for which insufficient data are available for classification.[13]
Cynara scolymus L. *C. cardunculus* L.	Artichoke Wild artichoke	No information based on the databases searched is available.
Cypripedium calceolus L.	Lady's slipper	Large dose causes psychedelic reaction, giddiness, restlessness. Dermatitis can develop after touching the glandular hairs.[6]

Scientific name	Common name	Toxicity *
Cytisus scoparius (L.) Link.	Broom, scoparium	Herbs for which significant data exist to recommend the following labeling: "To be used only under the supervision of an expert qualified in the appropriate use of this substance." Labeling must include proper use information: dosage, contraindications, potential adverse effects and drug interactions, and any other relevant information related to the safe use of this substance. The FDA classified Broom as an unsafe herb.[6, 13]
Daucus carota L.	Carrot	Following use restrictions apply, unless otherwise directed by an expert qualified in the use of the described substance: not to be used during pregnancy.[13]
Digitalis purpurea L.	Foxglove	Herbs for which significant data exist to recommend the following labeling: "To be used only under the supervision of an expert qualified in the appropriate use of this substance." Labeling must include proper use information: dosage, contraindications, potential adverse effects and drug interactions, and any other relevant information related to the safe use of this substance. This herb is on the poisonous plants list.[1, 5, 13]
Dioscorea oppositae Thunb. *D. villosa* L.	Chinese yam Mexican yam	Herbs that can be safely consumed when used appropriately.[13]

Scientific name	Common name	Toxicity *
Echinacea angustifolia DC. Cornq. *E. pallida* (Nutt.) Nutt. *E. pururea* (L.) Moench.	Narrowleaf echinacea Pale-flower echinacea Purple coneflower	Herbs that can be safely consumed when used appropriately, should not be consumed when pregnant or suffering from diabetes. Extracts from the German Commission E. Monograph: E. angustifolia/pallida not to be used by progressive conditions such as tuberculosis, leucopsis, collagenosis, multiple sclerosis, AIDS, HIV infection, and other autoimmune disorders.[3, 17]
Echinopanax horridus (Sm) Decne. & Planch. ex. H. A. T. Harms	Devil's club	See *Oplopanax horridus*
Elaeagnus angustifolia L.	Russian olive	Herbs for which insufficient data are available for classification.[13]
Eleutherococcus senticosus (Rupr. ex Maxim) Maxim.	Siberian ginseng	Herbs that can be safely consumed when used appropriately.[13]
Elymus repens L.	Couch grass	See *Agropyron repens*
Ephedra distachya L. *E. sinica* Stapf.	Ephedra Ma Huang	Following use restrictions apply, unless otherwise directed by an expert qualified in the use of the described substance: not to be used during pregnancy. Contraindicated in anorexia, bulimia, and glaucoma. Frequent use may result in nervousness and restlessness.[6, 13]
E. nevadensis Wats.	Mormon tea	Herbs that can be safely consumed when used appropriately.[13]
Epilobium angustifolium L.	Fireweed	It is not toxic, can be safely consumed when used appropriately.[13, 17]
Epilobium parviflorum Schreb	Small-flowered willow herb	Herbs that can be safely consumed when used appropriately.[13]

Scientific name	Common name	Toxicity *
Equisetum arvense L. *E. telmateia* Ehrh.	Horsetail	Herbs that can be safely consumed when used appropriately. This herb is on the poisonous plants list, poisonous to livestock.[1, 6, 13]
Erythroxylum coca Lam.	Coca	A narcotic euphoriant and hallucinogen.[6]
Eschscholtzia california Cham.	California poppy	Herbs for which insufficient data are available for classification.[13]
Eucalyptus citriodora Hool. *E. globulus* Labill	Gum tree Eucalyptus	Contraindicated in inflammatory diseases of the bile ducts and gastrointestinal tract and in severe liver diseases. Death is reported from ingestion of 4-24 ml of essential oil.[6, 13]
Eugenia caryophyllata L.	Clove	See *Syzygium aromaticum*
Eupatorium perfoliatum L.	Boneset	Following use restrictions apply, unless otherwise directed by an expert qualified in the use of the described substance: for external use only. Long-term use is not recommended. The FDA classified this herb as unidentified safety.[6, 13]
Euphrasia officinalis L.	Eyebright	Herbs that can be safely consumed when used appropriately.[13]
Fagopyrum tataricum L.	Buckwheat	This herb is on the poisonous plants list.[1]
Fagus grandifolia Ehrh. *F. sylvatica* L.	Beech European beech	Herbs for which insufficient data are available for classification.[13]
Filipendula ulmaria (L.) Maxim.	Meadow sweet	Herbs that can be safely consumed when used appropriately. FDA classified as undefined safety.[5, 6, 13]
Foeniculum vulgare Mill.	Fennel	Herbs that can be safely consumed when used appropriately.[13]

Scientific name	Common name	Toxicity *
Forsythia suspensa (Thunb.) Vahl.	Forsythia	Following use restrictions apply, unless otherwise directed by an expert qualified in the use of the described substance: not to be used during pregnancy.[13]
Fragaria vesca L.	Strawberry	Herbs that can be safely consumed when used appropriately, may cause allergy.[4, 13]
Fraxinus americana L. *F. excelsior* L.	White ash	Herbs that can be safely consumed when used appropriately.[13]
Fucus vesiculosus L.	Kelp	Not to be used during pregnancy. Therapeutic use is not recommended in hyperthyroidism. High sodium content may not be suitable for salt-restricted diets.[13, 17]
Galium aparine L.	Cleavers	Herbs that can be safely consumed when used appropriately.[13]
Gaultheria procumbens L.	Wintergreen	Herbs that can be safely consumed when used appropriately. Death from stomach inflammations have resulted from frequent and large doses of the oil.[6, 13]
Geranium macrorrhizum L.	Geranium	Herbs that can be safely consumed when used appropriately.[13]
Ginkgo biloba L.	Ginkgo	May potentiate pharmaceutical monoamine oxidase inhibitors, do not exceed recommended dose, not for long-term use. Severe dermatitis can result from handling broken or crushed fruit. Seeds contain ginkgotoxin which is a neurotoxic antivitamin B_6.[2]
Glechoma hederacea L.	Ground ivy	Poisonous to horses. Herbs for which insufficient data are available for classification.[1, 6, 13]

Scientific name	Common name	Toxicity *
Glycine max (L.) Merrill	Soybean	No information based on the databases searched is available.
Glycyrrhiza glabra L. *G. uralensis* Fisch ex DC.	Licorice Chinese licorice	Following use restrictions apply, unless otherwise directed by an expert qualified in the use of the described substance: not to be used during pregnancy, not for prolonged use or in high doses except under supervision of a qualified health practitioner, contraindicated for diabetics and in hypertension, liver disorder. May alleviate peptic ulcers.[6, 13]
Gossypium hirsutum L.	Cotton	Herbs for which insufficient data are available for classification.[13]
Grindelia robusta Nutt. *G. squarrosa* (Pursh) Donal	Gumweed	Large doses of selenium within the plant is toxic.[6]
Hamamelis virginiana L.	Witch hazel	Minor amount of toxic chemicals such as eugenol, acetalkdhyde, and carcinogen safrole and an internal dose of as little as 1 g can cause nausea, vomiting, and constipation. May be carcinogenic.[6, 17]
Harpagophytum procumbens DC. ex Meisn.	Devil's claw	Contraindicated in gastric and duodenal ulcers.[8]
Hedeoma pulegiodes (L.) Pers.	Pennyroyal	Following use restrictions apply, unless otherwise directed by an expert qualified in the use of the described substance: not to be used during pregnancy. Oil can cause dermatitis, abortion, and death.[6, 13]
Hedera helix L.	English ivy	Extract from leaf caused dermatitis, ingested leaf cause scarlatiniform eruption.[6]
Helianthus annuus L.	Sunflower	Herbs that can be safely consumed when used appropriately.[13]

Scientific name	Common name	Toxicity *
Helichrysum angustifolium (Lam) DC.	Curry plant	Herbs for which insufficient data are available for classification.[13]
Helonias dioica L.	False unicorn root	See *Chamaelirium luteum*
Heracleum maximum Bartr. *H. lanatum* Michx. *H. sphondylium* L.	Cow parsley Cow parsnip Cow parsnip	Juice of the plant was harmful to the skin when the skin was exposed to sunlight. Herbs for which insufficient data are available for classification.[6, 13]
Hibiscus sabdariffa L. *H. rosa-sinensis* L.	Hibiscus	Herbs that can be safely consumed when used appropriately.[13]
Hierochloe odorata (L) Beauv.	Sweet grass	Toxic in experimental animals, known to cause liver damage, retard growth, and cause cancer and testicular atrophy.[17]
Hippophae rhamnoides L	Sea buckthorn	Herbs for which insufficient data are available for classification.[13]
Humulus lupulus L.	Hops	Not considered toxic, some reports advise not to use in depression treatment. Hops dermatitis has been recognized.[6, 17]
Hydrangea arborescens L.	Hydrangea	Not for long-term use, do not exceed recommended dose. Illness is reported from ingestion of leaves or roots, hand dermatitis resulted from the repeated handling of this plant by a nurseryman.[6]
Hydrastis canadensis L.	Goldenseal	Daily dosage to be administered to achieve therapeutic effects without causing any side effect. Not to be used during pregnancy. External use may cause ulceration.[17]
Hypericum perforatum L.	St. John's wort	May potentiate pharmaceutical monoamine oxidase inhibitors. This herb is on the poisonous plants list.[1, 5]

Scientific name	Common name	Toxicity *
Hyssopus officinalis L.	Hyssop	Following use restrictions apply, unless otherwise directed by an expert qualified in the use of the described substance: not to be used during pregnancy.[4, 13]
Ilex aquifomium L.	English holly	Not recommended for excessive or long-term use; berries cause nausea, vomiting and diarrhea.[6]
Inula helenium L.	Elecampane	Following use restrictions apply, unless otherwise directed by an expert qualified in the use of the described substance: not to be used during pregnancy.[13]
Iris versicolor L.	Blue flag	Following use restrictions apply, unless otherwise directed by an expert qualified in the use of the described substance: not to be used during pregnancy. May cause nausea and vomiting. This herb is on the poisonous plants list.[1, 5, 13]
Jasminum grandiflorum L. *J. auriculatum* Vahl.	Jasmine	Herbs that can be safely consumed when used appropriately.[13]
Juglans regia L. *J. nigra* L.	Walnut Black walnut	Prolonged use is not advised because of the presence of significant quantities of juglone, a known mutagen in animals. Carcinogenic effects associated with the chronic external use of *J. regia* in humans have been observed.[6]
Juniperus communis L.	Juniper	Following use restrictions apply, unless otherwise directed by an expert qualified in the use of the described substance: not to be used during pregnancy, not for use exceeding 4-6 weeks in succession, contraindicated in inflammatory kidney disease.[6, 13]

Scientific name	Common name	Toxicity *
Laminaria digitata (Hudss.) Lamk. *L. saccharinā* (L.) Lamk.	Kelp	See *Fucus vesiculosus*
Lamium album L.	Nettle	No information based on the databases searched is available.
Larrea tridentata (Sesse. & Moc. ex DC.) Coville.	Chapatral	Not for use in large amounts by persons with kidney disease and liver conditions such as hepatitis and cirrhosis.[6]
Laurus nobilis L.	Bay laurel	Herbs that can be safely consumed when used appropriately, may be poisonous and cause stomatitis and cheilitis.[4, 6, 13]
Lavandula angustifolia Mill. *L. spica* L. *L. officinalis* L.	Lavender	Herbs that can be safely consumed when used appropriately. May be poisonous. Do not take more than two drops of undiluted oil internally.[4, 13]
Ledum latifolium Jacq.	Labrador tea	Herbs for which insufficient data are available for classification.[13]
Lemna minor L.	Duckweed	Herbs for which insufficient data are available for classification.[13]
Leonurus cardiaca L.	Motherwort	Following use restrictions apply, unless otherwise directed by an expert qualified in the use of the described substance: not to be used during pregnancy. Some individuals develop dermatitis after contact with leaves. The FDA classified as an undefined safety herb.[6, 13]
Levisticum officinale W. Koch. *L. levisticum* L.	Lovage	Following use restrictions apply, unless otherwise directed by an expert qualified in the use of the described substance: not to be used during pregnancy. Contraindicated in impaired kidney function of inflammation of the kidneys.[4, 13]

Scientific name	Common name	Toxicity *
Ligustrum lucidium W.T.Aiton. *L. vulgare* L.	Ligustrum fruit Privet	Herbs that can be safely consumed when used appropriately. Children have been poisoned after ingestion of berries, but reports of death are undocumented. Do not take internally. The fruits and leaves can cause gastrointestinal irritation, liver damage.[13]
Linum usitatissimum L.	Flax	Take with at least 150 ml of liquid, contraindicated in bowel obstruction. This herb is on the poisonous plants list.[1, 5]
Liquidambar styraciflua L. *L. orientalis* L.	Gum tree, storax Oriental sweet gum	Herbs for which insufficient data are available for classification.[13]
Lobelia inflata L. *L. siphilitica* L.	Lobelia Blue lobelia	Following use restrictions apply, unless otherwise directed by an expert qualified in the use of the described substance: not to be used during pregnancy. May cause nausea and vomiting, not to be taken in large doses, dose-dependent cardioactivity has been observed. This herb is on the poisonous plants list.[1, 5, 6, 8, 13]
L. pulmonaria L.	Lungwort	Classed as a narcotic. This herb is on the poisonous plants list.[1, 5, 6]
Lomatium dissectum (Nutt.) Math. & Const.	Lomatium	Following use restrictions apply, unless otherwise directed by an expert qualified in the use of the described substance: not to be used during pregnancy. When used internally, it can cause skin rashes.[13]
Lonicera caerulea L. *L. caprifolium* L.	Mountain fly honeysuckle Dutch honeysuckle	Herbs that can be safely consumed when used appropriately. Mild symptoms of feeling unwell and vomiting after ingestion have been reported.[13]

Scientific name	Common name	Toxicity *
Lycium barbarum L. *L. chinesis* Mill. *L. pallidum* L.	Lycium Chinese wolfberry Wolfberry	Following use restrictions apply, unless otherwise directed by an expert qualified in the use of the described substance: not to be used during pregnancy. Entire plant causes gastrointestinal upset.[8, 13]
Lycopodium clavatum L.	Clubmoss	Considerable toxicity especially on prolonged use.[6]
Lysimachia vulgaris L.	Loosestrife	No information based on the data bases searched is available.
Lythrum salicaria L.	Purple loosestrife	Herbs for which insufficient data are available for classification.[13]
Macropiper excelsum G. Forst	Kava-Kava	See *Piper methysticum*
Mahonia aquifolium (Pursh) Nutt.	Oregon grape	Following use restrictions apply, unless otherwise directed by an expert qualified in the use of the described substance: not to be used during pregnancy. It can cause intense pain when infiltrated into the eye.[6, 13]
Marrubium vulgare L.	Horehound	Following use restrictions apply, unless otherwise directed by an expert qualified in the use of the described substance: not to be used during pregnancy.[13]
Matricaria chamomilla L. *M. recutita* L.	Chamomile German chamomile	Herbs that can be safely consumed when used appropriately.[13]
Matteucia struthiopteris (L.) Tod.	Ostrich fern	Herbs for which insufficient data are available for classification.[13]
Medicago sativa L.	Alfalfa	Herbs that can be safely consumed when used appropriately. This herb is on the poisonous plants list.[1, 5, 13]

Scientific name	Common name	Toxicity *
Melaleuca alternifolia (Maid. & Bet.) Cheel	Melaleuca, medicinal tea tree	Herbs for which insufficient data are available for classification. Individuals with sensitive skin should dilute tea tree oil with an equal part of olive oil.[6, 13]
Melilotus officinalis (L.) Lamk. *M. arvensis* L.	Sweet clover Melilot	Herbs for which insufficient data are available for classification. This herb is on the poisonous plants list. Internal use should be avoided by those taking either coagulants or anticoagulants.[1, 4, 5, 13]
Melissa officinalis L.	Lemon balm Balm	Herbs that can be safely consumed when used appropriately.[13]
Mentha x piperita L.	Peppermint	Herbs that can be safely consumed when used appropriately. Oil contains pulegone is toxic.[6, 13]
M. spicata L.	Spearment	Herbs that can be safely consumed when used appropriately.[13]
M. pulegium L.	Pennyroyal	Following use restrictions apply, unless otherwise directed by an expert qualified in the use of the described substance: not to be used during pregnancy.[13]
Monarda odoratissima Bench. *M. punctata* T. J. Howell	Horsebalm Horsemint	Following use restrictions apply, unless otherwise directed by an expert qualified in the use of the described substance: not to be used during pregnancy.[13]
Myosotis scorpioides L.	Lily of the valley	No information based on the databases searched is available.
Myrica penxylvanica Lois.	Bayberry	Herbs that can be safely consumed when used appropriately.[13]

Scientific name	Common name	Toxicity *
Myristica fragrans Houtt.	Nutmeg	Following use restrictions apply, unless otherwise directed by an expert qualified in the use of the described substance: not to be used during pregnancy.[13]
Myrtus communis L.	Myrtle	Do not take the essential oil internally except with professional advice.[6, 16]
Narcissus pseudonarcissus L.	Daffodil	Listed as a poisonous plant. Accidental ingestion of bulbs has produced several hours of severe discomfort. Handling large quantities of bulbs causes dermatitis in some individuals.[1, 5, 6]
Nasturtium officinale L.	Watercress	Herbs for which the following use restrictions apply, unless otherwise directed by an expert qualified in the use of the described substance: not to be used during pregnancy. Contraindicated in cases of gastric and duodenal ulcers, inflammatory kidney disorders and for children under four years of age.[6, 13]
Nepeta cataria L.	Catnip, catmint	Following use restrictions apply, unless otherwise directed by an expert qualified in the use of the described substance: not to be used during pregnancy. Classified by the FDA as an undefined safety.[6, 13]
Nymphaea alba L.	White water lily	No information based on the databases searched is available.
Ocimum basilicum L.	Basil	Following use restrictions apply, unless otherwise directed by an expert qualified in the use of the described substance: not to be used during pregnancy. Not recommended for infants or toddlers or for extended periods of time.[13]
Ocotea bullata (Birth) E. May	Stinkwood	Herbs for which insufficient data are available for classification.[13]

Scientific name	Common name	Toxicity *
Oenothera biennis L.	Evening primrose	Herbs that can be safely consumed when used appropriately, may have side effect due to the gamma-linolenic acid content.[17]
Olea europaca L.	Olive	Herbs for which insufficient data are available for classification.[13]
Oplopanax horridus (Sm.) Miq.	Devil's club	Herbs that can be safely consumed when used appropriately, it may cause allergic reactions to the spines.[13, 17]
Opuntia fragilis (Nutt.) Haw.	Prickly pear	Herbs for which insufficient data are available for classification.[13]
Origanum majorana L.	Sweet marjoram	Herbs that can be safely consumed when used appropriately.[13]
Origanum vulgare L. subsp. *hirtum* (Link) Ietswaart.	Marjoram, Oregano	Herbs that can be safely consumed when used appropriately.[13]
Paeonia lactiflora Pall.	White peony (Bia shao)	Herbs that can be safely consumed when used appropriately.[13]
P. officinalis L.	European peony	Herbs that can be safely consumed when used appropriately.[13]
P. suffruiticosa Andr.	Tree peony (Mu Dan Pi)	Following use restrictions apply, unless otherwise directed by an expert qualified in the use of the described substance: not to be used during pregnancy.[13]
Panax ginseng C. A. Meyer. *P. notoginseng* (Burk.) F. H. Chen *P. quinquefolium* L.	Asian ginseng Tian Qi American ginseng	May cause hypertension.[12] The LD_{50} for mice is 1667 mg/kg.[6] Those with hay fever, asthma, emphysema and cardiac or blood clotting problems and pregnant women should limit consumption.[17]

Scientific name	Common name	Toxicity *
Papaver bracteatum L. *P. rhoens* L.	Iranian poppy Poppy, corn poppy	Some of the alkaloids are toxic. This herb is on the poisonous plants list. Toxic substances are present in foliage and pods.[1, 4, 5]
Passiflora incarnata L.	Passion flower	Herbs that can be safely consumed when used appropriately. It is narcotic.[6, 13]
Pelargonium capitatum (L.) L'Her *P. graveolens* L'Her. ex. Aiton *P. crispum* (L.) L'Her. ex.Aiton	Wild rose geranium Lemon geranium Rose geranium	Herbs that can be safely consumed when used appropriately.[13]
Peltigera canina L.	Ground liverwort (Dog lichen)	Herbs for which insufficient data are available for classification.[13]
Petroselinum crispum (Mill.) Nym.	Parsley	Following use restrictions apply, unless otherwise directed by an expert qualified in the use of the described substance: not to be used during pregnancy, contraindicated in inflammatory kidney disease.[6, 13]
Phataris canariensis L.	Canary creeper	See *Tropaeolum majus*
Phellodendron chinensis Schneid.	Chinese corktree (Huang Bai)	Following use restrictions apply, unless otherwise directed by an expert qualified in the use of the described substance: not to be used during pregnancy.[13]
Phoradendron leucarpum (Raf.) Rev. & M. C. Johnst. *P. flavescens* (Push.) Nutt.	Mistletoe American mistletoe	Contraindicated in protein hypersensitivity and chronic progressive infections such as tuberculosis and AIDS, do not exceed recommended dose. Poisoning after the ingestion of berries has been reported, but no serious poisonings are documented.[8]

Scientific name	Common name	Toxicity *
Phyllanthus emblica L.	Myrobalan, emblic	Herbs that can be safely consumed when used appropriately.[13]
Physalis alkekengi L.	Chinese lantern	Listed as a poisonous plant.[1, 5]
Phytolacca americana L.	Pokeweed	Herbs for which significant data exist to recommend the following labeling: "To be used only under the supervision of an expert qualified in the appropriate use of this substance." Labeling must include proper use information: dosage, contraindications, potential adverse effects and drug interactions, and any other relevant information related to the safe use of this substance. It was reported that this herb has caused severe poisoning when used as folk medicine. The FDA classified this as an herb of undefined safety.[6, 13]
Picea mariana (Mill) Black *P. glauca* (Moench) Voss. *P. ables* (L.) Karst	Black spruce White spruce Norway spruce	Herbs for which insufficient data are available for classification.[13]
Pimpinella anisum L.	Anise	Following use restrictions apply, unless otherwise directed by an expert qualified in the use of the described substance: not to be used during pregnancy. May cause dermatitis.[6, 13]
Pinus mugo Torra var. Pumilio *P. palustris* Mill. *P. strobus* L. *P. albicaulis* Engelm. *P. contorta* Dougl. ex. Loud.	Dwarf mountain pine Southern pitch pine White pine Lodgepole pine	Herbs for which insufficient data are available for classification.[13]

Scientific name	Common name	Toxicity *
Piper methysticum G. Forst	Kava-Kava	Following use restrictions apply, unless otherwise directed by an expert qualified in the use of the described substance: not to be used during pregnancy, do not exceed recommended dose. Continued use may cause inflammation of the body and eyes.[6, 13]
P. nigrum L.	English pepper	Herbs that can be safely consumed when used appropriately.[13]
Plantago asiatica L. *P. psyllium* L.	Psyllium	Take with at least 250 ml of liquid, contraindicated in bowel obstruction, take other drugs 1 hour prior to consumption of psyllium.[3]
P. lanceolata L. *P. major* L.	Plantains	Herbs that can be safely consumed when used appropriately.[13]
Podophyllum peltatum L.	Mayapple	This herb is on the poisonous plants list. One case is recorded of poisoning from young shoots; fruits can cause catharsis. Note: Mayapple is too toxic to attempt self-medication.[1, 5, 17]
Polygala senega L.	Seneca snakeroot	Not to be used during pregnancy. Severe irritant due to overdose, may cause nausea, dizziness, diarrhea, and vomiting.[13, 17]
Polygonum bistorta L. *Polygonum multiflorum* Thunb.	Bistort root Flowery knotweed (Fu-ti)	Herbs that can be safely consumed when used appropriately.[13]
Populus balsamifera L. *P. candicans* L. *P. tremuloides* Michx	Poplar, Balm of gilead American aspen	Herbs that can be safely consumed when used appropriately.[13]

Scientific name	Common name	Toxicity *
Primula vulgaris Huds. *P. veris* L.	Primrose Cowslip	Not given during pregnancy, or to patients sensitive to aspirin or taking anticoagulant drugs. Severe dermatitis occurs in some people, from a skin irritant in glandular hairs on flower stalks and calyx.[6, 13]
Prunella vulgaris L.	Heal all, selfheal	Herbs that can be safely consumed when used appropriately.[13]
Prunus africana L.	African prune	No information based on the databases searched is available.
P. dulcis (Mill) D. A. Webb	Almond	Herbs for which significant data exist to recommend the following labeling: “To be used only under the supervision of an expert qualified in the appropriate use of this substance.” Labeling must include proper use information: dosage, contraindications, potential adverse effects and drug interactions, and any other relevant information related to the safe use of this substance.[13]
P. mume Siebold & Zucc.	Japanese apricot	Herbs that can be safely consumed when used appropriately.[13]
P. serotina J. F. Ehrb	Black cherry, wild cherry	Not for long-term use, do not exceed recommended dose. Poisoning has occurred after the ingestion of seed in fruits, from chewing twigs and from making tea from leaves. This herb is on the poisonous plants list.[1]
P. virginiana L.	Chokecherry	Contain cyanic acid which is toxic. Poisoning and death are reported of children who ate large quantities of fruits without removing seeds. This herb is on the poisonous plants list.[1, 5]
Pueraria lobata (Willd) Ohwi	Kudzu	Herbs that can be safely consumed when used appropriately.[13]
Pygeum africanum L.	Pygeum	No information based on the databases searched is available.
Pyrethrum cinerarifolium L.	Pyrethrom	See *Chrysanthemum cinerarifolium*

Scientific name	Common name	Toxicity *
Quercus alba L.	White oak	Contraindicated for external use with extensive skin surface damage, full baths with a significant amount of the tea are contraindicated in the following conditions: weeping eczema, skin damage over a large area, febrile and infectious disorders, cardiac insufficiency stages. This herb is on the poisonous plants list.[1, 5]
Ranunculus occidentalis Nutt.	Buttercup	May be poison for children. This herb is on the poisonous plants list.[1, 5]
R. ficaria L.	Pilewort	Herbs for which insufficient data are available for classification.[13]
Raphanus sativus L.	Radish	Herbs for which insufficient data are available for classification.[13]
Rhamnus catharticus L. *R. frangula* L. *R. purshianus* L.	Purging buckthorn Alder buckthorn Cascara sagrada	Not to be used during pregnancy. Contraindicated in intestinal obstruction, abdominal pain of unknown origin, or any inflammatory condition of the intestines and in children less than 12 years of age, not use in excess of 8-10 days. Frequent use may result in loss of water and salts, deposition of pigment in the intestinal mucosa and red urine. May cause dermatitis.[6, 13, 17]
Rheum palmatum L. *R. officinal* Baill. *R. tanguticum* L.	Chinese rhubarb Rhubarb Rhubarb	Contraindicated in intestinal obstruction, abdominal pain of unknown origin, or any inflammatory conditions of the intestines and in children less than 12 years of age, not use in excess of 8-12 days, individuals with a history of kidney stones should use cautiously. Poisoning and death are reported after the ingestion of leaves.[6]
Rhodiola rosea (L.) Scop.	Roseroot	See *Sedum rosea*

Scientific name	Common name	Toxicity *
Rhus radicans L.	Poison ivy	Herbs that can be safely consumed when used appropriately. Sap from most plant parts produces an irritating dermatitis after an initial sensitization, may cause death.[13]
R. toxicodendron L.	Roseroot	Herbs for which insufficient data are available for classification.[13]
Ribes nigrum L. *R. lacustre* (Pers.) Poir.	Black currant Black gooseberry	Herbs that can be safely consumed when used appropriately.[13]
Robertium macrorrhizum Pic.	Geranium	See *Geranium macrorrhizum*
Rosa canina L. *R. damascena* Mill.	Dog rose Damask rose	Herbs that can be safely consumed when used appropriately.[13]
Rosmarinus officinalis L.	Rosemary	Following use restrictions apply, unless otherwise directed by an expert qualified in the use of the described substance: not to be used during pregnancy. Oil can cause erythema.[6, 13]
Rubus chamaemorus L.	Cloudberry	No information based on the databases searched is available.
R. fruiticosus L.	Blackberry	Herbs that can be safely consumed when used appropriately.[13]
R. idaeus L.	Raspberry	Herbs that can be safely consumed when used appropriately.[13]
Rumex acetosella L. *R. obtusifolia* L.	Sorrel Dock	Individuals with a history of kidney stones should use this herb cautiously. Overdoses of root extract may cause diarrhea. This herb is on the poisonous plants list.[1, 4, 5, 6]
Ruscus aculeatus L.	Butcher's broom	Herbs that can be safely consumed when used appropriately.[13]

Scientific name	Common name	Toxicity *
Ruta graveolens L.	Rue, herb of grace	Following use restrictions apply, unless otherwise directed by an expert qualified in the use of the described substance: not to be used during pregnancy, contraindicated in poor kidney function, avoid prolonged exposure to sunlight. An acronarcotic poison in excessive doses internally.[4, 6, 13]
Salix alba L. *S. discolour* Muhlenb *S. caprea* L.	White willow Pussy willow	Herbs that can be safely consumed when used appropriately.[13]
Salvia officinalis L.	Sage	Following use restrictions apply, unless otherwise directed by an expert qualified in the use of the described substance: not to be used during pregnancy, not for long-term use, do not exceed recommended dose. Sage should not be taken in large doses for a long period.[4, 13]
S. sclarea L.	Clary sage	Following use restrictions apply, unless otherwise directed by an expert qualified in the use of the described substance: not to be used during pregnancy, not for long term use, do not exceed recommended dose.[13]
Sambucus nigra L. *S. canadensis* L.	Elderberry	Herbs that can be safely consumed when used appropriately. Mild symptoms have been reported of feeling unwell and vomiting. This herb is on the poisonous plants list.[1, 13]
S. racemosa L.	Elder	Herbs for which insufficient data are available for classification.[13]

Scientific name	Common name	Toxicity *
Sanguinaria canadensis L.	Bloodroot	This herb is on the poisonous plants list. Following use restrictions apply, unless otherwise directed by an expert qualified in the use of the described substance: not to be used during pregnancy, may cause nausea and vomiting.[1, 5, 13, 17]
Sanicula marilandica L.	Black snakeroot	Herbs for which insufficient data are available for classification.[13]
Santalum album L.	Sandalwood	Contraindicated in diseases involving the parenchyma of the kidney, not for use beyond 6 weeks. Santalol can causes dermatitis.[8]
Saponaria officinalis L.	Soapwort	Liquid may cause severe eye irritation. This herb is on the poisonous plants list. This herb should not be taken internally.[1, 4, 5]
Sarothamnus scoparius L.	Broom	Herbs for which insufficient data are available for classification.[13]
Sassafras albidum (Nutt.) Nees	Sassafras	Not for long-term use, do not exceed recommended dose.[6, 346]
Satureja hortensis L. *S. montana* L.	Savory	Herbs that can be safely consumed when used appropriately.[13]
Schisandra chinensis (Turcz.) Ball.	Schizandra, wu wei zi	Herbs that can be safely consumed when used appropriately.[13]
Scutellaria baicalensis Georgi. *S. lateriflora* L. *S. galericulata* L.	Baical skullcap Virginia skullcap Skullcap	Herbs that can be safely consumed when used appropriately. Classified by FDA as an herb of unidentified safety.[6, 13]
Sedum acre L.	Stonecrop, small houseleek	Herbs for whieh insufficient data are available for classification.[13]

Scientific name	Common name	Toxicity *
S. rosea (L.) Scop. subsp. *integrifolium*	Roseroot	No toxic reactions reported.[17]
Sempervivum tectorum L.	Houseleek, hens and chicks	No information based on the databases searched is available.
Senecio vulgris L.	Groundsel	This herb is on the poisonous plants list.[1, 5]
Senna alexandrina L.	Alexandrina	See *Cassia angustifolia*
Serenoa repens (Bartr.) Small.	Saw palmetto	Herbs that can be safely consumed when used appropriately.[13]
Sesamum indicum L. *S. orientale* L.	Sesame	Herbs that can be safely consumed when used appropriately.[13]
Shepherdia canadensis (L.) Nutt.	Buffalo berry	No information based on the databases searched is available.
Silybum marianum (L.) Gaertn.	Milk thistle	Herbs that can be safely consumed when used appropriately.[13]
Simmondsia chinensis (Link) C. Schneid	Jojoba	Toxic to animals.[8]
Solidago virgaurea L.	Goldenrod	Classed as a narcotic hallucinogen, seeds and flowers are very poisonous. In chronic kidney disorders, a practitioner should be consulted.[6]
Sorbus aucuparia L.	Mountain ash	Herbs for which insufficient data are available for classification.[13]
Spirea ulmaria L.	Meadowsweet	See *Filipendula ulmaria*

Scientific name	Common name	Toxicity *
Stachys officinalis (L.) Trev.	Betony, woundwort	Herbs that can be safely consumed when used appropriately. Overdose may result in irritation of the stomach.[6, 13]
Stellaria media (L.) Vill.	Chickweed	Herbs that can be safely consumed when used appropriately.[13]
Stevia rebaudiana (Bertoni) Bertoni	Stevia, Sweet herb of Paraguay	Herbs that can be safely consumed when used appropriately.[13]
Symhytum officinale L.	Comfrey	Herbs for which the following use restrictions apply, unless otherwise directed by an expert qualified in the use of the described substance: for external use only. The plant has been shown to be toxic to animals.[4, 13]
Syringa vulgaris L.	Lilac	Herbs for which insufficient data are available for classification.[13]
Syzygium aromaticum (L.) Merr. & Perry	Clove	Herbs that can be safely consumed when used appropriately.[13]
Tagetes lucida Cav.	Mexican mint marigold	Herbs for which insufficient data are available for classification. It may be sensitive to some people.[6, 13]
Tanacetum parthenium (L.) Schultz.	Feverfew	See *Chrysanthemum parthenium*
T. cinerarifolium L.	Pyrethum	See *Chrysanthemum cinerarifolium*
T. vulgare L.	Tansy	Following use restrictions apply, unless otherwise directed by an expert qualified in the use of the described substance: not to be used during pregnancy. Potentially toxic, oil could be lethal.[4, 6, 13]

Scientific name	Common name	Toxicity *
Taraxacum officinale G. H. Weber ex Wigg.	Dandelion	Herbs that can be safely consumed when used appropriately. It may cause excessive urination. Avoid use if gallstones are present.[13, 17]
Taxus x media Rehd. *T. brevifolia* Nutt.	Yew	This herb is on the poisonous plants list. All parts except the fleshy arils (seed within the aril is deadly) can be poisonous.[1, 5, 17]
Thymus citriodorus (Pers.) Schreb.	Lemon thyme	Herbs that can be safely consumed when used appropriately.[13]
Thymus vulgaris L	Garden thyme	Herbs that can be safely consumed when used appropriately.[13]
Tillia cordata Mill. *T. europaea* L.	Linden, small-leaved lime	Herbs that can be safely consumed when used appropriately. Too frequent use of linden flower tea may result in heart damage.[6, 13]
Tribulus terrestris L.	Puncture vine	Listed as a poisonous plant in Pakistan.
Trifolium incarnatum L. *T. pratense* L.	Clover (crimson) Clover (red)	Following use restrictions apply, unless otherwise directed by an expert qualified in the use of the described substance: not to be used during pregnancy. This herb is on the poisonous plants list.[1, 5, 13]
Trigonella foenum-gracecum L.	Fenugreek	Following use restrictions apply, unless otherwise directed by an expert qualified in the use of the described substance: not to be used during pregnancy. Can be toxic in overdose.[6, 13]
Tropaeolum majus L.	Canary creeper, nasturtium	Herbs for which insufficient data are available for classification.[13]
Tussilago farfara L.	Colisfoot	Following use restrictions apply, unless otherwise directed by an expert qualified in the use of the described substance: not to be used during pregnancy.[13]

Scientific name	Common name	Toxicity *
Ulmus rubra Muhl. *U. procera* L.	Slippery elm, sweet elm	Herbs that can be safely consumed when used appropriately, pollen is allergenic.[6, 13]
Umbelluslaria california Nutt.	California laurel Bay laurel	Herbs for which insufficient data are available for classification.[13]
Uncaria tomentosa (Willd.) DC. *U. gulanensis* (Aubl.) Gmel.	Cat's claw	May be toxic based on the toxicological study.[11]
Urtica dioica L. *U. urens* L.	Stinging nettle	Herbs that can be safely consumed when used appropriately. It was reported that toxic liquid in hairs causes intense itching and pain. This herb is on the poisonous plants list.[1, 5, 13]
Vaccinium macrocarpon L. *V. vitis-idaea* L.	Cranberry Mountain cranberry cowberry, foxberry	Not considered toxic, it may cause diarrhea or gastrointestinal problems if too much cranberry fruit and juice is consumed.[17]
V. mrytilloides Michx.	Blueberry	Herbs that can be safely consumed when used appropriately.[13]
V. myrtillus L. *V. oreophilum* Rydb.	Bilberry	Not toxic, fresh berries may cause diarrhea, leaves (as tea) must not be consumed over a long period of time.[17]
Valeriana officinalis L.	Valerian	Herbs that can be safely consumed when used appropriately. After 2-3 weeks, stop taking valerian for a few days, may be narcotic.[4, 6, 13]
Verbascum thapsus L.	Mullein	Herbs that can be safely consumed when used appropriately. May be mildly toxic.[4, 13]
Veronica officinalis L.	Speedwell	Herbs that can be safely consumed when used appropriately.[13]

Scientific name	Common name	Toxicity *
Vetiveria zizaniodes L. (Nash.)	Vetiver	Following use restrictions apply, unless otherwise directed by an expert qualified in the use of the described substance: not to be used during pregnancy.[13]
Viburnum opulus L.	Highbush cranberry	Herbs that can be safely consumed when used appropriately. Mild symptoms are reported of feeling unwell and vomiting.[6, 13]
Vinca minor L.	Periwinkle	Herbs for which significant data exist to recommend the following labeling: "To be used only under the supervision of an expert qualified in the appropriate use of this substance." Labeling must include proper use information: dosage, contraindications, potential adverse effects and drug interactions, and any other relevant information related to the safe use of this substance. Contraindicated with low blood pressure and constipation. May be toxic.[6, 13, 20]
Viola tricolor L.	Pansy	Herbs that can be safely consumed when used appropriately.[13]
Viscum album L.	European mistletoe	Contraindicated in protein hypersensitivity and chronic progressive infections such as tuberculosis and AIDS, do not exceed recommended dose. Poisoning after the ingestion of berries has been reported, but no serious poisonings are documented. Some side effects such as shivering, high fever, headache, etc.[3]
Vitex agnus-castus L. *V. negundo* L.	Chaste tree Chinese chaste tree	Following use restrictions apply, unless otherwise directed by an expert qualified in the use of the described substance: not to be used during pregnancy, may counteract the effectiveness of birth control pills.[13]
Vitis labrusca L *V. vinifera* L	Grape	No information based on the databases searched is available.

Scientific name	Common name	Toxicity *
Withania somnifera Dunal.	Ashwagandha, Indian ginseng	Following use restrictions apply, unless otherwise directed by an expert qualified in the use of the described substance: not to be used during pregnancy, may potentiate the effects of barbiturates.[13]
Yucca aloifolia L.	Yucca	Herbs that can be safely consumed when used appropriately.[13]
Y. glauca L.	Soapweed	No information based on the databases searched is available.
Zanthoxylum americanum Mill.	Prickly ash, toothache tree	No information based on the databases searched is available.
Zea mays L.	Corn	Herbs that can be safely consumed when used appropriately.[13]
Zingiber officinale Roscoe	Ginger	Herbs that can be safely consumed when used appropriately.[13]

*The information contained herein is not intended to be a complete guide to the toxicity of the medicinal plants listed, or a recommendation.

References

1. Anonymous. 1990. Canadian Poisonous Plants Database. Health Canada.

2. Arenz, A., M. Klein, K. Fiehe, J. Gross, C. Drewke, T. Hemscheidt and E. Leistner. 1996. Occurrence of neurotoxic 4'-O-methylpyridoxine in *Ginkgo biloba* leaves, ginkgo medications and Japanese ginkgo food. Planta Medica 62: 548-551.

3. Bisset, N. G. 1994. Herbal Drugs and Phytopharmaceuticals. CRC Press, London. 566 p.

4. Bremness, L. 1994. The Complete Book of Herbs. Dorling Kindersley Ltd., London. 304 p.

5. Brown, D. 1998. Poisonous plants list. www.ansci.cornell.edu/plants/alpharest.html.

6. Duke, J. A. 1985. CRC Handbook of Medicinal Herbs. CRC Press, Inc., Boca Raton, FL. 677 p.

7. Duke, J. A. and J. L. duCellier. 1993. CRC Handbook of Alternative Cash Crops. CRC Press, London. 536 p.

8. Hedberg, I. 1987. Research on medicinal and poisonous plants of the tropics past, present and future. In: A. J. M. Leeuwenberg (ed.). Medicinal and Poisonous Plants of the Tropics. Pudoc. Wageningen, The Netherlands. p. 9-15.

9. Henry, M., A. M. Edy, P. Desmarest and J. du Manuir. 1991. *Glycyrrhiza glabra* L. (Licorice): cell culture, regeneration, and the production of glycyrrhizin. In: Bajaj, Y. P. S. (ed.). Biotechnology in Agriculture and Forestry 15. Medicinal and Aromatic Plants III. Springer-Verlag Co., New York. p. 270-282.

10. Keeler, R. F. and A. T. Tu. 1983. Handbook of Natural Toxins, Vol. 1: Plant and Fungal Toxins. Marcel Dekker, New York.

11. Keplinger, K., G. Laus, M. Wurm, M. P. Dierich and H. Teppner. 1999. *Uncaria tomentosa* (Willd.) DC., ethnomedicinal use and new pharmacological, toxicological and botanical results. J. Ethnopharmacology 64: 23-34.

12. Li, T. S. C. and L. C. H. Wang. 1998. Physiological components and health effects of ginseng, echinacea and sea buckhtorn. In: G. Mazza (ed.). Functional Foods, Biochemical and Processing Aspects. Technomic Publishing Co.. Inc., Lancaster, PA. 460 p.

13. McGuffin, M., C. Hobbs, R. Upton and A. Goldberg. 1997. Botanical Safety Handbook. CRC Press, New York.

14. Rizvi, M. A., G. R. Sarwar and A. Mansoor. 1998. Poisonous plants of medicinal use growing around Madinat al-Hikmah. Hamdard Medicus 41: 88-95.

15. Sheikh, N. M., R. M. Philen and L.A. Love. 1997. Chaparral-associated hepatotoxicity. Arch. Med. 157: 913-919.

16. Small, E. (ed.). 1997. Culinary Herbs. NRC Research Press, Ottawa. 710 p.

17. Small, E. and P. M. Catling (eds.). 1999. Canadian Medicinal Crops. NRC Research Press, Ottawa. 240 p.

18. Srivastava, K. C. and N. Malhotra. 1991. Acetyl eugenol, a component of oil of cloves (*Syzygium aromaticum* L.) inhibits aggregation and alters arachidonic acid metabolism in human blood platelets. Prostaglandins Leukot Essent Fatty Acid 42: 73-81.

19. Tyler, V. E. 1993. The Honest Herbal. A Sensible Guide to the Use of Herbs and Related Remedies. Pharmaceutical Products Press, New York. 375 p.

20. Yokoyama, M. and S. Inomata. 1998. *Catharanthus roseus* (Periwinkle): in vitro culture, and high-level production of arbutin by biotransformation. In: Bajaj, Y. P. S. (ed.). Biotechnology in Agriculture and Forest 41. Medicinal and Aromatic Plants X. Springer-Verlag Co., New York. p. 67-80.

CHAPTER 3

Essential Oils and Fractions from Medicinal Plants

For practical purposes, essential or volatile oils may be defined as odoriferous bodies of an oily nature obtained from plant leaves, flowers, fruit, seed, wood, resin, bark, or roots. They are obtained generally in liquid, semisolid, or solid form, and usually can be steam distilled and the solvent extracted, including supercritical carbon dioxide. Essential oils do not dissolve in water or dissolve only very poorly, evaporate at room temperature without residue, and often have characteristic strong odors and tastes.

An essential oil is not a simple compound; it is a mixture of various compounds, mainly terpenes and terpene derivatives, aldehydes, or esters. In cosmetics, household chemicals, and food flavoring preparations such as for soft drinks, certain components of essential oils can be replaced with synthetic chemicals. However, the characteristics of essential oils, especially natural essence, cannot be replaced. They are the initial basic materials for producing new fragrances for perfumes used by the cosmetic industry.

Because of the lipophilic character of compounds in oil, essential oils are also used for air purification, in aroma therapy, bathing products, car fresheners, compresses, inhalants, perfumes, room fresheners, and skin creams. They are used in saunas and for relief of stress insomnia. It is important to understand the strength and power of essential oils, because small amounts of some taken internally can be lethal. For this reason, one should never consume any essential oil internally without consulting a physician or pharmacist.

Table 3. Essential oil and its fractions from medicinal plants.

Scientific name	Common name	Essential oil	
		Major components	Source
Abies balsamea (L.) Mill.	Balsam fir	Monoterpenes, alpha-pinene, phyllandrene[101]	Leaf
Achillea millefolium L.	Yarrow	Azulenes, lactones, linalool, camphor, sabinene, bernyl acetate, limonene chamazulene, sesquiterpene, germacrene, pinene, sabinene, achilline, cineole[34, 60, 110, 125]	Flowering stems
Aconitum napellus L.	Monkshood	Volatile oil, flower essence[110]	Flower
Acorus calamus L. var. Americanus Wolff. *A. tatarinowii* L.	Calamus, sweet flag Shi Chang Pu	Sesquiterpenes, phenylpropanes, and monoterpenes. Beta-asarone, acorin, calamenenol, beta-farnesene, methyl eugenol, cadinol, linalool, azulene, shyobunone, camphor, epishyobunone, cineole[30, 34, 101, 110, 125, 137]	Leaf, rhizome, root
Actaea alba L. *A. rubra* (Ait.) Wild.	White baneberry Red baneberry	Monoterpenes, flower essences[110]	Floral fragrances
Aesculus hippocastanum L.	Horse chestnut	Polyine, acthusin, aethusanol A and B, acetylenes[68]	Seed
Agastache foeniculum L. *A. anethrodora* L.	Aniseed	Methyl chavicol (benzene, 1-methyoxy-4-(2-propenyl), estragole), limonene, γ-elemene, (E)-anethole, phyllandrene[65, 91, 110]	Seed
Agrimonia cuparoria L.	Agrimony	Volatile oil[68]	Seed
Agropyron repens (L.) Beauv.	Couch grass	Agropyrene, carvacrol, trans-anethole, carvone, flower essences[110]	Seed, root, flower

Scientific name	Common name	Essential oil	
		Major components	Source
Alchemilla vulgaris L. *A. xanthochlora* Rothm.	Lady's mantle	No information based on the database searched is available.	
Allium cepa L.	Onion	Dipropyl disulphide, methylalliin, cycloalliin, dihydroalliin, dipropyl trisulphide[101]	Bulb
A. sativum L.	Garlic	Alliin, allicin, allypropyl, disulphide[34, 101]	Cloves
A. schoenoprasum L.	Chives	Volatile oil, cycloalliin, allicin[101]	Bulb
Alnus crispus (Ait.) Pursh *A. incana* (L.) Moench. subsp. *tenufolia* *A. glutinosa* (L.) Gaertn.	Alder	Oleoresins/essential oils, alder flower essence[110]	Bark, flower
Aloe vera (L.) Burm. *A. barbadensis* L.	Aloe	Flower essences[45, 111]	Flower
Althaea officinalis L.	Marshmallow	No information based on the database searched is available.	
Amelanchier alnifolia Nutt.	Saskatoon	Saskatoon flower essence[110, 112]	Flower
Ananas comosus (L.) Merr.	Pineapple	Methanol, ethanol, propanol, isobutanol, pentanol, capronate, lactate, diacetyl[42]	Fruit
Anaphalis margaritacea (L.) Bench & Hook	Everlastings	Hexyl, phenethyl, heptyl-2-methyl butyrates[110]	Leaf

Scientific name	Common name	Essential oil	
		Major components	Source
Anethum graveolens L.	Dill	Carvone, limonene, phyllandrene, eugenol, pinene, 3,9-epoxy-p-menth-1-one, 4,5-dimethoxy-6-(2-propenyl)-1,3-benzodioxole[30, 42]	Seed
Angelica archangelica L.	Angelica	Beta-phyllandrene, limonene, sesquiterpenes, pinene, limonene, linalool, borneol, camphoric, anisic, and azelaic acids, benzene, dicarbonic acid, myristic acid, carvacrol, ligustilide, alpha-pinene, allo-ocimene, trans-beta-farnescene, beta-ocimene[42, 101]	Root, seed
A. sinensis (Oliv.) Diels	Dong quai	Butylidine, phthalide, ligustilide, carvacrol, sesquiterpenes, phyllandrene, limonene[34]	Root
Antennaria magellanica Schultz	Everlastings	See *Anaphalis margaritacea*	
Anthemis nobilis L.	Chamomile	See *Chamaemelum nobile*	
Apium graveolens L.	Celery	Limonene, phthalides, beta-selinene, selinene, apiol, santalol, sedanolide, lsedanic acid, citric, isocitric, fumaric, malic, and tartaric acids From seed oil: oleic, palmitic, paliloleic, petroselinic, petroselaidic, stearic, myristic, myristoleic acid. From celery tissue: citric, isocitric, fumaric acid, malic, tartaric acid, selinene, pinene, elemene, humulene, cymene, caryophyllene, cymol, terpineol, ionone. From root: malic, citric, pyruvic, malonic, glycolic acid[34, 42, 101]	Seed, aerial part, root
Aralia racemosa L. *A. nudicaulis* L.	Spikenard	Araliene, flower essences[110].	Root

Scientific name	Common name	Essential oil	
		Major components	Source
Arctium lappa L.	Burdock	Sesquiterpenes, inulin, volatile oil[34]	Root
Arctostaphylos uva-ursi (L) Spreng	Bearberry, Uva-ursi	Flower essences[110]	Flower
Armoracia rusticana Gaertn, Mey & Scherb.	Horseradish	Allyl phenylethyl isothiocyanate, 2-phenylethyl isothiocynate[124]	Root
Arnica latifolia Bong. *A. montana* L. *A. chamissonis* subsp. *foliosa* *A. condifolia* Hook *A. fulgens* Pursh *A. sororia* Greene	Arnica	Thymol-hydroquinone dimethyl ether, mucilage, isobutyric ester of phlorol, arnidiol, foradiol[86, 101, 110]	Root, flower, leaf
Artemisia absinthium L.	Wormwood	Phelladrene, thujone, pinene, artabsin, anabsinthin, azulenes, thujyl alcohol[34, 42, 110]	Aerial part
A. annua L.	Chinese wormwood, Qing Hao	Abrotamine, beta-bourbonene, artemisia ketone[34, 101]	Aerial part
A. dracunculus L.	Tarragon	Estragole, ocimene, methyl chavicol, alpha- and beta-pinene, camphene, limonene, nerol, sabinene, myrcene, menthol, trans-anethole, anisole, anisic acid[42, 101]	Aerial part

Scientific name	Common name	Essential oil	
		Major components	Source
A. tridentata Nutt.	Sagebrush	1,8-cineole, palmitic acid, n-undecylic acid[110]	Aerial part
A. vulgaris L.	Mugwort	Cineole, thujone, quebrachitol, tauremisin, sitosterol, tetracosanol, delta-hydromatricaria-ester[42, 101, 110]	Aerial part
Asarum canadensis L.	Wild ginger	Diasarone, asarone, methylisoeugenol, pinene, linalool, borneol, terpineol, geraniol, eugenol[101]	Root
Asparagus officinalis L. *A. racemosus* Willd.	Asparagus	Asparagus aroma vetiver oil[42, 43]	Rhizomes
Astragalus americana Bunge. *A. membranaceus* Bunge *A. sinensis* L.	Astragalus Huang Qi	Methyl palmitate, dimethyl azelate, methy levulinate[94]	Root
Avena sativa L.	Oat	Palmitic acid, oleic acid, alpha-linoleic acid, beta-linoleic acid[110]	Seed
Baptisia tinctora (L.) R. Br. ex. Ait. f.	Wild indigo, clover broom	No information based on the database searched is available.	
Bellis perennis L.	Daisy	Volatile oil[110]	Flower
Berberis aquifloium L.	Oregon grape	See *Mahonia aquifolium*	
B. vulgaris L.	Barberry	No information based on the database searched is available.	

Scientific name	Common name	Essential oil	
		Major components	Source
Beta vulgaris L.	Beet root	Monoterpenes[109, 128]	Seed
Betula lenta L.	Birch	Birch bark: creosol, carbonic acid, dimethyl phenols, cresol, xylenol gualacol, pyrocatechol, and pyrobetulin Birch bud oil: caryophyllene, alpha-betulenolacetate, betulol, betulenol, betulene and other sesquiterpenes Birch oil: methyl salicylate[110]	Leaf, bark, bud
Borago officinalis L.	Borage	Gamma-linolenic acid, pyrrolizidine alkaloid, and volatile oil from flower[92, 110]	Seed
Boswellia sacra Roxb. ex Colebr. *B. carteri* Birdwood	Frankincense	Verbenol, pinene, dipentene, limonene, thujone, phyllandrene, cymene, myrcene, terpinene[101]	Gum
Brassica nigra (L.) Kock.	Black mustard	Allyl isothiocyanate (obtained from seeds only by macerating in warm water during production)[20, 101]	Seed
Buxus sempervirens L.	Boxwood	4-mercapto-4-methylpentan-2-one, alpha-tocopherol acetate[139]	Root, bark
Calamintha nepeta (L.) Savi *C. ascendens* L. *C. officinalis* L.	Basil thyme Calamint Mountain mint	Pulegone, menthone, isomenthone[101, 124]	Leaf
Calendula officinalis L.	Marigold	Calendulin, citral, nerol, citronellol, limonene, geraniol, erucic acid[42]	Flower

Scientific name	Common name	Essential oil	
		Major components	Source
Camellia sinensis (L.) Kunize	Green tea	Monoterpene aldehydes and alcohols, hexenal, hexenol, butylaldehyde, isovaleraldehyde, phenols, cresol[24]	Aerial part
Cananga odorata (Lam) J. D. Hook & T. Thompson	Ylang-Ylang	Benzyl bemzoate, linalool, safrole, eugenol, geraniol, methyl salicylate, benzyl acetate, geranyl acetate, germacrene, terpineol, famesene, delta-cadinene, p-cresyl, methyl benzoate, pinene, safrole, valeric acid, beta-caryophyllene, para-methylanisole, humulene[101]	Fresh flower
Cannabis sativa L.	Hemp	Sesquiterpenes, terpenes, alpha-pinene, beta-pinene, camphene, limonene, beta-phyllandrene, beta-caryophyllene, allo-aromadendrene, terpenic oxides[101, 110]	Seed, flower
Capsella bursa-pastoris (L.) Medik.	Shepherd's purse	Phyllandrene, cis-3-hexen-1-ol, 3-octanol, camphor, limonene, n-decane, geraniol, thymol, carvacrol, flower essences[110]	Leaf, root, flower
Capsicum annuum L. var. Annuum *C. annum* L. var. Grossum *C. frutescens* L. *C. annum* L. var. Acuminatum	Cayenne pepper Sweet pepper Chilli pepper	Capsaicin, volatile oil[124, 147]	Fruit
Carbenia benedicta L.	Blessed thistle	See *Cnicus benedictus*	
Carduus benedita L.	Blessed thistle	See *Cnicus benedictus*	

Scientific name	Common name	Essential oil	
		Major components	Source
Carica papaya L.	Papaya	Palmitic, stearic, arachic acid, oleic, linoleic acid[96]	Seed
Carthamus tinctorius L.	Safflower	Linoleic acid and alpha-tocopherol from seed oil, carthamin from flower. Others such as erucic acid, ubiquinone-9, fatty acid[58, 68]	Seed, flower
Carum carvi L.	Caraway	Carvone, limonene, carveol, phyllandrene, dihydrocarveol, dihydrocarvone, pinene, terpene[46 51, 53]	Fruit
Cassia senns L. *C. angustifolia* Vahl.	Tinnevelly senna Alexandria senna	Carvone, limonene[34, 101]	Aerial part
Castanea sative Mill	Chestnut	No information based on the database searched is available.	
Catharanthus roseus (L.) Don	Periwinkle	See *Vinca minor*	
Caulophyllum thalietroides (L.) Michx. *C. giganteum* (F.) Loconte & W. H. Blackwell	Blue cohosh	No information based on the database searched is available.	
Cedrus libani Riich subsp. *atlantica*	Cedarwood	Himachalenes, atlantones, cis- and trans-10-11 dihydroatlantones, camphor, longipinene, alpha-ylangene, sativene, cedrene, 8,9-dehydroeoisolongifolene[29, 34, 101]	Leaf, wood
Centaurea calcitrapa L.	Star thistle	Carlinoxide[15]	Leaf, flower

Scientific name	Common name	Essential oil	
		Major components	Source
Centella asiatica (L.) Urb.	Gotu Kola	Sesquiterpenoids, beta-caryophyllene, alpha-humulene, germacrene D. Alpha-pinene, beta-pinene, alpha-copaene, beta-caryophyllene, trans-beta-farnesene, alpha-humulene[73, 152]	Aerial part
Chamaelirium luteum (L.) A. Gray	False unicorn root, fairywand	No information based on the database searched is available.	Root
Chamaemelum nobile (L.) All.	Roman Chamomile	Azulene, proazulenes, farnesene, alpha-bisabolol, spiroether, tiglic, angelic acid esters, chamazulene[42, 101]	Flower, root
Chamaenerion angustifolium (L.) Scop.	Firewood	See *Epilobium angustifolium*	
Chenopodium album L.	Lamb's quarters	Ascaridole, cymeme, terpenes[110]	Leaf, seed
C. ambrosioides L.	Wormseed	Ascaridole, geraniol, methyl salicylate[42]	Seed
Chrysanthemum cinerarifolium (Trevir.) Vis.	Pyrethrum	L-camphor, L-borneol, terpenes, esters, beta-cyclopyrethrosin, chrysanolide, chrysanin, beta-amyrin, limonene, cineole[101, 157]	Flower, shoot
C. parthenium (L.) Burm.	Feverfew	Camphor (chamomile camphor), dihydrocarvone, borneol, thujone, alpha-pinene, monoterpernoids, sesquiterpene[34, 42]	Flower, leaf
C. vulgare L.	Tansy	See *Tanacetum vulgare*	
Cichorium intybus L.	Chicory	Inulin, sesquiterpene lactones[68]	Root

Scientific name	Common name	Essential oil	
		Major components	Source
Cimicifuga racemosa (L.) Nutt.	Black cohosh	No information based on the database searched is available.	
Cinnamomum verum J. Presl.	Cinnamon	Leaf: eugenol, eugenol acetate, benzyl bezoate, linalool, safrol, cinnamaldehyde. Bark: cinnamaldehyde, eugenol, benzaldehyde, cuminaldehyde, pinene, cineol, phyellandrene[34, 42]	Leaf, bark
C. cassia (Nees) Nees & Eberm.	Cassia	Cinnamic aldehyde with methyl eugenol, salicylaldehyde, methylsalicylaldehyde[101]	Leaf, bark
C. camphora (L.) Nees & Eberm	Camphor	Camphor, safrole, terpene alcohols, linalool sesquiterpenes. cineole, aldehyde, camphene, dipentene, limonene, phyllandrene, pinene[101]	Leaf
Citrus limon (L.) Burm.	Lemon	Limonene, gamma-terpinene, beta-pinene, sabinene, alpha-terpinene, alpha-pinene, citral[34, 101]	Crushed fruit
C. bergamia Risso & Poit	Bergamot orange	d-Limonene, linalool, linalyl acetate, bergapten, diterpene[101]	Peel
C. reticulata Blanco	Mandarin	Limonene, gamma-terpinene, myrcene, p-cymene, nerol, thujene, pinene, selinene[61]	Peel
C. aurantifolia Swingle	Lime	Limonene, gamma-terpinene, alpha-terpinene, linalool, nonanol, citronellal[101]	Crushed fruit
C. aurantium L.	Bitter orange	Linalool, linalyl acetate, geraniol, hesperidin, citric acid, malic acid, beta-pinene, sabinene, mycene, terpinene[61]	Leaf

Scientific name	Common name	Essential oil	
		Major components	Source
C. paradisi Macfad	Grapefruit	d-Limonene, gamma-terpinene, palmitic, stearic, linoleic, oleic acids[101]	Pulp
Cnicus benedictus L.	Blessed thistle	Sesquiterpene lactones, volatile oil[68]	Leaf; flower
Cochlearia amorucia L.	Horseradish	See *Armoracia rusticana*	
Codonopsis pilosula (Franch) Nannfeldt. *C. tangshen* Oliver	Codonopsis, Dang Shen, bellflower	From *C. clematidea* Schrenk, 0.3% essential oil were extracted. From *C. lanceolata*, trans-2-hexanal, 1-hexanol, cis-3-hexanol, trans-2-hexanol[85]	Aerial part
Coffea arabica L.	Coffee	Palmitic, stearic, oleic, linoleic acids, 3, 4-benzopyrene, lanosterol, coffeasterol, cycloartenol, cycloartanol[4, 16, 52]	Green beans
Colchicum autumrnale L.	Crocus Meadow saffron	No information based on the database searched is available.	
Collinsonia canadensis L.	Pilewort	Germacrene, caryophyllene, elemicin, beta-elemene, delta-cadinene, alpha-naginatene, unidentified furano-monoterpene	Root
Commiphora molmol Engl. ex Tschirch *C. myrrha* (Nees) Engl.	Myrrh	Heerabolene, eugenol, furanosesquiterpenes[101]	Gum, resin
Convallaria majalis L.	Lily of the valley	Alkylmethoxy pyrazines, (2E, 6Z)-nonadienal, (2E, 6Z)-nonadienol, (Z)-3-hexenol, (Z)-3-hexenyl acetate, phenylacetic aldehyde[68]	Flower

Scientific name	Common name	Essential oil	
		Major components	Source
Convolvulus arvensis L. *C. sepium* L.	Bindweed	Unidentified (oil from seed contains iodine), flower essences[101, 110, 111]	Seed, flower
Coriandrum sativum L.	Coriander (fruit) Cilantro (leaf)	Linalool (coriandrol), alpha-pinene, terpinene, cymene, decylaldehyde, borneol, geraniol, carvone, anethole[101]	Fruit, leaf
Cornus canadensis L. *C. florida* L.	Bunchberry Dogwood	Palmitic acid, oleic acid, oil (not edible) used for soap, flower essences[79, 110]	Seed, bark flower
Corylus cornuta Marsh. *C. rostrata* Marsh. *C. avellana* L.	Hazelnut	Palmitic acid, paraffin, volatile oil, hazelnut oil may used as carrier oil[110]	Leaf, nut, flower
Crataegus monogyna Jacq. *C. pinnatifida* Bge.	English hawthorn Chinese hawthorn	Triterpenoid acid, corosolic acid[2, 121]	Leaf
Crocus sativus L.	Saffron	Pinene, safranal, cincole[124]	Flower, bulb
Cryptotaenia japonica Hassk.	Japanese parsley, Mitsuba	Cryptotaenene, mitsubene, (2E,4E)-2,4-decadienal, (2E, 4Z)-2,4-decadienal, (2E, 4Z)-2,4-heptadienal, palmitic acid, linoleic acid, ethyl palmitate, ethyl linoleate, ethyl linolenate, abietatriene, ferruginol[71]	Leaf
Cucurbita pepo L.	Pumpkin	Unsaturated fixed oil including linoleic and oleic fatty acids[45]	Seed

Scientific name	Common name	Essential oil	
		Major components	Source
Cuminum cyminum L.	Cumin	Aldehydes, cumin ester, limonene, pinene, alpha-terpineol, cymene, phyllandrene, myrecene, camphene, borneol (Essential oil is fungistatic at low doses and also repellent to insect *Allocophora foveicollis.*)[30, 101]	Seed
Curcuma domestica L. *C. xanthorrhiza* L. *C. longa* L.	Curcuma, Jiang Huang	Sesquiterpene, zingiberen, turmeron, p-cymene, 1,8-cineole, alpha-phyllandrene, sabinene, borneol, ar-turmerone, alpha-atalantone, gamma-atlantone[4, 101, 149]	Root
Cuscuta epithymum Murr. *C. chinensis* Lam.	Dodder	Volatile oil, flower essence[110]	Flower
Cymbopogon citratus (DC ex Nees) Stapf.	Lemon grass	Geraniol, neral, myrcene, beta-ionone, methyl ionone, citronellal, dipentene, linaloolm nerol, methylheptenone, geraniol, cymbopogonol, citral, alpha-terpineol, farnesol, aldehydes[101, 132, 142]	Leaf
C. martinii (Roxb) Wats	Palmorosa, gingergrass	Geraniol, geranyl acetate, linalool, gamma-terpinene, phyllandrene, citronellol, citral[101]	Leaf
C. nardus L. *C. winterianus* Jowitt	Citronella	Geraniol, aldehydes, citronellol, citronellal, limonene, camphene[101]	Leaf

Scientific name	Common name	Essential oil	
		Major components	Source
Cynara scolymus L. *C. cardunculus* L.	Artichoke Wild artichoke	No information based on the database searched is available.	
Cypripedium calceolus L.	Lady's slipper	Volatile oil[42]	Flower
Cytisus scoparius (L.) Link.	Broom, scoparium	Volatile oil[42]	Flower
Daucus carota L.	Carrot	Seed: sabinene, daucene, beta-bisabolene, beta-caryophyllene, geraniol, linalool, carotol, geraniol acetate, asarone, daucol. Leaf: sabinene, linalyl acetate, carvone, carotol. Root: beta-bisabolene, cis- and trans-asarone, asarone aldehyde, eugenol, 2-hydroxy-4-methoxyacetophenone, vanillin[30, 101]	Seed, leaf, root
Digitalis purpurea L.	Foxglove	Linoleic acid, oleic acid[22, 68, 124]	Leaf
Dioscorea oppositae Thunb. *D. villosa* L.	Chinese yam Mexican yam	No information based on the database searched is available.	

Scientific name	Common name	Essential oil	
		Major components	Source
Echinacea angustifolia DC. Cornq. *E. pallida* (Nutt.) Nutt. *E. pururea* (L.) Moench.	Narrowleaf echinacea Pale-flower echinacea Purple coneflower	Tetrahydrosesquiterpen\e hydrocarbon, 2-methyltetradeca-5,12-diene, 2-methyltetradeca-6.12-diene, penta-(1,8Z)-diene, 1-pentadecene, (E)-10-hydroxy-4,10-dimethyl-4,11-dodecadien-2-one (echinolone), humulene, caryophyllene, caryophyllene eposide, pentadeca-(8Z)-en-2-one, geranylisobutyrate, dodeca-2,4-dien-1-yl-isovalerate, palmitic, linolenic acid, germacrene D, vanillin, p-hydroxycinnamic acid methyl ester, borneol, bornylacetate, pentadeca-8-en-2-one, caryophyllene, beta-farnesene, myrcene, limonene, carvomenthene, epishiobunol, 1,8-pentadecadiene. Volatile components from: Flower: beta-myrcene, alpha-pinene, limonene, camphene, beta-pinene, trans-ocimene, 3-hexen-1-ol, 2-methyl-4-pentenal. Root: alpha-phyllandrene, dimethyl sulfide, 2-methyl-butanal, 3-methyl-butanal, 2-methyl-propanal, acetaldehyde, camphene, 2-propanal, limonene. Flower and stems: alpha- and beta-pinene, beta-myrcene, ocimene, limonene, camphene, terpinene[90]	Flower, root, stem
Echinopanax horridus (Sm) Decne. & Planch. ex. H. A. T. Harms	Devil's club	See *Oplopanax horridus*	
Elaeagnus angustifolia L.	Russian olive	Volatile oil, flower essence. Oil from leaf for making wax[56, 110]	Flower, leaf

Scientific name	Common name	Essential oil	
		Major components	Source
Eleutherococcus senticosus (Rupr. ex Maxim) Maxim.	Siberian ginseng	Unidentified essential oil in roots (0.8%), fruit (0.5%), leaves (0.31%), and stem (0.26%)[42]	Root, fruit, leaf
Elymus repens L.	Couch grass	See *Agropyron repens*	
Ephedra distachya L. *E. sinica* Stapf. *E. nevadensis* Wats.	Ephedra Ma Huang Mormon tea	Volatile oil[34]	Leaf
Epilobium angustifolium L.	Fireweed	Volatile oil, also sun infused oil from flower bud[41]	Flower
Epilobium parviflorum Schreb	Small-flowered willow herb	No information based on the database searched is available.	
Equisetum arvense L. *E. telmateia* Ehrh.	Horsetail	cis-3-Hexenal, trans-2-hexenal[110]	Leaf
Erythroxylum coca Lam.	Coca	Methyl salicylate[101]	Leaf
Eschscholtzia californi Cham.	California poppy	No information based on the database searched is available.	
Eucalyptus citriodora Hool. *E. globulus* Labill	Gum tree Eucalyptus	Citronellal, isopulegol, neoisopulegol, eucalyptol, pinene, limonene, alpha-terpineol, linalool, geraniol, pinocarvone, myrtenal, carvone, cuminaldehyde, citral, aromadendrene, globulol, eudesmol, eudesmyl acetate, cineole[34, 44, 55, 101]	Leaf

Scientific name	Common name	Essential oil	
		Major components	Source
Eugenia caryophyllata L.	Clove	See *Syzygium aromaticum*	
Eupatorium perfoliatum L.	Boneset	Volatile oil[42] From *E. foeniculacum:* dimethyl ether of thymolydroquinone, trace of aldehydes, ketones[101]	Aerial part
Euphrasia officinalis L.	Eyebright	Volatile oil[42]	Aerial part
Fagopyrum tataricum L.	Buckwheat	Bornyl acetate, alpha terpineol, alpha thujene, alpha-pinene[116]	Leaf
Fagus grandifolia Ehrh. *F. sylvatica* L.	Beech European beech	Volatile oil[104]	Seed
Filipendula ulmaria (L.) Maxim.	Meadow sweet	Salicin, gaultherin, spiraein, spiracoside, isosalicin, salicylic acid, salicylaldehyde[34, 42]	Leaf
Foeniculum vulgare Mill.	Fennel	Anethole, fenchone, methyl chavicol, limonene, phyllandrene, pinene, anisic acid, camphene, palmitic, oleic, linoleic, petroselinic acids[42, 101]	Seed
Forsythia suspensa (Thunb.) Vahl.	Forsythia	Gamma-terpinene, alpha-pinene, beta-phellendrene, beta-pinene, myrcene, campheae, p-cymene, beta-ocimene, linalool, terpinene-4-ol, carene, nor-lapachol[87, 89]	Leaf, seed, bark
Fragaria vesca L.	Strawberry	Methyl salicylate, borneol, linalool, nonanal, anarin, linolenic acid, oleic acids[110]	Fruit seed

Scientific name	Common name	Essential oil	
		Major components	Source
Fraxinus americana L. *F. excelsior* L.	White ash	Volatile oil	Leaf, bark
Fucus vesiculosus L.	Kelp	Seaweed absolute (deep green or greenish-brown liquid extracted by petroleum ethel). It is an extremely powerful perfume material[5, 6, 97, 145]	Leaf
Galium aparine L.	Cleavers	Volatile oil (also unidentified oil with vanilla like perfume from leaf)[68]	Flower
Gaultheria procumbens L.	Wintergreen	Methyl salicylate, formaldehyde, gaultheriline[42, 101]	Leaf, fruit
Geranium macrorrhizum L.	Geranium	Monoterpenes, sesquiterpenes[101]	Leaf, root
Ginkgo biloba L.	Ginkgo	14-methylthexadecanoic (14-MHD), anteiso-methyl branched fatty acid, 5,9-octadecadienoic acid, 5,9,12-octadecatrienoic acid[1, 67, 153]	Seed
Glechoma hederacea L.	Ground ivy	Germacrene, cis-ocimene, beta-elemene 1,8-cineole, myrcene, beta-pinene, 3-octanone, epsilon-bulgarene, sabinene, terpinene-4-ol, 1-octen-3-ol, gamma-elemene, alpha-terpineol, beta-bourbonene, limonene, epsilon-muurolene, pinocamphone, menthone[42]	Flower
Glycine max (L.) Merrill	Soybean	Soybean oil may be used as carrier oil[34]	Seed

Scientific name	Common name	Essential oil	
		Major components	Source
Glycyrrhiza glabra L. *G. uralensis* Fisch ex DC.	Licorice Chinese licorice	Volatile oil[68]	Flower
Gossypium hirsutum L.	Cotton	n-Amyl alcohol, gossypol[124]	Seed
Grindelia robusta Nutt. *G. squarrosa* (Pursh) Donal	Gumweed	Alpha-pinene, borneol, terpineol, beta-pinene, flower essences[42, 110]	Flower head, seed
Hamamelis virginiana L.	Witch hazel	Volatile oil, sesquiterpene, eugenol, safrole, hamamelin[34, 42]	Leaf
Harpagophytum procumbens DC. ex Meis.	Devil's claw	No information based on the database searched is available.	
Hedeoma pulegiodes (L) Pers.	Pennyroyal	Pulegone, n-butyric acid, d-1-methyl-3-cyclohexanone, n-caprylic acid, n-capric acid[42, 101]	Aerial part
Hedera helix L.	English ivy	Palmitic, petroselinic, oleic, linoleic acids[42]	Seed
Helianthus annuus L.	Sunflower	Alpha-tocopherol, linolenic, oleic, myristic, palmitic, palmitoleic, margaric, stearic, arachidic, linoleic, and eicosenic acids[35, 58]	Seed
Helichrysum angustifolium (Lam) DC.	Curry plant	Nerol, neryl acetate, geraniol, pinene, linalool, isovaleric aldehyde, sesquiterpenes, furfural[72]	Leaf, flower top
Helonias dioica L.	False unicorn root	See *Chamaelirium luteum*	

Scientific name	Common name	Essential oil	
		Major components	Source
Heracleum maximum Bartr *H. lanatum* Michx. H. sphondylium L.	Cow parsley Cow parsnip Cow parsnip	Ethyl butyrates, caproic esters, acetic ester, bergaterpeme, octyl alcohol, pimpinellin, iso-pimpinellin, sprondrin, lauric acid ester of octyl alcohol, flower essence[101, 110]	Seed, root
Hibiscus sabdariffa L. *H. rosa-sinensis* L.	Hibiscus	Ambrettolide, alcohol farnesol, palmitic acid, cholesterol, ergosterol, beta-sitosterol, campesterol[101]	Seed
Hierochloe odorata (L.) Beauv.	Sweet grass	Coumarin, 3-methylbutanal, 3-methylbutanol, massoilactone[110, 125, 143]	Whole plant
Hippophae rhamnoides L	Sea buckthorn	Ethyl hexanoate, 3-methylbutyl benzoate, 3-methylbutyl, dipentene, farnesene, octanoate, aliphatic esters, tocopherol, terpenes, linoleic, linolenic, oleic, palmitic acid[18, 100, 129]	Seed, pulp
Humulus lupulus L.	Hops	Humulene, mycrene, alpha-humulene, beta-caryophyllene, farnesene, valerianic acid, monoterpenes[42, 101, 110, 125, 131]	Flower, leaf
Hydrangea arborescens L.	Hydrangea	Volatile oil[42]	Root
Hydrastis canadensis L.	Goldenseal	Volatile oil[34, 44]	Seed
Hypericum perforatum L.	St. John's wort	Carophyllene, pinene, terpinene, germacrene, menthy-2-octane, hymulene, octanol, caryophyllene, isotridecane, ethyl hexanoate, undecane, cineole, dodocanol, ketones, ishwarane, 5 and 6-methylheptan-2,4-dione, 3-methybutylbenzoate, 3-methylbutyloctanoate[34, 42, 110]	Flower, seed

Scientific name	Common name	Essential oil	
		Major components	Source
Hyssopus officinalis L.	Hyssop	Pinene, camphor, limonene and pinocamphene, beta-pinene, isopinocamphone, estragole[34, 101]	Seed
Ilex aquifomium L.	English holly	Maté absolute (a semisolid, sticky very dark green mass prepared by extraction of the dry leaves with petroleum ether), used by perfumers[42, 45]	Leaf
Inula helenium L.	Elecampane	Alantoi, sesquiterpene lactones, isolactone, dihydralantolactone, dihydroisalantolactone[34, 101]	Flower, seed
Iris versicolor L.	Blue flag	Furfural[42, 101]	Rhizome
Jasminum grandiflorum L. *J. auriculatum* Vahl.	Jasmine	Benzyl acetate, benzyl acetate, linalool, linalyl acetate, phytol, isophytol, phenylacetic acid, cis-jasmone, methyl jasmonate, farnesol, sesquiterpene, benzylic alcohol, eugenol, phytolester, methylpalmitate, indole, benzylbenzoate[30, 75, 101]	Flower
Juglans regia L. *J. nigra* L.	Walnut Black walnut	Aliphatic hydrocarbons, hexadecane, 3-hexen-1-ol. Volatile oil from leaf: caryophyllene, (E)-beta-ocimene, beta-pinene, limonene[26, 28]	Fruit coat, leaf
Juniperus communis L.	Juniper	Berry: 1-terpinen 4-ol, alpha pinene, camphene, cadinene, monoterpenes. Leaf: delta-sabinene, alpha-pinene, limonene, myrcene, cadinenes, beta-elemene, spathulenol, muurolene, elemen-7-ol, cineole, cymene, terpinene, thujene[42, 101, 110]	Berry, leaf

Scientific name	Common name	Essential oil	
		Major components	Source
Laminaria digitata (Huds) Lamk. *L. saccharina* (L.) Lamk.	Kelp	See *Fucus vesiculosus*	
Lamium album L.	Nettle	Eucalyptol, citronellal, isoeugenol[113]	Aerial parts
Larrea tridentata (Sesse. & Moc. ex DC.) Coville.	Chapatral	No information based on the database searched is available.	
Laurus nobilis L.	Bay laurel	Monoterpenoid, acetates, cineole, benzenoides, linalool, alpha-pinene, alpha-terpineol[30, 42, 101]	Leaf
Lavandula angustifolia Mill. *L. spica* L. *L. officinalis* L.	Lavender	Linalyl acetate, linalool, camphor, borneol, cineole, nerol, amyl alcohol, geraniol, lavadulol, acetic, butyric, valeric acid, caproic acids[34, 101, 120]	Leaf, flower
Ledum latifolium Jacq.	Labrador tea	Monocarboxylic acid, cineol, camphor, pinene, sabinene, ledene, terpinene, selinene, sellnadiene, ledol, borneol, lepalox[17, 101, 110]	Leaf
Lemna minor L.	Duckweed	Volatile oil, flower essence[110]	Flower
Leonurus cardiaca L.	Motherwort	Oleic, stearic, linoleic acid, linolenic acids[68]	Herb seed

Scientific name	Common name	Essential oil	
		Major components	Source
Levisticum officinale W. Koch. *L. levisticum* L.	Lovage	Phthalides such as butylidene, dihydrobutylinene, volatile acids, butylpythalides, ligostilides, terpenoid, coumarins[34, 101]	Leaf
Ligustrum lucidi W. T. Aiton. *L. vulgare* L.	Ligustrum fruit Privet	Dillapiole, sedanonic acid, n-butylidene phthafide[124]	Seed
Linum usitatissimum L.	Flax	Fatty oil (linseed oil) with esters of linoleic, linolenic, stearic, oleic, and erucic acid[13]	Seed
Liquidambar styraciflua L. *L. orientalis* L.	Gum tree, storax Oriental sweet gum	Styrene, vanillin, phenylpropyl alcohol, cinnamic alcohol, benzyl alcohol, ethyl alcohol[101]	Bark
Lobelia inflata L. *L. siphilitica* L.	Lobelia Blue lobelia	No information based on the database searched is available.	
L. pulmonaria L.	Lungwort	Volatile oil[110]	Aerial part
Lomatium dissectum (Nutt.) Math. & Const.	Lomatium	Terpenes, sesquiterpenes, also volatile oil from flower[110]	Root, seed
Lonicera caerulea L. *L. caprifolium* L.	Mountain fly honeysuckle Dutch honeysuckle	Flower essences[110]	Flower

Scientific name	Common name	Essential oil	
		Major components	Source
Lycium barbarum L. *L. chinesis* Mill. *L. pallidum* L.	Lycium Chinese wolfberry Wolfberry	From *L. turcomanicum,* palmitic acid, linoleic acid, oleic acid, linolenic acid, palmitoleic acid[11, 34]	Berry
Lycopodium clavatum L.	Clubmoss	Volatile oil, flower essence[110]	Spore
Lysimachia vulgaris L.	Loosestrife	Hexadecanoic acid, heptadecanoic acid, methy ester of eicosa-11, 14,17-trienoic acid, fluoranthene[161]	Aerial part
Lythrum salicaria L.	Purple loosestrife	Volatile oil, linoleic acid, triacylglycerols[34, 118]	Aerial part
Macropiper excelsum G. Forst	Kava-Kava	See *Piper methysticum*	
Mahonia aquifolium (Pursh) Nutt.	Oregon grape	Essential oil (with antibacterial and fungicidal properties related to skin athogens) used in perfumery and cosmetic industry[77, 138]	Fruit
Marrubium vulgare L.	Horehound	Volatile oil[68]	Leaf
Matricaria chamomilla L. *M. recutita* L.	Chamomile German chamomile	Chamazulene, farnesene, bisabolol oxide, apigene, apigetrin, apiin, rutin, luteolin, quercimeritrin, proazulenes, spiroether[13, 34, 42, 101]	Herb
Matteucia struthiopteris (L.) Tod.	Ostrich fern	Hexyl and octyl esters, butyric acid, pelargonic acid[110]	Herb

Scientific name	Common name	Essential oil	
		Major components	Source
Medicago sativa L.	Alfalfa	Acetone, butanone, propanal, pentanal, flower essences[42, 110]	Seed, flower
Melaleuca alternifolia (Maid. & Bet.) Cheel	Melaleuca, medicinal tea tree	Terpinen-4-ol, gamma-terpinene, beta-pinene, alpha-terpinene, cymene, sesquiterpene, cineol[34, 101]	Leaf
Melilotus officinalis (L.) Lamk. *M. arvensis* L.	Sweet clover Melilot	Citral, caryophyllene oxide, linalool, citronellal, coumarins, melilotic acid, orthocoumaric acid, menthol, flower essences[110]	Leaf, flower
Melissa officinalis L.	Lemon balm, Balm	Citral, linalool, geraniol, citronellal, piperitone, methone, caryophyllene oxide[34]	Leaf
Mentha x piperita L.	Peppermint	Menthol, menthone, acetaldehyde, dimethyl sulfide, isovaleric aldehyde, pinene, limonene, terpinene, beta-caryophyllene, neomenthol, 2,5-trans-p-methanediol, methyl acetate, methyl ethers, isomenthone, piperitone, pulegone[30, 34, 42]	Leaf
M. spicata L.	Spearmint	Linalool, cineole, limonene, L-carvone, dihydrocarvone, phyllandrene, menthone[30, 101]	Leaf
M. pulegium L.	Pennyroyal	Menthol, pinene, limonene, pulegone, isomenthone, piperitone[101]	Leaf
Monarda odoratissima Bench. *M. punctata* T. J. Howell	Horsebalm Horsemint	Thymol, limonene, cymene, carvacrol, linalool, geraniol, neral, terpinene[101, 110]	Aerial part
Myosotis scorpioides L.	Lily of the valley	See *Convallaria majalis*	

Scientific name	Common name	Essential oil	
		Major components	Source
Myrica penxylvanič̆a Lois.	Bayberry	d-α-Phellandrene, cineo, sesquiterpene, caryophyllene[101]	Leaf
Myristica fragrans Houtt.	Nutmeg	Oleoresin, alpha-, beta-pinene, alpha-, beta-terpinene, sabinene, myristicin, elincin, safrole, camphene, cymene, eugenol, linalool, pinene, safrole, terpineol[34, 101]	Crushed nut
Myrtus communis L.	Myrtle	Alpha-pinene, cineole, myrtenol (used in perfume, soap, skin care products), limonene, linalool, cymene, camphene, beta-pinene, delta-3-carene[34, 101]	Leaf
Narcissus pseudonarcissus L.	Daffodil	Narcissine, benzyl alcohol, cinnamaldehyde, indole[16, 42]	Root, seed
Nasturtium officinale L.	Watercress	Gluconasturtiin decomposes into a pungent essential oil[101]	Aerial part
Nepeta cataria L.	Catnip, catmint	Carvacrol, thymol, alpha-, beta-nepetalactone, citronellol, geraniol[34, 42, 101]	Whole plant
Nymphaea alba L.	White water lily	(from *N. hybrida*) Henicosane, nonadecane, triconsane, fatty acids, hydrocarbons, terpenoidal compounds (isocalamendiol and beta-caryophyllens)[114]	Flowering aerial part
Ocimum basilicum L.	Basil	Linalool, methy chavicol, fenchyl alcohol, eugenol, methy cinnamate, cineole, ocimene[14, 30, 42]	Leaf
Ocotea bullata (Birth) E. May	Stinkwood	Safrole, alpha-pinene, eugenol, futural, cineole. From *O. usambarensis:* nerolidol, alpha-terpineol, alpha-copaen, sesquiterpene alcohol[147, 148]	Bark, leaf

Scientific name	Common name	Essential oil	
		Major components	Source
Oenothera biennis L.	Evening primrose	Gamma-linolenic acid, beta-sitosterol, citrostadienol, alpha-tocopherol, linoleic acid, gamolenic acid, cis 6, cis 9, cis 12-octadecatrienoic acid (Oeparol, a natural drug from evening primrose essential oil)[63, 123]	Seed
Olea europaca L.	Olive	Hydrocarbons, aliphatic alcohols, oleic acid, aldehydes and acids, aliphatic esters, methy phenyl esters, titerpenoids. Olive oil may be used as carrier oil[3, 80]	Fruit
Oplopanax horridus (Sm.) Miq.	Devil's club	Nerolidol, torreyol, dodinene, equinopanacol, bulnesol, dodecenol, cadenene, cedrol, equinopanacene[110, 141]	Root
Opuntia fragilis (Nutt.) Haw.	Prickly pear	Flower essence[110]	
Origanum majorana L.	Sweet marjoram	Terpineol, sabinene hydrate, sabinene, linalool, carvacrol[34]	Aerial part
Origanum vulgare L. subsp. *hirtum* (Link) Ietswaart.	Marjoram Oregano	Thymol, carvacrol, beta-borneol, pinene, dipentene, cymene, caryophyllene, bisabolene[34]	Aerial part
Paeonia lactiflora Pall. *P. officinalis* L. *P. suffruiticosa* Andr.	White peony (Bia shao) European peony Tree peony, Mu Dan Pi	Benzoic acid, paeonol From *P. emodi* root, n-alkanes, beta-amyrin, butyrospermol, cycloarienol, lupeol, lauric, myristic, myristoleic, palmitic, palmitoleic, stearic, oleic, linoleic acid. From *P. anomala,* methyl salicylate, p-hydroxybenzaldehyde[9, 76, 96, 101, 154]	Root

Scientific name	Common name	Essential oil	
		Major components	Source
Panax ginseng C. A. Meyer *P. notoginseng* (Burk.) F. H. Chen *P. quinquefolium* L.	Asian ginseng Tian Qi American ginseng	Sesquiterpenes, monoterpenes, monoenes, linoleic, linolenic[42, 100]	Seed
Papaver bracteatum L. *P. rhoens* L.	Iranian poppy Poppy, corn poppy	Palmitic, stearic, oleic, linoleic acid[42]	Seed
Passiflora incarnata L.	Passion flower	Hexanal, benzyl alcohol, linalool, 2-phenylethyl alcohol, 2-hydroxybenzoic acid methyl ester, carvone, trans-anethole, eugenol, isoeugenol, beta-ionone, alpha-bergamotol, phytol, palmitic, oleic acid[27]	Seed, fruit
Pelargonium capitatum (L.) L'Her *P. graveolens* L'Her. ex. Aiton *P. crispum* L'Her. ex. Aiton	Wild rose geranium Lemon geranium Rose geranium	Citronellol, geraniol, alpha-pinene, alpha-phyllandrene, p-cymene, alpha- and gamma-terpinene, beta-carophyllene, myrcene, linalool, isomenthone, cadinene, nerol, cis-rose oxides, trans-rose oxides, d-terpineol, alpha-pinene[32, 38, 101]	Leaf
Peltigera canina L.	Ground liverwort (Dog lichen)	Volatile oil[110]	Whole plant
Petroselinum crispum (Mill.) Nym.	Parsley	Apiole, myristicin, pinene, tetramethozyally benzene, apiol, phyllandrene, terpinolene[34]	Root, seed

Scientific name	Common name	Essential oil	
		Major components	Source
Phataris canariensis L.	Canary creeper	No information based on the database searched is available.	
Phellodendron chinensis Schneid.	Chinese corktree (Huang Bai)	Tetracyclic triterpenes, mycene, methylketones, monoterpenoids[64]	Fruit
Phoradendron leucarpum (Raf.) Rev. & M. C. Johnst. *P. flavescens* (Push.) Nutt.	Mistletoe American mistletoe	No information based on the database searched is available.	
Phyllanthus emblica L.	Myrobalan, emblic	Cineol, cymene, linalool[101]	Aerial part
Physalis alkekengi L.	Chinese lantern	1,2-Benzenedicarboxylic acid, dibutyl ester[155]	Berry
Phytolacca americana L.	Pokeweed	Saturated acids, mono-acids, di-acids[42]	Seed
Picea mariana (Mill) Black *P. glauca* (Moench) Voss. *P. ables* (L.) Karst	Black spruce White spruce Norway spruce	Camphene, pinene, 3-carene, longifolene, borneoal acetate, longiborneal, sesquiterphenol, flower essences[54, 110]	Bark, flower
Pimpinella anisum L.	Anise	Trans-anethole, methyl chavicol, cis-anethole, p-anisic acid, carvone, estragole, limonene, eugenol, anisaldehyde, alpha- and beta-pinene, camphene, sabinene, saffrol, myrcene, linool, cis-anethol[48]	Seed, leaf, root

Scientific name	Common name	Essential oil	
		Major components	Source
Pinus mugo Torra var. Pumilio *P. palustris* Mill. *P. strobus* L. *P. albicaulis* Engelm. *P. contorta* Dougl. ex. Loud.	Dwarf mountain pine Southern pitch pine White pine Lodgepole pine	L-camphene, beta-pinene, dipentene, borneol, bornyl acetate, cadinene, monoterpene Bishomophinolenic, 1,4-methylhexadecanoic acid, hydrocarbons, limonene, pinene 14-Methylhexadecanoic[54, 101, 110, 151]	Needle, seed
Piper methysticum G. Forst	Kava-Kava	Camphor, borneol, azulene, cineole kavalacetone such as methystirin + dihydromethysticia, kavain, dihydrokavain, desmethozyyangenin, yangonin[101]	Leaf, bark
P. nigrum L.	English pepper	Limonene, sabinene, camphene, beta-bisabolene, beta-caryephyllene, beta-pinene, piperine, piperidine, piperettine, monoterpenes, thujene[42, 101]	Berry powder
Plantago asiatica L. *P. psyllium* L.	Psyllium	Linoleic, oleic, palmitic fatty acid[34]	Seed
P. lanceolata L. *P. major* L.	Plantains	Adenine, planteose trisaccharides, aucubin, iso-ricinoleic acid, choline, linoleic acid, linolenic acid, campesterol, sitosterol, stigmasterol, cholesterol, oleic acid, linoleic and linolenic acid, carvacrol, linalool, geranyl acetate, flower essences[42, 110]	Seed Flower, root
Podophyllum peltatum L.	Mayapple	Podophyllin[54]	Root

Scientific name	Common name	Essential oil	
		Major components	Source
Polygala senega L.	Seneca snakeroot	Methyl salicylate, carboxylic acids, hexanoic acid, n-hexanal, o-cresol[101, 110]	Root
Polygonum bistorta L. *Polygonum multiflorum* Thunb.	Bistort root Flowery knotweed (Fu-ti)	Acetic, butyric, persicariol (volatile oil from flower)[110]	Root
Populus balsamifera L. *P. candicans* L. *P. tremuloides* Michx	Poplar, Balm of gilead American aspen	Populene, alpha-caryophyllene[110]	Leaf bud
Primula vulgaris Huds. *P. veris* L.	Primrose Cowslip	Volatile oil, primeverin, primulaverin[101]	Flower
Prunella vulgaris L.	Heal all, selfheal	d-Camphor, d-fenchone, hexadecanoic acid, flavanoids, coumarines, sesquiterpene lactones, isoquercitin[110]	Leaf
Prunus africana L.	African prune	β-Sitosterol, 3-0-glucoside, β-sitostenone, triterpenic acids, oleanolic, ursolic and crataegolic acid. Fatty acids mainly palmitic acid, long-chain fatty alcohols (n-docosanol, n-tetracosanol) assoicated with their trans-ferutic acid esters[12]	Bark
P. dulcis (Mill) D. A. Webb	Almond	Lycopine, quercitin, benzyl alcohol, linalool, isobutyric acid, capronic acid, lactone, may be used as a carrier oil[42]	Fruit

Scientific name	Common name	Essential oil	
		Major components	Source
P. mume Siebold & Zucc.	Japanese apricot	Benzaldehyde, hexanal, (E)-2-hexenal, butyl acetate, hexyl acetate, gamma decalactone, lycopine, isoquercitin, linalool, capronic acid, phenylethyl alcohol, eugenol, benzoic acid, gamma jasmolactone[33, 156]	Fruit
P. serotina J. F. Ehrb	Black cherry, wild cherry	Hydrocyanic acid, benzaldehyde	Bark
P. virginiana L.	Chokecherry	Hydrocyanic acid, benzaldehyde[101, 110]	Bark
Pueraria lobata (Willd) Ohwi	Kudzu	Methyl palmitate[95]	Root
Pygeum africanum L.	Pygeum	No information based on the database searched is available.	
Pyrethrum cinerarifolium L.	Pyrethrum	See *Chrysanthemum cinerarifolium*	
Quercus alba L.	White oak	Hexenyl derivatives and acetals, terpenes including monoterpenes, sesquiterpenes, and diterpenes, alkane derivative, benzyl alcohol[47]	Bark
Ranunculus occidentalis Nutt.	Buttercup	Protocanemonin, anemonin, flower essence[101, 110]	Root
R. ficaria L.	Pilewort	Paeonol[101]	Root, bark
Raphanus sativus L.	Radish	Volatile oil, flower essence[110]	Flower

Scientific name	Common name	Essential oil	
		Major components	Source
Rhamnus catharticus L. *R. frangula* L. *R. purshianus* L.	Purging buckthorn Alder buckthorn Cascara sagrada	Unidentified oil (13.4-56.9%)[42]	Seed
Rheum palmatum L. *R. officinal* Baill. *R. tanguticum* L.	Chinese rhubarb Rhubarb Rhubarb	Chrysophanic acid, anthraquinones, biisobutyl phthalate, ferulic acid, cinnamic acid, phenylpropionic acid, rhein anthrones[42]	Whole plant
Rhodiola rosea (L.) Scop.	Roseroot	See *Sedum rosea*	
Rhus radicans L.	Poison ivy	Urushiol[110]	Leaf
R. toxicodendron L.	Roseroot	Terpenes, beta-pinene, limonene, depentene[101]	Root
Ribes nigrum L. *R. lacustre* (Pers.) Poir.	Black currant Black gooseberry	Nitribine, quinic acid, terpenes, norpinene, sabinene, d-cadeine, caryophyllene, 4-methoxy-2-methyl-2-mercaptobutane, flower essences[110]	Flower bud, leaf
Robertium macrorrhizum Pic.	Geranium	See *Geranium macrorrhizum*	
Rosa canina L.	Dog rose	Citronellol, geraniol and neral, methyl eugenol and C_{18} paraffin, volatile oil[7, 101, 110]	Flower
R. damascena Mill.	Damask rose	Geraniol, nerol, stearopten, phenyl ethanol, farnesol, linoleic acid, linolenic acid[8, 30, 101]	Rosehip

Scientific name	Common name	Essential oil	
		Major components	Source
Rosmarinus officinalis L.	Rosemary	1,8-Cineole, camphor, camphene, borneol, alpha-pinene, olefinic terpenes, sesquiterpene, santene, tricyclene, thujene, fenchene, sabinene, myrecene, cerene, phyllandrene, limonene, terpinene, cymene, bornyl acetate[7, 8, 42, 101, 128, 134]	Leaf
Rubus chamaemorus L.	Cloudberry	Aromatic compounds constitued about 53% of essential oil, and 1.5% of terpene[69, 88, 110]	Flower
R. fruiticosus L.	Blackberry	No information based on the database searched is available.	
R. idaeus L.	Raspberry	Tocopherol, linalool, linoleic acid, linolenic acid, oleic acid, palmitic acid, isoamyl alcohol, flower essences[110]	Fruit, seed, flower
Rumex acetosella L. *R. obtusifolia* L.	Sorrel Dock	Flower essences Sun infused dock root oil[110]	Flower, root, seed
Ruscus aculeatus L.	Butcher's broom	No information based on the database searched is available.	
Ruta graveolens L.	Rue, herb of grace	Methyl nonyl ketones, 2-undecanone, pinene, limonene, cineole. Volatile oil, leaf; 2-nonanone, 2-undecylacetate, 2-nonylacetate; fruit; 2-undecanone, root; elemol, myrcene, pregeijerene, geijerene[42, 101]	Aerial part, root
Salix alba L. *S. discolour* Muhlenb *S. caprea* L.	White willow Pussy willow	Volatile oil, flower essence (the bracket fungus from willow, *Trametes suaveolens*, produces amanitol)[110]	Flower

Scientific name	Common name	Essential oil	
		Major components	Source
Salvia officinalis L.	Sage	Alpha- and beta-thujone, ocimene, borneol, cineole, camphor, linalool, linolenic acid[42, 128]	Leaf
S. sclarea L.	Clary sage	Linalyl acetate, linalool, pinene, myrcene, phyllandrene[61, 68]	Flower top, leaves
Sambucus nigra L. *S. canadensis* L.	Elderberry	Volatile oil[68]	Berry
S. racemosa L.	Elder	Trans-3,7-dimethyll, 3,7-octatrien-3-ol, palmitic, linalool, cis-hexanl, cis-trans-rose oxides, linoleic acid, linolenic acid[110]	Berry
Sanguinaria canadensis L.	Bloodroot	No information based on the database searched is available.	
Sanicula marilandica L.	Black snakeroot	Volatile oil, resin, flower essence[110]	Aerial part
Santalum album L.	Sandalwood	Alpha- and beta-santalols, sesquiterpene hydrocarbon, santene, teresantol, borneol, santalone, tri-cyclo-ekasantalal, nor-alpha-trans-bergamotenone, cyclo- and epicyclosantalal, palmitic acid, oleic acid, linoleic acid[42, 101]	Root, flower
Saponaria officinalis L.	Soapwort	Trace of volatile oil[59]	Root
Sarothamnus scoparius L.	Broom	Volatile oil, 4-mercapto-4-methylpentan-2-one[139]	Flower

Scientific name	Common name	Essential oil	
		Major components	Source
Sassafras albidum (Nutt.) Nees	Sassafras	Safrole, phyllandrene, pinene, asarone, camphor, thujone, myristicin, menthone[42, 101]	Inner bark of root
Satureja hortensis L. *S. montana* L.	Savory	Carvacrol, cymene, thymol, gamma-terpinene, linalool[42, 101]	Flower
Schisandra chinensis (Tur.) Ball	Schizandra ,wu wei zi	Volatile oil[34]	Fruit
Scutellaria baicalensis Georg. *S. lateriflora* L. *S. galericulata* L.	Baical skullcap Virginia skullcap Skullcap	Scutellarin, acetophenone, 1-phenyl-1, 3-butanedione, palmitic acid, oleic acid, tannins, lignan, resin[42, 110]	Root
Sedum acre L.	Stonecrop (small houseleek)	Volatile oil, flower essence[110, 130]	Flower
S. rosea (L.) Scop. subsp. *integrifolium*	Roseroot	Volatile oil[130]	Root, flower
Sempervivum tectorum L.	Houseleek, hens and chicks	Flavonoid aglycones[130]	
Senecio vulgris L.	Groundsel	Monoterpene hydrocarbons, oxygenated monoterpenes, alpha-pinene, sesquiterpene hydrocarbons, oxygenated sesquiterpenes, allo-aromadendrene[92, 144]	Aerial part
Senna alexandrina L.	Alexandrina	See *Cassia angustifolia*	

Scientific name	Common name	Essential oil	
		Major components	Source
Serenoa repens (Bartr.) Small.	Saw palmetto	Fatty acids such as caproic, caprylic, capric, lauric, linoleic, linolenic, palmitic, oleic, myristic and stearic acids. Sterols such as campesterol, beta-sitosterol, stigmasterol[12, 119]	Leaf, seed
Sesamum indicum L. *S. orientale* L.	Sesame	Phospholipids, triacylglycerols, oleic, stearic and linoleic acids, saponin, lignans (cyclolignans, tetrahydronaphthalene), butane, butanolide, 3-7-dioxabicyclo (3,3,0)-octane derivatives, tetrahydrofuran. Sesame oil may be used as a carrier oil[101]	Seed
Shepherdia canadensis (L) Nutt.	Buffalo berry	Oleoresins/essential oil[108]	Stem
Silybum marianum (L) Gaertn.	Milk thistle	Palmitic, oleic, linoleic, linolenic and behenic acids, tocopherol, erucic acid, flavonolignan[68, 99, 133]	Root, leaf, seed
Simmondsia chinensis (Link) C. Schneid	Jojoba	Jojoba oil may used as a carrier oil and in India, for medicinal purposes[68]	Seed
Solidago virgaurea L.	Goldenrod	Pinene, phyllandrene, dipentene, canadensis curlone, germacrene, methyl chavicol, mycrecene, borneol, bornyl acetate, cadinene, limonene, beta-sesquiphyllandrene, musilage, saponin[42, 101]	Root, flower
Sorbus aucuparia L.	Mountain ash	Malic acid, benzaldehyde, hydrocyanic acid, volatile oil, flower essences[110]	Berry, flower
Spirea ulmaria L.	Meadowsweet	See *Filipendula ulmaria*	
Stachys officinalis (L.) Trev.	Betony, woundwort	Unidentified trace of essential oil[42]	Herb

Scientific name	Common name	Essential oil	
		Major components	Source
Stellaria media (L.) Vill.	Chickweed	Unidentified 4.8 to 5.9% oil[42]	Seed
Stevia rebaudiana (Bertoni) Bertoni	Stevia, Sweet herb of Paraguay	Monoterpenols, beta-selinene, beta-caryophyllene, alpha-muurolene[45]	Leaf
Symhytum officinale L.	Comfrey	No information based on the database searched is available.	
Syringa vulgaris L.	Lilac	Flower essences[31, 74, 111]	Flower
Syzygium aromaticum (L.) Merr. & Perry	Clove	Eugenol, caryophyllene, alpha-humulene, alpha-terpinyl acetate, eugenyl, methyl eugenol, acetyl eugenol, naphthalene, chavicol, heptanone, sesquiterpenes, acetyl eugenol, methyl salicylate, pinene, vanillian[34, 42, 127, 158]	Dried buds
Tagetes lucida Cav.	Mexican mint marigold	Estragole, methyleugenol, anethole, linalool[20, 30, 124]	Leaf
Tanacetum parthenium (L.) Schultz.	Feverfew	See *Chrysanthemum parthenium*	
T. cinerarifolium L.	Pyrethum	See *Chrysanthemum cinerarifolium*	
T. vulgare L.	Tansy	Beta-thujone, camphor, borneol, limonene, chamazulene, dihydrochamazulene, artemisia ketone, gamma-terpinene[34, 42, 101]	Leaf
Taraxacum officinale G. H. Weber ex Wigg.	Dandelion	Unidentified dandelion oil, fixed oil from seeds, volatile oil from flower[81, 82, 110]	Flower head, seed

Scientific name	Common name	Essential oil	
		Major components	Source
Taxus x media Rehd. *T. brevifolia* Nutt.	Yew	Aliphatic alcohol, 1-octen-3-ol, 3,5-dimethoxyphenol[74]	Leaf
Thymus citriodorus (Pers.) Schreb.	Lemon thyme	Thymol, carvacrol, hexenol, terpene, pinene[101]	Aerial part
Thymus vulgaris L	Thyme	Thymol, carvacrol, methylchavicol, cineole, borneol, cymene, terpinene, camphene, pinene, myrcene[13, 34, 42, 110]	Aerial part
Tillia cordata Mill. *T. europaea* L.	Linden Small-leaved lime	Farnesol, docosane, eucosine, eugenol, nerol, linolool, hexacosane[42]	Flower
Tribulus terrestris L.	Puncture vine	Essential oil (used against *Meloidogyne incognita* larvae)[121]	Seed
Trifolium incarnatum L. *T. pratense* L.	Clover (crimson) Clover (red)	Methyl salicylate, benzyl alcohol, palmitic, oleic, linoleic, and stearic acids[101, 110]	Flower, seed
Trigonella foenum-gracecum L.	Fenugreek	Volatile oil, linolenic acid, oleic acid, palmitic acid[23, 78, 124]	Seed
Tropaeolum majus L.	Canary creeper, nasturtium	Benzyl isothiocyanate (benzyl mustart oil), benzyl cyanide[101]	Seed
Tussilago farfara L.	Coltsfoot	Tannin, mucilage, inulin, flower essences[42]	Leaf, flower
Ulmus rubra Muhl. *U. procera* L.	Slippery elm, sweet elm	No information based on the database searched is available.	

Scientific name	Common name	Essential oil	
		Major components	Source
Umbelluslaria california Nutt.	California laurel Bay laurel	Umbellulone, anethole, dodecanoid acid, monoterpenoid, sabinene, 1,8 cineole, pinocarvone. From seeds: lauric acid, triglyceride, dilaurocaprin, trilaurin, dicaprolaurin, tricaprin[57, 98, 101, 107]	Leaf, seed
Uncaria tomentosa (Willd) DC. *U. gulanensis* (Aubl) Gmel.	Cat's claw	No information based on the database searched is available.	
Urtica dioica L. *U. urens* L.	Stinging nettle	Alpha-pinene, sabinene, oleic acid, linoleic acid, saturated acids, glycerol, linolenic acid[42]	Fresh seed head
Vaccinium macrocarpon L. *V. vitis-idaea* L.	Cranberry, Mountain cranberry cowberry, foxberry	Flavonoids, alkaloids, volatile oil[39]	Leaf, fruit
V. mrytilloides Michx. *V. myrtillus* L. *V. oreophilum* Rydb.	Blueberry Bilberry	Volatile oil, flavonoids[21, 37, 110]	Flower
Valeriana officinalis L.	Valerian	Bornyl acetate, beta-caryophyllene, isovalerate, valepotriates, acelic acid, formic acid, valeric acid[13, 34, 40, 42]	Aerial part
Verbascum thapsus L.	Mullein	Volatile oil, flower esence[110]	Flower
Veronica officinalis L.	Speedwell	No information based on the database searched is available.	

Scientific name	Common name	Essential oil	
		Major components	Source
Vetiveria zizaniodes L. (Nash.)	Vetiver	Vetiverol, vetivene, alpha-and beta-vetivone, khusimol, elemol, vetiselinenol, beta-eudesmol, terpenes, zizanoic acid, vanillin, hydrocarbons, sesquiterpenes, alcohols, ketones[10, 30, 45, 101]	Root
Viburnum opulus L.	Highbush cranberry	Volatile oil, from Japanese species, chavicol[110]	Berry
Vinca minor L.	Periwinkle	Aldehydes, sesquiterpenes[68]	Flower
Viola tricolor L.	Pansy	Alkylmethoxy pyrazines, (2E, 6Z)-nonadinal, (2E, 6Z)-nonadienol, (Z)-3-hexenol, (Z)-3-hexenyl acetate, phenylacetic aldehyde[68, 124]	Flower
Viscum album L.	European mistletoe	See *Phoradendron leucarpum*	
Vitex agnus-castus L. *V. negundo* L.	Chaste tree Chinese chaste tree	Cineole, sesquiterpenes, beta-chryophyllene, germacrene B, monoterpenes, alpha-pinene, sabinene, beta-caryophyllene[44, 150]	Seed
Vitis labrusca L *V. vinifera* L	Grape	Methyl anthranilate, grape seed oil may used as a carrier oil[44, 150]	Seed
Withania somnifera Dunal.	Ashwagandha, Indian ginseng	Triglycerols, monoacylglycerols, digalactosylglycerol, acylated sterylglucoside, sterylglucoside, phosphatidylcholine, palmitic, and oleic acids[19]	Seed

Scientific name	Common name	Essential oil	
		Major components	Source
Yucca aloifolia L.	Yucca	Pinene, camphene, 1-p-menthene, limonene, p-cymene, terpinolene, alpha-copaene, gamma-gurjunene, aromadendrene, murolene, cadinene, n-heptadecane, cis-8-heptadecene, n-nonadecane, cis-9-nonadecene, n-hexanol, 3-hexen-1-ol,3-octanol, 1-octen-3-ol, linalool, alpha-terpineol, citronellol, allphatic hydrocarbons[146]	Flower
Y. glauca L.	Soapweed	No information based on the database searched is available.	
Zanthoxylum americanum Mill.	Prickly ash, toothache tree	Sesquiterpenes (germacrene B, methy linoleate, gamma-elemene and germacrone), monoterpenes, nerolidol[36]	Bark
Zea mays L.	Corn	Volatile oil, corn oil may used as carrier oil[34]	Seed
Zingiber officinale Roscoe	Ginger	Sesquiterpenoid hydrocarbons, zingiberene, ar-curcumene, farnesene, alpha- and beta-selinene, camphene, neral, nerol, beta-sesquiphyllandrene, oxygenated monoterpenoids, 1, 8-cineole, beta-bisabolene, geranial, geraniol, geranyl acetate, alpha-copaene[23, 117]	Rhizome

References

1. Ahlemeyer, B. and J. Krieglstein. 1998. Neuroprotective effects of *Ginkgo biloba* extract. In: Lawson, L. D. and R. Bauer (eds.). Phytomedicines of Europe, Chemistry and Biological Activity. Am. Chem. Soc., Washington, DC. p. 210-220.

2. Ahn, K., M. Hahm, E. Park and H. Lee. 1998. Corosolie acid isolated from the fruit of *Crataegus pinnatifida* var. psilosa is a protein kinase C inhibitor as well as a cytotoxic agent. Planta Medica 64: 468-470.

3. Akasbi, M., D. W. Shoeman and A. Saari-Csallany. 1993. High-performance liquid chromatography of selected phenolic compounds in olive oils. J. Am. Oil Chemists Soc. 70: 367-370.

4. Alcaide, A., M. Devys and M. Barber. 1971. Triterpenes and sterols of coffee oil. Phytochemistry 10: 209.

5. Ando, Y. 1953. Studies on essential oil of sea weeds. II. On the essential oil contents of various kinds of sea weeds. Bull. Jpn. Soc. Sci. Fish. 19: 713-716.

6. Ando, Y. 1953. Studies on essential oil of sea weeds. III. The constituents of the oil of *Dictyota dichotoma*. Bull. Jpn. Soc. Sci. Fish. 19: 717-721.

7. Arctander, S. 1960. Perfume and Flavor Materials of Natural Origin. Published by the author, Elizabeth, NJ. 736 p.

8. Arctander, S. 1969. Perfume and Flavor Chemicals. Vol. 1 and 2. Published by the author, Elizabeth, NJ.

9. Asaf, M., M. Shamim Ahmad, A. Mannan, T. Itoh and T. Matsumoto. 1983. Analysis of *Paeonia emodi* root oil. J. Am. Oil Chem. Soc. 60: 581-583.

10. Ashour, F. M. 1980. Physicochemical properties and chemical composition of vetiver oil. Ann. Agri. Sci. Moshtohor. 12: 183.

11. Aslanov, S. M. and M. E. Mamedova. 1986. Fatty acid composition of the oil of the seeds and flesh with feel of the fruit of *Lycium turcomanicum*. V. L. Komarov Inst. Botany, Academy of Sci. Azerbaidzhan, SSR, Baku. No. 6: 835.

12. Awang, D. 1997. Saw palmetto, african prune and stinging nettle for benign prostatic hyperplasic (BPH). Can. Pharmaceutical J. Nov., 1997. p. 37-40, 43-44, 60.

13. Bajaj, Y. P. S., M. Furmanowa and O. Olszowska. 1988. Biotechnology of the micropropagation of medicinal and aromatic plants. In: Bajaj, Y. P. S. (ed.). Biotechnology in Agriculture and Forestry 4. Medicinal and Aromatic Plants I. Springer-Verlag Co., New York. p. 60-103.

14. Baranyk, P., V. Zeleny, H. Zukalova and P. Horejs. 1995. Oil content of some species of alternative oil plants. Rostlinna-Vyroba. 41: 433-438.

15. Baresova, H. 1988. *Centaurium erythraea* Rafn: micropropagation and the production of secoiridoid glucosides. In: Bajaj, Y. P. S. (ed.). Biotechnology in Agriculture and Forestry 4. Medicinal and Aromatic Plants I. Springer-Verlag Co., New York. p. 350-366.

16. Bastida, J., S. Bergonon, F. Viladomat and C. Codina. 1994. Alkaloids from *Narcissus primigenius.* Planta Medica 60: 95-96.

17. Belousova, N. I. and Yu. G. Slizhov. Monocarboxylic acids of the essential oils of Siberian and Far Eastern species of *Ledum*. Chem. Nat. Compd. 22: 474-475.

18. Beveridge, T., T. S. C. Li, B. D. Oomah and A. Smith. 1999. Sea buckthorn products: manufacture and composition. J. Agric. Food Chem. 47: 3480-3488.

19. Bhakare, H. A., R. R. Khotpal and A. S. Kulkarni. 1993. Lipid composition of *Withania somnifera, Phoenix sylvestris* and *Indigofera enualphylla* seeds of central India. J. Food Sci. & Technol. Mysore 30: 382-384.

20. Bicchi, C., M. Fresia, P. Rubiolo, D. Monti, C. Franz and I. Goehler. 1997. Constituents of *Tagetes lucida* Cav. spp. *lucida* essential oil. Flavour Fragrance J. 12: 47-52.

21. Bisset, N. G. 1994. Herbal Drugs and Phytopharmaceuticals. CRC Press, London. 566 p.

22. Bogovac, P., M. Miric, and M. Gorunovic. 1992. Fats from foxglove seeds (*Digitalis lanata* Ehrh.). Internat. Symp. Medicinal and Aromatic Plants. Budapest, Hungary. Sept. 4-6, 1990. Acta Hort. 306: 233-238.

23. Bordia, A., S. K. Verma and K. C. Srivastava. 1997. Effect of ginger (*Zingiber officinale* Rosc.) and fenugreek (*Trigonella foenumgraecum* L.) on blood lipids, blood sugar and platelet aggregation in patients with coronary artery diseases. Prostaglandins Leukot. Essent. Fatty·Acids 56: 379-384.

24. Bottoni, D. 1998. Tea tree oil. Int. J. Pharmaceutical Compounding 2: 284-285.

25. Brunke, E. J., G. Schmaus and K. H. C. Baser. 1995. Trace constituents of sensory importance—recent results. Proc.13th International Cong. Flavours, Fragrances and Essential

Oils. Istanbul, Turkey. 15-19 October 1995. Chemistry-Technology, Sensory-Biological 3: 186-215.

26. Buchbauer, G. and L. Jirovetz. 1992. Volatile constituents of the essential oil of the peels of *Juglans nigra* L. J. Essential Oil Res. 4: 539-541.

27. Buchbauer, G. and L. Jirovetz. 1992. Volatile constituents of the essential oil of *Passiflora incarnata* L. J. Essential Oil Res. 4: 329-334.

28. Buttery, R. G., R. A. Flath, T. R. Mon and L. C. Ling. 1986. Identification of germacrene D in walnut and fig leaf volatiles. J. Agric. Food Chem. 34: 820-822

29. Canard, D., O. Perru, V. Tauzin, C. Devillard and J. P. Bonhoure. 1997. Terpene composition variations in diverse provenances of *Cedrus libani* (A.) Rich. and *Cedrus atlantica* Manet. Trees Structure and Function 11: 504-510.

30. Chandra, G. 1985. Investigations on essential oils and isolates of potential value at H B T I, Kanpur. Indian Perfumer 29: 23-30.

31. Chapple, C. C. S. and B. E. Ellis. 1991. *Syringa vulgaris* L. (Common lilac): in vitro culture and the occurrence and biosynthesis of phenylpropanoid glycosides. In: Bajaj, Y. P. S. (ed.). Biotechnology in Agriculture and Forestry 15. Medicinal and Aromatic Plants III. Springer-Verlag Co., New York. p. 478-497.

32. Charlwood, B. V. and K. A. Charlwood. 1991. *Pelargonium* spp. (Geranium): in vitro culture and the production of aromatic compounds. In: Bajaj, Y. P. S. (ed.). Biotechnology in Agriculture and Forestry 15. Medicinal and Aromatic Plants III. Springer-Verlag Co., New York. p. 339-352.

33. Chen, H. M. and T. T. Fang. 1993. Studies on the preparation of mei juice VI. Distribution pattern of volatile compounds of mei fruit juice. Memoirs College Agric. National Taiwan Univ. 33: 163-179.

34. Chevallier, A. 1996. The Encyclopedia of Medicinal Plants. Dorling Kindersley Ltd. London. 336 p.

35. Conte, L., A. Zazzerini and L. Tosi. 1989. Changes in composition of sunflower oil extracted from achenes of *Sclerotium bataticola* infected plant. J. Agric. Food Chem. 37: 36-38.

36. Craveiro, A. A., J. W. Alencar, F. J. A. Matos and M. I. L. Machado. 1991. Essential oil of *Zanthoxylum gardneri* Engl. A new source of nerolidol. J. Essential Oil Res. 3: 371-372.

37. Cunio, L. 1994. *Vaccinium myrtillus*. Australian J. Medical Herbalism 5: 81-85.

38. Demame, F. E., A. M. Viljoen and J. J. A. Van der Walt. 1993. A study of the variation in the essential oil and morphology of *Pelargonium capitatum* (L.) L'Herit. (Geraniaceae). Part I. The composition of the oil. J. Essential Oil Res. 5: 493-499.

39. Dube, S., P. D. Upadhyay and S. C. Tripathi. 1991. Fungitoxic and insect repellent efficacy of some spices. Indian Phytopathol. 44: 1, 101-105.

40. Dudas, P., B. Galmbosi and G. Bujaki. 1985. Aphids damaging umbellate volatile oil plants. Novenyvedelem 21: 5, 196-198.

41. Dobelism, I. N. 1990. Magic and Medicine of Plants (5th Edition). The Reader's Digest Assoc., Inc., NewYork. 464 p.

42. Duke, J. A. 1985. CRC Handbook of Medicinal Herbs. CRC Press, Inc., Boca Raton, FL. 677 p.

43. Duke, J. A. 1992. Handbook of Biologically Active Phytochemicals and Their Activities. CRC Press, Boca Raton, FL.

44. Duke, J. A. 1992. Handbook of Phytochemical Constituents in GRAS Herbs, Plant Foods and Medicinal Plants. CRC Press, Boca Raton, FL.

45. Duke, J. A. and J. L. duCellier. 1993. CRC Handbook of Alternative Cash Crops. CRC Press, London. 536 p.

46. Embong, M. B., D. Hadziyer and S. Molmar. 1977. Essential oils from species grown in Alberta—caraway oil (*Carum carvi*). Can. J. Plant Sci. 57: 829-837.

47. Engel, R., P. G. Gutz, T. Herrmann and A. Nahrstedt. 1993. Glandular trichomes and the volatiles obtained by steam distilllation of *Quercus robur* leaves. Seitschrift-fur-Naturforschung-Section-C, Biosciences 48: 9-10, 736-744.

48. Ernst, D. 1989. *Pimpinella anisum* L. (Anise): cell culture, somatic embryogenesis, and the production of anise oil. In: Bajaj, Y. P. S. (ed.). Biotechnology in Agriculture and Forestry 7. Medicinal and Aromatic Plants II. Springer-Verlag Co., New York. p. 381-397.

49. Ference, D. 1989. Economic opportunities for Canada in essential oils and medicinal crops. Don Ference and Assoicates Ltd., Vancouver, BC. 287 p.

50. Ferlay, V., G. Mallet, A. Masson, E. Ucciani and M. Gruber. 1993. Fatty Acid Composition of Seed Oils from Wild Species of the South-East Mediterranean. Oleagineux, Paris. 48: 91-97.

51. Fisher, G., J. DeAngelis and D. M. Burgett. 1993. Pacific NorthWest Insect Control Handbook. Washington State Univ., Pullman, WA. 352 p.

52. Folstar, P. , W. Pilnik, J. G. Heus, and H. C. Van der Plas. 1975. The composition of the fatty acids in coffee oil and coffee wax. Lebensmittel-wissenschaft Technologie 8: 286-288.

53. Furmanowa, M., D. Sowinska and A. Pietrosiuk. 1991. *Carum carvi* L. (Caraway): in vitro culture, embryogenesis, and the production of aromatic compounds. In: Bajaj, Y. P. S. (ed.). Biotechnology in Agriculture and Forestry 15. Medicinal and Aromatic Plants III. Springer-Verlag Co., New York. p. 177-192.

54. Goldsmith, S. K. 1987. Resource distribution and its effect on the mating system of a longhorned beetle, *Perarthrus linsleyi* (Coleoptera: Cerambycidae). Oecologia 73: 317-320.

55. Golfari, L. 1963. Observations on *Cephisus siccifolius* on Eucalyptus trees in Misiones. Idia 189: 9-14.

56. Goncharova, N. P. and A. I. Glushenkova. 1990. Lipids of *Elaeagnus* fruit. Hort. Abst. 61: 3566.

57. Goralka, R. J. L. and J. H. Langenheim. 1995. Analysis of foliar monoterpenoid content in the California bay tree, *Umbellularia california*, among populations across the distribution of the species. Biochem. Syst. and Ecol. 23: 439-448.

58. Gour, T.B., T. V. K. Singh, A. Sathe and S. N. Pasha. 1991. *Stemmatophora fuscibasalis* Snellen—a new record as a pest of citronella. Indian J. Plant Protection 19: 220.

59. Grzbowska, T. 1986. Chemical control of brown leaf spot of lovage (*Levisticum officinale* Koch.) Herba-Polonica 32: 3-4, 225-231.

60. Guedon, D., P. Abbe and J. L. Lamaison. 1993. Leaf and flower head flavonoids of *Achillea millefolium* L. subspecies. Biochem. Syst. Ecol. 21: 607-611.

61. Guenther, E. and D. Althausen (eds.). 1949. The Essential Oils. Vol. 1, 2, and 3. D. Van Nostrand Co. Inc., New York. 427 p, 852 p, and 777 p.

62. Gunstone, F. D. and R. L. Wolff. 1996. Conifer seed oils: distribution of DELTA5 acids between alpha and beta chains by 13C nuclear magnetic resonance spectroscropy. 1996. J. Am. Oil Chem. Soc. 73: 1611-1613.

63. Hanczakowski, P., B. Szymezyk and T. Wolski. 1993. The nutritive value of the residues remaining after oil extraction from seeds of evening primrose (*Oenothera biennis* L.) J. Sci. Food Agric. 63: 375-376.

64. Heinrich, G. and W. Schultze. 1985. Composition and site of biosynthesis of the essential oil in fruits of *Phellodendron amurense* Rupr. Isr. J. Bot. 34: 205-217.

65. Henning, J. 1995. Aromatic *Agastache*. American Nurseryman 181: 59-67.

66. Hethelyl, E., P. Tetenyi, B. Danos and I. Koczka. 1992. GC/MS investigation of essential oils. Internat. Symp. Medicinal Aromatic Plants, Budapest, Hungary, 4-6 Sept. 1990. Acta Hort. 306: 302-312.

67. Hierro, M. T. G., G. Robertson, W. W. Christle, Y. Joh and Y. G. Joh. 1996. The fatty acid composition of the seeds of *Ginkgo biloba*. J. Am. Oil Chem. Soc. 73: 575-579.

68. Hornok, L. 1992. Cultivation and Processing of Medicinal Plants. John Wiley & Sons, New York. 338 p.

69. Hunkanen, E. and T. Pyysalo. 1976. The aroma of cloudberries (*Rubus chamaemorus* L.). Zeitschrift fur Lebensmittel Untersuchung und Forschung 160: 393-400.

70. Husain, A. 1994. Essential Oil Plants and Their Cultivation. Central Inst. Medicinal Aromatic Plants, Lucknow, India. 292 p.

71. Ishikura, N. 1989. *Crytomeria japonica* Don (Japanese cedar): in vitro production of volatile oils. In: Bajaj, Y. P. S. (ed.). Biotechnology in Agriculture and Forestry 7. Medicinal and Aromatic Plants II. Springer-Verlag Co., New York. p. 129-134.

72. Jakupovic, J., V. P. Pathak, F. Bohlmann, R. M. King, and H. Robinson. 1987. Obliquin derivatives and other constituents from Australian *Helichrysum* species. Phytochemistry 26: 803-807.

73. Jayatilake, G. S. and A. J. MacLeod. 1987. Volatile constituents of *Centella asiatica*. In: Martens, M., G. A. Dalen and H. Russwurm Jr. (eds.). Flavour Science and Technology. John Wiley & Sons Ltd., New York. p. 79-82.

74. Jean, F. I., F. X. Garneau, G. J. Collin, M. Bouhajib and L. O. Zamir. 1993. The essential oil and glycosidically bound volatile compounds of *Taxus canadensis* Marsh. J. Essential Oil Res. 5: 7-11.

75. Jonard, R. 1989. *Jasminum* spp. (Jasmine): micropropagation and the production of essential oils. In: Bajaj, Y. P. S. (ed.). Biotechnology in Agriculture and Forestry 7. Medicinal and Aromatic Plants II. Springer-Verlag Co., New York. p. 315-331.

76. Kalinkina, G. I., A. D. Dembitskii, E. Sh. Bergaliev and L. A. Zarubina. 1993. An investigation of the essential oil of the roots of *Paeonia anomala*. Chem. Natural

Compounds. 29: 905-906.

77. Katsarska, P. and D. Desev. 1976. Study on the essential oil (concrete and absolute) of *Mahonia aquifolia* with a view to its utilization in the perfumery and cosmetic industry. Perfumer Flavorist 2: 62.

78. Kaushalya, G., K. K. Thakral, S. K. Arora, M. L. Chowdhary and K. Gupta. 1996. Structural carbohydrate and mineral contents of fenugreek seeds. Indian Cocoa, Arecanut and Spices J. 20: 120-124.

79. Kelen, A., G. Y. Hangay, Z. S. Kernoczi and J. D. Voonar. 1992. Sports cosmetics containing medicinal herb extracts. Acta Hort. 306: 281-289.

80. Kubo, A., C. S. Lunde and I. Kubo. 1995. Antimicrobial activity of the olive oil flavor compounds. J. Agric. Food Chem. 43: 1629-1633.

81. Kuusi, T., K. Hardh and H. Kanon. 1984. Experiments on the cultivation of dandelion for salads use. 1. Study of cultivation methods and their influence on yield and sensory quality. J. Agric. Sci. Finland 56: 9-22.

82. Kuusi, T., K. Hardh and H. Kanon. 1984. Experiments on the cultivation of dandelion for salads use. II. The nutritive value and intrinsic quality of dandelion leaves. J. Agric. Sci. Finland 56: 23-31.

83. Lawless, J. 1995. The Illustrated Encyclopedia of Essential Oils. Element Books Ltd., England. 256 p.

84. Lawrence, B. M. 1981. Essential Oils 1979-1980. Allured Publishing Corp., Wheaton, IL. 292 p.

85. Lee, S., S. Kim. G. Min, J. Cho, B. Choi and S. Lee. 1996. Agronomic characteristics and aromatic compositions of Korea wild *Codonopsis lanceolata* collections cultivated in the field. Korean J. Crop Sci. 41: 188-199.

86. Levin, W. and G. Willuhn. 1987. Sesquiterpene lactones from *Arnica chamissonis* Less. VI. Identification and quantitative determination by high performance liquid and gas chromatography. J. Chromatogr. 41: 329-342.

87. Liu, G. S., J. X. Liu, H. J. Fang and Q. Hu. 1984. Chemical study on the essential oil from the seeds of *Forsytha suspensa* Vahl. Acta Bot. Sin. Peking 26: 672-674.

88. Manninen, P., J. Pakarinen and H. Kallio. 1997. Large-scale supercritical carbon dioxide extraction and supercritical carbon dioxide countercurrent extraction of cloudberry seed oil.

J. Agric. Food Chem. 45: 2533-2538.

89. Marion, C., Y. Pelissier, R. Sabatier, C. Andary and J. M. Bessiere. 1994. Calculation of essential oil yield without prior extraction—application to the genus *Forsythia* Vahl. J. Essential Oil Res. 6: 379-387.

90. Mazza, G. and T. Cottrell. 1999. Volatile components of roots, stems, leaves, and flowers of *Echinacea* species. J. Agric. Food Chem. 47: 3081-3085.

91. Mazza, G. and F. A. Kiehn. 1992. Essential oil of *Agastache foeniculum*, a potential source of methyl chavicol. J. Essential Oil Res. 4: 295-299.

92. Mengi, N., S. N. Garg, S. K. Agarwal and C. S. Mathela. 1995. The occurrence of beta-thujone and a new p-menthane derivative in *Senecio chrysanthemoides* leaf oil. J. Essential Oil Res. 7: 511-514.

93. Mierendorff, H. J. 1995. Determination of pyrrolizidine alkaloids by thin layer chromatography in the oil of seeds of *Borage officinalis* L. Fett-Wissenschaft-Technologie 97: 33-37.

94. Miyazawa, M. and H. Kameoka 1987. Volatile flavor components of *Astragali radix* (*Astragalus membranaceus* Bunge.). Agric. and Biol. Chem. 51: 3153-3154.

95. Miyazawa, M. and H. Kameoka. 1988. Volatile flavor components of *Puerariae radix* (*Pueraria lobata* Ohwi). Agric. Biol. Chem. 52: 1053-1055.

96. Miyazawa, M., H. Maruyama, and H. Kameoka. 1984. Essential oil constituents of *Paeoniae radix, Paeonia lactiflora* Pall. Agric. Biological Chem. 48: 2847-2849.

97. Muller, D., A. Carnat and J. L. Lamaison. 1991. *Fucus*: comparative study of *Fucus vesiculosus* L., *Fucus serratus* L. and *Ascophyllum nodosum* Le Jolis. Plantes Medicinales et Phytotherapie 25: 194-201.

98. Neville, H. A., B. A. Bohm. 1994. Flavonoids of *Umbellularia california*. Phytochemistry 36: 1229-1231.

99. Omer, E. A., A. M. Refaat, S. S. Ahmed, A. Kamel and F. M. Hammouda. 1993. Effect of spacing and fertilization on the yield and active constituents of milk thistle, *Silybum marianum.* J. Herbs Spices Medicinal Plants 1: 17-23.

100. Oomah, D., L. Stephanie and D. V. Godfrey. 1999. Properties of sea buckthorn (*Hippophae rhamnoides* L.) and ginseng (*Panax quinquefolium* L.) seed oils. Proc. Canadian Inst. Food Sci. Technology Annual Conf. Kelowna, BC. p. 55.

101. Parry, E. J. 1922. The Chemistry of Essential Oils and Artificial Perfumes Vol. I. Scott, Greenwood and Son, London. 365 p.

102. Parry, E. J. 1922. The Chemistry of Essential Oils and Artificial Perfumes Vol. II. Scott, Greenwood and Son, London. 549 p.

103. Pisano, R. C. 1986. The future of natural essential oils in the flavour and fragrancies industry. Perfumer and Flavorist 11: 35-42.

104. Polle, A. and B. Morawe. 1995. Seasonal changes of the antioxidative systems in foliar buds and leaves of field-grown beech tree (*Fagus sylvatica* L.) in a stressful climate. Botanica Acta 108: 314-320.

105. Puri, D., N. Baral and B. P. Upadhyaya. 1997. Indigenous plant remedies in Nepal used in heart diseases. J. Nepal Medical Assoc. 36: 123, 334-337.

106. Rawlinson, C. J. and P. A. Dover. 1986. Pests and diseases of some new and potential alternative arable crops for the United Kingdom. Brighton Crop Prot. Conf. Pests Dis. Surrey: British Crop Protection Council 1986. 2: 721-732.

107. Reynolds, T., J. V. Dring and C. Hughes. 1991. Lauric acid containing triglycerides in seeds of *Umbellularia californica* Nutt. (Lauraceae). J. Am. Oil Chemists Soc. 68: 976-977.

108. Ritch-Krc, E.M., N. J. Turner and G. H. N. Towers. 1996. Carrier herbal medicine: an evaluation of the antimicrobial and anticancer activity in some frequently used remedies. J. Ethnopharmacol. 52: 151-156.

109. Rodov, V. S., P. S. Bugorskii, R. G. Butenko and A. S. Popov. 1988. The capacity of tissue cultures of higher plants for biotransformation of monoterpenes. Fiziologiya-Rastenli. 35: 526-533.

110. Rogers, R. D. 1997. Sundew, Moonwort, Medicinal Plants of the Prairies. Vol. 1 and 2. Edmonton, Alberta. 282 p.

111. Rogers, R. and L. Szott-Rogers. 1997. Prairie Deva, Flower Essences. 40 p.

112. Rogiers, S. Y. and N. R. Knowles. 1997. Physical and chemical changes during growth, maturation, and ripening of saskatoon. Can. J. Bot. 75: 1215-1225.

113. Roussis, V., I. Chinou, D. Perdetzoglou and A. Loukis. 1996. Identification and bacteriostatic activity of the essential oil of *Lamium garganicum* L. ssp. *leevigatum arcangeli*. J. Essential Oil Res. 8: 291-293.

114. Saeed, A., G. M. Wassel and El-S. Omer. 1995. The essential oil of *Nymphaea hybrida* Tach. V. and *Nelumbo nucifera* Gaertn. Egypt. J. Pharm. Sci. 36: 353-359.

115. Sakurai, N. and M. Nagai. 1996. Chemical constitutents of original plants of *Cimicifuga rhizoma* in Chinese medicine. Yakugaku Zasshi 116: 850-865.

116. Samalya, G. C. and V. K. Saxena. 1986. T. L. C. and G. L. C. studies of the essential oil from *Fagopyrum tataricum* leaves. Indian Perfumer 30: 299-303.

117. Sakamura, F. and T. Suga. 1989. *Zingiber officinale* Roscoe (Ginger): in vitro propagation and the production of volatile constituents. In: Bajaj, Y. P. S. (ed.). Biotechnology in Agriculture and Forestry 7. Medicinal and Aromatic Plants II. Springer-Verlag Co., New York. p. 524-538.

118. Schmelzer, K. R., M. Gippert, M. Weisenfels, and L. Beczner. 1975. Symptoms of host plants of broad bean wilt virus. I. Communication. Zentralblatt Bakteriol. Parasitenkde Infektionskr. Hygiene 130: 696-703.

119. Schreiner, O. 1900. Saw palmetto oil. Pharmaceut. Rev. 1900. p. 217-222.

120. Segura, J. and M. C. Calvo. 1991. *Lavandula* spp. (Lavender): in vitro culture, regeneration of plants and the formation of essential oils and pigments. In: Bajaj, Y. P. S. (ed.). Biotechnology in Agriculture and Forestry 15. Medicinal and Aromatic Plants III. Springer-Verlag Co., New York. p. 283-310.

121. Shahat, A. A., S. I. Ismail, F. N. Hammouda and S. A. Azzam. 1998. Anti-HIV activity of flavonoids and proanthocyanidins from *Crataegus sinalica*. Phytomedicine 5: 133-136.

122. Singh, R. P., S. S. Tomar, C. Devakumar, B. K. Goswami and D. B. Saxena. 1991. Nematicidal efficacy of some essential oils against *Meloidogyne incognita*. Indian Perfumer 35: 35-37.

123. Skrzypczak, L., B. Thiem and M. Wesolowska. 1998. *Oenothera* species (Evening primrose): in vitro regeneration, production of flavonoids, fatty acids and other secondary metabolites. In: Bajaj, Y. P. S. (ed.). Biotechnology in Agriculture and Forestry 41. Medicinal and Aromatic Plants X. Springer-Verlag Co., New York. p. 286-304.

124. Small, E. (ed.). 1997. Culinary Herbs. NRC Research Press, Ottawa. 710 p.

125. Small, E. and P. M. Catling (eds.). 1999. Canadian Medicinal Crops. NRC Research Press, Ottawa. 240 p.

126. Solet, J. M., A. Simon-Ramiasa, L. Cosson and J. L. Guignard. 1998. *Centella asiatica* (L.)

Urban. (Pennywort): cell culture, production of terpenoids, and biotransformation capacity. In: Bajaj, Y. P. S. (ed.). Biotechnology in Agriculture and Forestry 41. Medicinal and Aromatic Plants X. Springer-Verlag Co., New York. p. 81-96.

127. Srivastava, K. C. and N. Malhotra. 1991. Acetyl eugenol, a component of oil of cloves (*Syzygium aromaticum* L.) inhibits aggregation and alters arachidonic acid metabolism in human blood platelets. Prostaglandins Leukot Essent. Fatty Acids 42: 73-81.

128. Standard, S. A., P. Vaux and C. M. Bray. 1985. High-performance liquid chromatography of nucleotides and nucleotide sugars extracted from wheat embryo and vegetable seed. J. Chromatogr. 318: 433-439.

129. Stastova, J., M. Bartlova and H. Sovova. 1996. Rate of the vegetable oil extraction with supercritical CO_2. III. Extraction from sea buckthorn. Chem. Eng. Sci. 51: 4347-4352.

130. Stevens, J. F., H. Hart, E. T. Elema, A. Bolck and H. Hart. 1996. Flavonoid variation in Eurasian *Sedum* and *Sempervivum*. Phytochemistry 41: 503-512.

131. Stevens, R. 1967. The chemistry of hop constituents. Chem. Rev. 67: 19-71.

132. Strenath, H. L. and K. S. Jagadishchandra. 1991. *Cymbopogon* Spreng. (Aromatic grasses): in vitro culture, regeneration, and the production of essential oils. In: Bajaj, Y. P. S. (ed.). Biotechnology in Agriculture and Forestry 15. Medicinal and Aromatic Plants III. Springer-Verlag Co., New York. p. 211-236.

133. Szentimihalyi, K., M. Then, V. Illes, S. Perneczky, Z. Sandor, B. Lakatos and P. Vinkler. 1998. Phytochemical examination oi oils obtained from the fruit of milk thistle (*Silybum marianum* L. Gaertner) by supercritical fluid extraction. Zeitschrift-fur-Naturforschung Sec. C-Biosciences 53: 9-10, 779-784.

134. Tawfik, A. A., P. E. Read and S. L. Cuppett. 1998. *Rosmarinus officinalis* L. (Rosemary): in vitro culture, regeneration of plants, and the level of essential oil and monoterpenoid constituents. In: Bajaj, Y. P. S. (ed.). Biotechnology in Agriculture and Forestry 41. Medicinal and Aromatic Plants X. Springer-Verlag Co., New York. p. 349-365.

135. Thapa, V. K. 1989. Some higher Himalayan Typhlocybine leafhopper (Homoptera, Cicadellidae) of Nepal. Insecta Marsumurana 42: 93-110.

136. Teobald, F. V. 1928. Aphididae from Italian Somaliland and Eritrea. Bull. Ent. Res. xix. pt. 2: 177-180.

137. Todorova, M. N., I. V. Ognyanov and S. Shatar. 1995. Chemical composition of essential oil from Mongolian *Acorus calamus* L. rhizomes. J. Essent. Oil Res. 7: 191-193.

138. Toleva, P., L. Tolev, P. Katsarska and D. Desev. 1977. Study on the essential oil (concrete and absolute) of *Mahonia aquifolia* with a view to its utilization in the perfumery and cosmetic industry. Higher Inst. Food Industry, Higher Inst. Medicine 26, Lenin Boul, 4000 Plovdiv, Bulgaria (from Perfumer and Flavorist. Vol. 2 (Oct. 1977), # 52).

139. Tominaga, T. and D. Dubourdieu. 1997. Identification of 4-mercapto-4-methylpentan-2-one from the box tree (*Buxus sempervirens* L.) and broom [*Sarothamnus scoparius* (L.) Koch.]. Flavour and Fragrance J. 12: 373-376.

140. Tsvetkova, N. 1994. Content of carbohydrates in the assimilation apparatus of some broadleaved woody species in conditions of industrial pollution. Nauka-za-Gorata 31: 26-32.

141. Turner, N. J. 1982. Traditional use of devil's-club (*Oplopanax horridus*; Araliaceae) by native peoples in western North America. J. Ethnobiol. 2: 17-38.

142. Tyler, V. E. 1993. The Honest Herbal. A Sensible Guide to the Use of Herbs and Related Remedies. Pharmaceutical Products Press, New York. 375 p.

143. Ueyama, Y., T. Arai and S. Hashimoto. 1991. Volatile constituents of ethanol extracts of *Hierochloe odorata* var. pubescens Kryl. Flavour Fragrance J. 6: 63-68.

144. Vera, R. R., S. J. Laurent and D. J. Fraisse. 1994. Chemical composition of the essential oil of *Senecio ambavilla* (Bory) Pers. from Reunion Island. J. Essential Oil Res. 6: 21-36.

145. Walkiw, O. and D. E. Douglas. 1975. Health food supplements prepared from kelp—a source of elevated urinary arsenic. Clin. Toxicol. 8: 325-331.

146. Wang, C. P. and H. Kamecka. 1977. The constituents of the essential oil from the flower of *Yucca gloriosa*. Nippon Nogeikagaku Kaishi 51: 649-653.

147. Weiss, E. A. 1997. Essential Oil Crops. CAB International, NewYork. 590 p.

148. Wichtl, M. 1998. Curcuma (Tumeric): biological activity and active compounds. In: Lawson, L. D. and R. Bauer (eds.). Phytomedicines of Europe, Chemistry and Biological Activity. Am. Chem. Soc., Washington, DC. p. 133-139.

149. Wichtl, M. and H. D. V. Prendergast 1998. Quality control and efficacy evaluation of phytopharmaceuticals. Proc. Soc. Economic Botany, London, UK. 1-6 July, 1996. p. 309-316.

150. Winterhoff, H. 1998. *Vitex agnus-castus* (Chastle tree): pharmacological and clinical data. In: Lawson, L. D. and R. Bauer (eds.). Phytomedicines of Europe, Chemistry and Biological Activity. Am. Chem. Soc., Washington, DC. p. 299-307.

151. Wolff, R. L., W. W. Christie and D. Coakley. 1997. The unusual occurrence of 14-methylhexadecanoic acid in Pinaceae seed oil among plants. Lipids 32: 971-973.

152. Wong, K. C. and G. L. Tan. 1994. Essential oil of *Centella asiatica* (L.) Urb. J. Essent. Oil Res. 6: 307-309.

153. Xing, S., G. Huangpu, Y. Zhang, J. Hou, X. Sun, F. Han and J. Yang. 1997. Analysis of the nutritional components of the seeds of promising ginkgo cultivars. J. Fruit Sci, 14: 39-41.

154. Yamamoto, H. 1988. *Paeonia* spp.: in vitro culture and the production of paeoniflorin. In: Bajaj, Y. P. S. (ed.). Biotechnology in Agriculture and Forestry 4. Medicinal and Aromatic Plants I. Springer-Verlag Co., New York. p. 464-483.

155. Yang, M., X. Pan, X. Zhao, J. Wang, M. Yang and X. Pan. 1996. Study on the chemical constituents of the essential oil of berries of *Physalis pubescene*. J. NorthEast Forestry Univ. 24: 94-98.

156. Yukawa, C., H. Iwabuchi and K. H. C. Baser. 1995. Volatile constituents of Japanese apricot (ume, *Prunus mume* sieb. Et. Zucc.). Proc. 13th International Cong. Flavours, Fragrances and Essential Oils. Istanbul, Turkey. Vol. 2. Plenary Lectures, Analytical Composition, Trade-Industry Agriculture Botany. p. 276-282.

157. Yuknyavichene, G. K., A. V. Morkunas and N. A. Stankyavichene. 1973. Some biological characterististics and essential oil content in *Pyrethrum*. Ploeznye Rasterniya Pribaltiskikh Respubik i Belorussii 1973, p. 299-303.

158. Zheng, G. Q., P. M. Kenney and L. K. T. Lam. 1992. Sesquiterpenes from clove (*Eugenia caryophyllata*) as potential anticarcinogenic agents. J. Natural Products 55: 999-1003.

159. Zheng, M. S., G. Z. Zhao and A. L. Hu. 1987. Studies on the bionomics of *Plebejus argus* L. and its control. Insect Knowledge 24: 232-234.

160. Zhang, Z., J. Lin, Z. Chen, Y. Zhang and Z. J. Zhang. 1997. The distribution of tannin in vegetative organs of three important tannin plants. J. Tropical and Subtropical Botany 5: 89-92.

161. Zhu, K., Q. Wang, X. Nie, K. Zhu, Q. L. Wang and X. Nie. 1995. Study on the chemical components of the essential oil of *Lysimachia foenum-graecum* Hance. Chemistry and Industry of Foresent Products 15: 73-76.

CHAPTER 4

Value-Added Products and Possible Usage Derived from Medicinal Plants

Traditionally, medicinal plants were not consumed in their fresh state. Certain parts (leaves, bark, root, or fruit) were picked from the wild, dried, and used for treating ailments. With the advance of modern technology, the parts of plants that contain active ingredients are now being processed in various forms, such as extracts and powders, for use in value-added products.

Recently, nutraceuticals and functional foods have emerged as major trends in the food and health supplements industries, for the purpose of delivering specific non-nutritive physiological benefits that may enhance health. These applications require quantities of plant-based materials and increased understanding of the nature of medicinal plants. The use of herbs is no longer only for flavoring, seasoning, and coloring, or in the form of infusions and decoctions, as in traditional Chinese medicine. More research is needed to discover the best means of extracting the maximum amounts of active ingredients without reducing their quality. In addition, stability, optimum storage conditions, and shelf life also need to be identified.

Table 4. Value-added products and possible usage derived from medicinal plants.

Scientific name	Common name	Source	Uses	Value-added products
*Abies balsamea (*L.) Mill.	Balsam fir	Leaf	Resin from leaf	Essential oil (turpentine), residue "rosin oil" is used in the manufacture of varnishes, lacquers, and carbon black (for pigment and ink), flavor, perfume; Canada balsam oil
Achillea millefolium L.	Yarrow	Flower, leaf	Infusion for cosmetic, decoration	Tincture and from fresh or dried flowering stems, and fresh root, essential oil Food and ornamental usage
Aconitum napellus L.	Monkshood	Root	Flower essence, liniment	Tincture from fresh herb
Acorus calamus L. var. Americanus Wolff. *A. tatarinowii* L.	Calamus, sweet flag Shi Chang Pu	Dried or fresh rhizome, root	Tea from dried leaf, decoction, powder	Tincture from fresh or dried rhizomes, leaf and root for perfumery, natural pesticide, antifungal and antibacterial agent; essential oil extracted from herbage is used against *Meloidogyne incognita* larvae
Actaea alba L. *A. rubra* (Ait.) Wild.	White baneberry Red baneberry	Root, berry	Infusion, resin	Tincture from dry root
Aesculus hippocastanum L.	Horse chestnut	Leaf, bark, seed	Used in cosmetics, fruits are ground for fodder	Tincture from fruit, bark, and seeds
Agastache foeniculum L. *A. anethrodora* L.	Aniseed	Seed, leaf, flower	Infusion from dried seeds or herb, salad	Dried seed

Scientific name	Common name	Source	Uses	Value-added products
Agrimonia cuparoria L.	Agrimony	Aerial parts	Flower top used as dye for butter yellow. Infusion	Liniment for wounds
Agropyron repens (L.) Beauvois	Couch grass	Rhizomes	Decoration	Combines well with *Agathosma* spp. for cystitis and *Hydrangea arborescens* for prostates
Alchemilla vulgaris L. *A. xanthochlora* Rothm	Lady's mantle	Leaf, root	Infusion from leaf, leaves used as source of green dye	Dried powder, tincture from root
Allium cepa L.	Onion	Bulb	Skin used as dye for orange and brassy yellow, cooking	Tincture from fresh bulb
A. sativum L.	Garlic	Bulb	Syrup, cooking, garlic oil	Tincture, oil, juice, and extracts from fresh bulb, capsules, and tablets from powers
A. schoenoprasum L.	Chives	Leaf, bulb, flower	Cooking, salad, soft cheeses, omelettes, sauce	Dried leaf
Alnus crispus (Ait.) Pursh *A. incana* (L.) Moench. subsp. *tenufolia* *A. glutinos* (L.) Guertn	Alder	Bark, leaf	Decoction, bark as coloring agent for red and yellowish gray dyes	Powder from bark, tincture from fresh leaf

Scientific name	Common name	Source	Uses	Value-added products
Aloe vera (L.) Burm. *A. barbadensis* L.	Aloe	Leaf, inner bark	Juice, infusion from bark, decoction from unripe female cones	Tincture, lotion, pill, leaf juice, aloe gel
Althaea officinalis L.	Marshmallow	Leaf	Liquid extracts, syrups ointments	Capsules and tincture from dried juice Althaea root syrup (cough prep.)
Amelanchier alnifolia Nutt.	Saskatoon	Berry, flower, bark	Decoction, dried or fresh of unripe or ripe fruit for tea, cooking, jam	Flower essence, frozen berries
Ananas comosus (L.) Merr.	Pineapple	Fruit, leaf	Juice from unripe fruit for medicinal use, infusion from leaf	Juice from unripe and ripe fruit, tincture or powder of leaf for medicinal use
Anaphalis margaritacea (L.) Bench & Hook	Everlastings	Leaf, flower, root	Infusion from aerial part	Tincture from fresh root
Anethum graveolens L.	Dill	Aerial part	Infusion	Tincture, essential oil for medicinal and perfume, dill oil
Angelica archangelica L. *A. sinensis* (Oliv.) Diels	Angelica Dong quai	Root, seed, flower stem	Infusion and decoction	Tincture and capsule from dried fruit and flowering stems, liqueurs
Antennaria margaritacea Schult.	Everlastings	See *Anaphalis margaritacea*		

Scientific name	Common name	Source	Uses	Value-added products
Anthemis nobilis L.	Chamomile	See *Chamaemelum nobile*		
Apium graveolens L.	Celery	Seed, root	Dried seeds for infusion, liquid extract and powder	Tincture from fresh or dried roots, seeds for tincture, capsule, and oil
Aralia racemosa L. *A. nudicaulis* L.	Spikenard	Root	Dried root for use in liquid extracts, infusion, decoctions, poultices	Capsules and tincture from roots, compound syrup of spikenard
Arctium lappa L.	Burdock	Seed, stem, root	Dried root for decoction, liquid extracts	Tincture, capsule, and tablets from dried root
Arctostaphylos uva-ursi (L.) Spreng	Bearberry, Uva-ursi	Leaf, fruit	Dry leaf used as dye, infusion, liquid extracts	Tincture from fresh or dried leaf, tablets, tea. A drug product "Belladonne extract" is derived from this plant; fruits used in jelly, jam and sauces
Armoracia rusticana Gaertn, Mey & Scherb	Horseradish	Leaf, rhizome	Salads, cooking, sauces, dried leaves used as a source of yellow dye	Essential oil

Scientific name	Common name	Source	Uses	Value-added products
Arnica latifolia Bong. *A. montana* L. *A. chamissonis* L. subsp. *foliosa* *A. condifolia* Hook *A. fulgens* Pursh *A. sororia* Greene	Arnica	Leaf, root, flower	Dried roots of 1-yr-old, and leaf collected before flowering for infusion, ointments	Tincture, fluid extracts from fresh root and dried flower, dried flowers are also used in creams Oil used in perfumery
Artemisia absinthium L.	Wormwood	Leaf	Leaf is used in sachets and powders to repel moths and fleas. Dried plant for decoction, infusion	Tincture from dried and fresh leaves, dry leaf powder, oil extraction, flavoring
A. annua L.	Chinese wormwood (Quing Hao)	Whole plant	Whole plant dried for decoction, infusion	Tincture from fresh or dried rhizomes and flower heads (without involucre and receptacle)
A. dracunculus L.	Tarragon	Leaf, flower	Dried flowering stems and herb for infusion, leaf for cooking	Oil from leaf is used in commercial flavorings, perfumery and detergents, flavoring
A. tridentata Nutt.	Sagebrush	Whole plant	Fresh and dry leaves for infusion, oil	Tincture
A. vulgaris L.	Mugwort	Whole plant	Cooking, decoction, infusion	Dried flowering stems and herb for capsules

Scientific name	Common name	Source	Uses	Value-added products
Asarum canadensis L.	Wild ginger	Rhizomes	Dried rhizomes for powder, decoction, liquid extract	Tincture
Asparagus officinalis L. *A. racemosus* Willd.	Asparagus	Young shoot, rhizomes	Cooking, juice, rhizomes boiled and dried for decoction	Juice for medicinal use, powder from roots, oil
Astragalus americana Bunge. *A. membranaceus* Bunge. *A. sinensis* L.	Astragalus Huang Qi	Leaf, flower, root	Infusion from herb, root for decoction	Tincture and powder from dried roots
Avena sativa L.	Oat	Whole plant	Tea, infusion of leaf and flower	Liquid extracts, tincture, a type of drug product "oatmeal concentrate" is derived from this plant
Baptisia tinctora (L.) R. Br. ex Ait.f.	Wild indigo	Root	Decoction, liquid extracts	Tincture
Bellis perennis L.	Flower	Root	Infusion from sliced root, fluid extract	Tincture
Berberis aquifolium L.	Oregon grape	See *Mahonia aquifolium*		

Scientific name	Common name	Source	Uses	Value-added products
Berberis vulgaris L.	Barberry	Seed, dry stalks, berry	Dried stalks for infusion, dried bark and root for decoction, liquid extraction	Tincture from unripe seeds, capsule from dried roots
Beta vulgaris L.	Beet root	Root	Decoction from dried root	Tincture from root or whole plant
Betula lenta L.	Birch	Flower bud, leaf, bark	Infusion, addition to tea mixture	Tincture from young leaves, buds, oil from bark, perfumery, flavoring, counterirritant Birch oil
Borago officinalis L.	Borage	Leaf, flower, seed, root	Infusion and liquid extracts from young leaves	Tincture and capsules from bark and root (mild doses only), fresh, dried ripe berries
Boswellia sacra Roxb. ex Colobr. *B. carteri* Birdwood	Frankincense	Gum resin	Fresh or dried gum resin collected for decoction	Gum resin for oil, and powder
Brassica nigra (L.) Kock.	Black mustard	Bark, twig	Leaf and flower add pungency to salads, seeds are ground to make mustard	Volatile oil from seed, bark, and twig
Buxus sempervirens L.	Buxwood	Leaf, bark	Infusion, wood is used in engraving	Often included in preparations to stimulate hair growth

Scientific name	Common name	Source	Uses	Value-added products
Calamintha nepeta (L.) Savi. *C. ascendens* L. *C. officinalis* L.	Basil thyme Calamint Mountain mint	Whole plant	Fresh or dried flowering plant and leaf for infusion	Tincture
Calendula officinalis L.	Calendula, Marigold	Whole plant	Petal used as dye, plant for infusion	Dried or fresh flowering stem and leaf for powder. A drug product "calendule oil" is derived from this plant
Camellia sinensis (L.) Kunize	Green tea	Leaf	Dried leaf for infusion	Tea, essential oil from dried leaf is used for perfumes, hair, and commercial food flavoring
Cananga odorata (Lam) J. D. Hook & T. Thompson	Ylang-Ylang	Seed	Crushed seeds and dried flowers for infusion	Capsules from seed powder, dried flower distilled for oil (an important perfumery raw material)
Cannabis sativa L.	Hemp	Whole plant	Fiber, seed oil, and flower extracts for ointments	Fresh or dried flower heads for extracts, tinctures. Pigments for pharmaceutical use, 41-46% oil from seed
Capsella bursa-pastoris (L.) Medik.	Shepherd's purse	Leaf	Infusion, decoction and liquid extracts	Dry (wither) young leaves for culinary use

Scientific name	Common name	Source	Uses	Value-added products
Capsicum annuum L. var. Annuum *C. annum* L. var.. Grossum *C. frutescens* L. *C. annum* L. var. Acuminatum	Cayenne pepper Sweet pepper Chilli pepper	Flower	Fruits for pickle, cooking or dried, decoction, ointments	Essential oil from flower, powder, tincture, tablets and oleoresin from fruits. A drug product "Capsicum oleoresin" is derived from this plant
Carbenia benedicta L.	Blessed thistle	See *Cnicus benedictus*		
Carduus benedita L.	Blessed thistle	See *Cnicus benedictus*		
Carica papaya L.	Papaya	Whole plant, seed	Papaia is extracted from unripe fruit and sap and dried to a powder	Tincture from fresh or dried flowering stem of female plants, essential oil from seeds
Carthamus tinctorius L.	Safflower	Whole plant	Flower used as dye (a red pigment, carthamin), flower for infusion	Tincture from fresh or dried flowering stems (large doses are poisonous), oil extracted from ovaries. Dietary supplement in hyper - cholesterol problems
Carum carvi L.	Caraway	Leaf, root, seed	As vegetable, seed for infusion	Fruits for extracts, tinctures, ointments, and plasters, oil from seed for flavoring and pharmaceutical products
Cassia senns L. *C. angustifolia* Vahl.	Tinnevelly senna Alexandrian senna	Leaf, pod	Infusion and decoction from root, dried leaf and pod	Fruit juice, seed for vermifuge medicine, latex for medicinal use (25% fatty oil in seeds)

Scientific name	Common name	Source	Uses	Value-added products
Castanea sative Mill.	Chestnut	Leaf, seed	Dried leaf for infusion and liquid extracts, seeds for food	Tincture from leaf and seeds
Catharanthus roseus (L.) Don	Periwinkle	See *Vinca minor*		
Caulophyllum thalietroides (L) Michx. *C. giganteum* (F.) Loconte & W. H. Blackwell	Blue cohosh	Seed, root rhizome,	Decoction, liquid extract, fresh root and young leaves used as vegetable or salads and soups	Seed for oil, root and rhizome for tincture
Cedrus libani Riich. subsp. *atlantica*	Cedarwood	Leaf, branches	Dried branches for decoction, infusion from leaf	Branches are chipped for oil distillation
Centaurea calcitrapa L.	Star thistle	Whole plant	Dried leaf for infusion, liquid extracts	Tincture and powder from leaf
Centella asiatica (L.) Urb.	Gotu Kola	Whole plant	Infusion, milk and decoction	Powder, ghee or medicated oil, extracts added to cosmetic masks and creams to increase collagen and firm the skin
Chamaelirium luteum (L.) A. Gray	False unicorn root, fairywand	Root	Decoction	Tincture, tablets

Scientific name	Common name	Source	Uses	Value-added products
Chamaemelum nobile (L.) All.	Roman chamomile	Wood shaving, flower	Flower used as dye, used as an addition of chamomile tea	Oil from bark, wood shaving, and sawdust, infusion to prevent damping off of seedlings, perfumery, aromatic bitter, emetic
Chamaenerion angustifolium (L) Scop.	Firewood	See *Epilobium angustifolium*		
Chenopodium album L.	Lamb's quarters	Leaf, seed	Liquid extracts, powder	Essential oil from plant, used as a fumigant against mosquitoes and in fertilizer to inhibit insect larvae
C. ambrosioides L.	Wormseed	Leaf, flower	Used as a fumigant against mosquitoes and in fertilizer to inhibit larvae	Tincture of roots, chenopodium oil, wormseed oil
Chrysanthemum cinerarifolium (Trev) Vis.	Pyrethrum	Flower, whole palnt	Infusion, liquid extract, dried flowers are used in insecticides and fumigants	Power and tincture of whole plant; a drug product "Pyrethrins" is derived from this plant
C. parthenium (L.) Berhn.	Feverfew	Flower, leaf, fruit	Infusion and liquid extract, fresh or dried leaves for herbal mixtures, nonwoody flowering tops for tea	Tincture from root and fresh flowering stems, oil from seeds

Scientific name	Common name	Source	Uses	Value-added products
C. vulgare L.	Tansy	See *Tanacetum vulgare*		
Cichorium intybus L.	Chicory	Seed, leaf, root	Root can be roasted and used as coffee additive	Tincture of dried flowering stems and seeds (poisonous in large doses)
Cimicifuga racemosa (L.) Nutt.	Black cohosh	Flower, rhizome	Dried root for use in decoction, liquid extract	Fresh root for tincture
Cinnamomum verum J. Presl. *C. cassia* (Nees) Nees & Eberm	Cinnamon	Inner bark, leaf, fruit	Bark for infusion, branches and leaf for oil, unripe fruits dried as cassia buds	Powder and tincture from bark, flavor, perfume
C. camphora (L.) Nees & Ebern.	Camphor	Leaf, bark	Wood is boiled to extract camphor, or infusion, ointments	Wood powder for capsule, tablet and tincture. Industrial solvent, perfume for soap, detergents, insect repellent
Citrus limon (L.)Burm. *C. bergamia* Risso & Poit *C. reticulata* Blanco *C. aurantifolia* Swingle *C. aurantium* L. *C. paradisi* Macfad.	Lemon Bergamot orange Mandarin Lime Bitter orange Grapefruit	Root, fruit, leaf	Leaf for infusion, decoction of dried roots	Oil from foliage and fruit, whole or in parts of fruit for processing, flavoring, perfume
Cnicus benedictus L.	Blessed thistle	Root, whole plant	Dried whole plant for infusion, liquid extract	Fresh and dried roots for tincture and capsules

Scientific name	Common name	Source	Uses	Value-added products
Cochlearia amorucia L.	Horseradish	See *Armoracia rusticana*		
Codonopsis pilosula (Franch.) Nannfeldt. *C. tangshen* Oliver	Codonopsis, Dang Shen bellflower	Inner bark, root	Dry inner bark, twig and root	Substitute for *Panax ginseng*
Coffea arabica L.	Coffee	Seed	Beverage	Caffeine extracted from seed
Colchicum autumnale L.	Meadow	Root	Root used as dye, liquid extract	Tincture, dry powder. A drug product "Colchicine" is derived from this plant
Collinsonia canadensis L.	Pilewort	Rhizome	Dried for decoction, liquid extracts	Tincture
Commiphora molmol Engl. ex. Tschirch. *C. myrrha* (Nees) Engl.	Myrrh	Branches for gum, resin	Gum collected from cutting the branches	Tablets
Convallaria majalis L.	Lily of the valley	Whole plant	Infusion (lily of the valley should never be collected and used for self-medication), leaves used as a source of green dye	Aromatic extracts from flowers are used in cosmetic and perfumery industries
Convolvulus arvensis L. *C. sepium* L.	Bindweed	Whole plant	Dried flowering stems and leaves for infusion	Tincture (side effects on large doses)

Scientific name	Common name	Source	Uses	Value-added products
Coriandrum sativum L.	Coriander (seed) Cilantro (leaf)	Leaf, seed	Young leaves used fresh, seed for culinary purpose	Powdered seeds, oil for perfumery
Cornus canadensis L. *C. florida* L.	Bunchberry Dogwood	Flower, berry, root, bark	Infusion, decoction, syrup	Tincture, oil from seeds
Corylus cornuta Marsh *C. rostrata* Marsh. *C. avellana* L.	Hazelnut	Nut, bud, stem, root, leaf	Hazelnut milk, decoction	Flower essences, essential oil from leaf
Crataegus monogyna Jacq. *C. pinnatifida* Bge.	English hawthorn Chinese howthorn	Fruit, leaf	Fruit for decoction, liquid extracts	Tincture from herb and fruit
Crocus sativus L.	Saffron	Leaf, seed, flower stigmas	As a flavoring and used as color for cakes and liqueurs	Extract from fruits for use internally, externally used in ointments
Cryptotaenia japonica Hassk.	Japanese parsley, Mitsuba	Leaf	Young leaves used for celery-like flavor, infusion, decoction	Tincture, syrup
Cucurbita pepo L.	Pumpkin	Seed, flower, leaf, fruit	Dehusked seeds are eaten raw or roasted as a snack, or added to bread	Dried flowers, leaves and fruits in the form of an infusion, decoction, tincture, tablet and tea mixtures

Scientific name	Common name	Source	Uses	Value-added products
Cuminum cyminum L.	Cumin	Seed	Ground seeds for culinary use	Oil from seed for flavoring and veterinary use. Flavoring (Indian curry powder)
Curcuma domestica L. *C. xanthorrhiza* L. *C. longa* L.	Curcuma, Jiang Hueng	Rhizomes	Grinding rhizomes for use in decoction	Essential oil extraction, dried roots powder for tablets
Cuscuta epithymum Murr. *C. chinesis* Lam.	Dodder	Seed	Decoction	N/A
Cymbopogon citratus (DC. ex Nees) Stapf. *C. nardus* L. *C. winterianus* Jowitt *C. martinii* (Roxb) Wats.	Lemon grass Citronella Palmorosa, gingergrass	Seed, leaf	Fresh or dried stems for infusion	Dried seeds oil extracts, raw fruit juice (up to 50% fatty oil from seeds) Insect repellent. Lemon grass oil for aerosol, deodorants, floor polishes, household detergents
Cynara scolymus L. *C. cardunculus* L.	Artichoke Wild artichoke	Leaf, root, flower head	Liquid extracts, syrup, flower head for vegetable	Tablets from dried leaf, use flowers to produce cheese
Cypripedium calceolus L.	Lady's slipper	Rhizomes	Infusion, liquid extract	Dried root powder, tincture
Cytisus scoparius (L.) Link.	Broom	Whole plant	Shoots for decoction, infusion, liquid extract	Tincture

Scientific name	Common name	Source	Uses	Value-added products
Daucus carota L.	Carrot	Whole plant	Infusion and liquid extract	Seed for oil
Digitalis purpurea L.	Foxglove	Leaf	Tea, powder from tea	Dried leaf for alkaloids extraction. Drug products "Digoxin," "lanetoside C" and "acetylgitaxin" are derived from this plant
Dioscorea oppositae Thunb. *D. villosa* L.	Chinese yam (Shen yao) Wild or Mexican yam	Tubers	Liquid extract from dried tubers, fresh for homeopathic preparations	Tincture, dried powder
Echinacea angustifolia DC. Cornq *E. pallida* (Nutt.) Nutt. *E. pururea* (L.)Moench.	Narrowleaf echinacea Pale-flower echinacea Purple coneflower	Root, flower	Decoction, infusion, liquid extracts	Tincture, capsule, tablets, oil from seeds; insecticides
Echinopanax horridus (Sm) Decne. & Planch. ex. H. A. T. Harms	Devil's club	See *Oplopanax horridus*		
Elaeagnus angustifolia L.	Russian olive	Leaf, fruit	Infusion from leaf, flower	Tincture from fresh or dried leaves, oil from leaf
Eleutherococcus senticosus (Rupr. ex. Maxim) Maxim.	Siberian ginseng	Root	Decoction, tea, infusion and fluid extract from fresh flowers	Tincture from dried or fresh roots and flowers, dried powders

Scientific name	Common name	Source	Uses	Value-added products
Elymus repens L.	Couch grass	See *Agropyron repens*		
Ephedra distachya L. *E. sinica* Stapf.	Ephedra Ma Huang	Stem	Decoction and liquid extracts	Powder and tincture from stem
E. nevadensis Wats.	Mormon tea	Stem	Decoction, infusion	Powder and tincture from stem
Epilobium angustifolium L. *Epilobium parviflorum* Schreb.	Fireweed Small-flowered willow herb	Whole plant	Infusion, decoction, ointment	Tincture, oil from young flower head and buds, flower essences
Equisetum arvense L. *E. telmatcia* Ehrb.	Horsetail	Stem	Fresh stem used as dye, leaf for infusion and liquid extract	Tincture and powder
Erythroxylum coca Lam.	Coca	Leaf, root	Infusion	Tincture from fresh or dried leaf, roots Cocaine
Eschscholtzia california Cham.	California poppy	Whole plant	Infusion	Tincture and dried powder of herb
Eucalyptus citriodora Hool. *E. globulus* Labill	Gum tree Eucalyptus	Leaf	Decoction and infusion, liquid from leaf used as spot removers for oil and grease	Used as a flavoring in pharmaceutical products
Eugenia caryophyllata L.	Clove	See *Syzygium aromaticum*		

Scientific name	Common name	Source	Uses	Value-added products
Eupatorium perfoliatum L.	Boneset	Whole plant	Infusion, liquid extract	Powder and tincture of herb
Euphrasia officinalis L.	Eyebright	Whole plant	Infusion and liquid extracts	Tincture
Fagopyrum tataricum L.	Buckwheat	Leaf, flower, seed	Leaf and flower for infusion, dried seeds for use whole or ground	Dried leaf and flower for tincture, tablets. A drug product "Rutin" is derived from this plant
Fagus grandifolia Ehrb. *F. sylvatica* L.	Beech European beech	Wood, seed	Seed oil for salads, nuts for pig food	Extract creosote from wood, and oil from (nuts) seeds
Filipendula ulmaria (L.) Maxim.	Meadow sweet	Whole plant	Infusion, liquid extracts	Tincture, tablets
Foeniculum vulgare Mill.	Fennel	Whole plant	Cooking, root for decoction, fruit for spices	Dried root powder, oil from seeds, aromatic compound anthol, fennel oil
Forsythia suspensa (Thunb.) Vahl.	Forsythia, Lian qiao˙	Fruit	Decoction from dried fruit	Tincture, often combined with honeysuckle
Fragaria vesca L.	Strawberry	Leaf, root, fruit	Infusion from leaves, root for decoction, fruit eat fresh, jam	Fresh or dry herb for tincture

Scientific name	Common name	Source	Uses	Value-added products
Fraxinus americana L. *F. excelsior* L.	White ash	Sap	Used as a sweetener and anticaking agent	Nature sweetener
Fucus vesiculosus L.	Kelp	Whole plant	Used for fertilizer, livestock feed, mineral supplements	Algin for the food, textile, cosmetic, and pharmaceutical industries; as a soil amendment for micronutrients and nitrogen
Galium aparine L.	Cleavers	Leaf, seed	Fresh leaf for juice or oil for external use	Leaf used as yellow dye for coloring food such as cheese and butter, root provides red dye
Gaultheria procumbens L.	Wintergreen	Leaf, fruit	Leaf for infusion and liquid extract	Fresh leaf for tincture and oil, which is used in rubbing oils, inhalers, liniment, and ointments. A drug product "Wintergreen oil" is derived from this plant
Geranium macrorrhizum L.	Geranium	Leaf, root	Decoction, infusion and liquid extract	Tincture, powder
Ginkgo biloba L.	Ginkgo	Leaf, seed	Dried leaves for infusion, seeds for decoction	Extracts, tincture and powder from leaves and flower heads
Glechoma hederacea L.	Ground ivy	Leaf, flower	Infusion, liquid extract	Tincture

Scientific name	Common name	Source	Uses	Value-added products
Glycine max (L.) Merrill	Soybean	Seed	Cooking	Oil and snack products from whole or crushed seeds, tinctures from flower heads, young fresh leaves for cooking. A drug product "Sitosterols" is derived from this plant
Glycyrrhiza glabra L. *G. uralensis* Fisch ex DC	Licorice Chinese licorice	Root, stolon	Dried roots for decoction, liquid extracts, dried leaves for tea, fermented leaves as tea substitute	Dried root for lozenges and powder Liquorice extract is derived from this plant
Gossypium hirsutum L.	Cotton	Leaf, root bark, seed, flower	Root and seeds for decoction, liquid extracts	Leaves for lotions, root for tincture, seeds for oil extraction (cottonseed oil)
Grindelia robusta Nutt. *G. squarrosa* (Pursh) Donal	Gumweed	Flower bud, aerial parts	Infusion, liquid extract	Dried herb for capsule, tincture, fresh herb for poultices
Hamamelis virginiana L.	Witch hazel	Leaf, bark branches,	Liquid extracts and ointment from leaf	Branches for tincture, twigs used in distilled extracts. Poultices and infusions of the leaves and bark. Cosmetic products from leaf
Harpagophytum procumbens DC. ex Meisn.	Devil's claw	Thallus	Decoction, ointment	Tincture and powder

Scientific name	Common name	Source	Uses	Value-added products
Hedeoma pulegiodes (L.) Pers.	Pennyroyal	Whole plant	Infusion and liquid extract	Tincture and capsule from leaves, oil for insect repellants and cleaning products
Hedera helix L.	English ivy	Leaf, seed	Leaf for decoction, liquid extract, ointments, poultices	Tincture, oil and milk from seeds
Helianthus annuus L.	Sunflower	Whole plant, seed	Liquid extract from fresh plant, seed can be roasted for food, flower buds used as a source of yellow dye	Tincture, seed is pressed for oil for manufacture of margarine, residue for animal feeds
Helichrysum angustifolium (Lam) DC.	Curry plant	Leaf	Infusion	Tea, essential oil
Helonias dioica L.	False unicorn root	See *Chamaelirium luteum*		
Heracleum maximum Barx. *H. lanatum* Michx. *H. sphondylium* L.	Cow parsley Cow parsnip Cow parsnip	Leaf, fruit, root, seed	Fruit infusion, leaf extracts Fresh flower stem as vegetable	Tincture; leaves used as a mild insect repellent
Hibiscus sabdariffa L. *H. rosa-sinensis* L.	Hibiscus	Leaf, flower	Leaf and flower for infusion, decoction	Tincture and capsule from dried roots and underground stolons, stem is a source of fiber, calyces are used to give color and flavor to herbal tea

Scientific name	Common name	Source	Uses	Value-added products
Hierochloe odorata (L.) Beauv.	Sweet grass	Whole plant	As an incense, mix with tobacco, tea	Essential oil from leaves used in perfumes and food, flower essence and leaf used for flavoring, widely used in weaving
Hippophae rhamnoides L.	Sea buckthorn	Leaf, berry, seed	Juice, jam, animal feed, oil	Essential oil from seed and pulp used in cosmetic and pharmaceutical industries
Humulus lupulus L.	Hops	Leaf, flower, bark	Fresh or dried flowers for infusion, liquid extracts, young shoots for culinary use, decoction from bark	Tincture from fresh twigs and leaves, hops are the main flavoring in beers, oil extracted for used in food flavoring and soft drinks Hop extracts and oil used in flavored tobacco, yeast, beverages, frozen dairy desserts, candy, baked goods, etc.
Hydrangea arborescens L.	Hydrangea	Root	Decoction, liquid extract	Capsules and tincture from roots
Hydrastis canadensis L.	Goldenseal	Rhizome, leaf	Dried root for decoction, liquid extracts	Tincture and tablets from dried roots, a yellow dye used for fabrics
Hypericum perforatum L.	St. John's wort	Leaf, flower, seed	Flower use as a dye, dried herb used in creams, infusion, liquid extracts	Dried young leaves and flower for tincture and oil
Hyssopus officinalis L.	Hyssop	Whole plant	Infusion and liquid extract from calyx, leaf, and fruit	Oil extracted and used to flavor liqueurs, perfumery

Scientific name	Common name	Source	Uses	Value-added products
Ilex aquifolium L.	English holly	Leaf, bark	Infusion and liquid extract, bark for decoction and liquid extract	Fresh fruits, dried leaves and seed oil extracts
Inula helenium L.	Elecampane	Leaf, flower, root	Fresh root for liquid extracts and syrup, decoction, young leaves and shoots for salads	Tincture from fresh cones or strobiles, essential oil used in perfumes, flavoring for candy and liqueurs
Iris versicolor L.	Blue flag	Rhizome	Decoction, liquid extract	Tincture from fresh roots, powder from dried roots
Jasminum grandiflorum L. *J. auriculatum* Vahl.	Jasmine	Root, flower, leaf	Decoction, infusion, pastes, tea	Fresh or dried leaf and root for tincture, dried root capsule
Juglans regia L. *J. nigra* L.	Walnut Black walnut	Fruit, leaf, seed	Leaf, green husks and shells used as brown dye	Fresh flowering stems for powders and tincture (less active from dry material)
Juniperus communis L.	Juniper	Fruit	Fresh and dried crushed berry used as dye	Dried flowering stems for powders and infusion, flavor, perfumery, preservative
Laminaria digitata (Hudss.) Lamk. *L. saccharina* (L.) Lamk.	Kelp	See *Fucus vesiculosus*		

Scientific name	Common name	Source	Uses	Value-added products
Lamium album L.	Nettle	Whole plant	Dried whole plant for infusion	Tincture
Larrea tridentata (Sesse. & Moc. ex DC.) Coville.	Chapatral	Root	Decoction	Tincture
Laurus nobilis L.	Bay laurel	Leaf	Leaf as spice, decorative garland	Essential oil for perfume, flavoring to commercial condiments, meat products and liqueurs
Lavandula angustifolia Mill. *L. spica* L. *L. officinalis* L.	Lavender	Root, flower	Decoction from fresh or dried roots	Dried roots for powder, fresh roots for essential oil and is used in perfumery, flavor, pharmaceutical preparation Lavender oil Essential oil is used as pigment, insect repellent
Ledum latifolium Jacq.	Labrador tea	Flower, whole plant	Infusion	Dried roots for powder (rarely used, because it causes vomiting and nausea), tincture from roots
Lemna minor L.	Duckweed	Whole plant	Infusion	Oil from flower
Leonurus cardiaca L.	Motherwort	Whole plant	Infusion, decoction	Fresh pericarp and leaflets for tincture, dried for powder

Scientific name	Common name	Source	Uses	Value-added products
Levisticum officinale W. Koch. *L. levisticum* L.	Lovage	Seed, leaf, stem	Infusion, tea mixture, salad from leaves, seeds used in bread, cheese crackers	Crushed dried berries in tincture, oil is used in perfumery
Ligustrum lucidium W. T. Aiton. *L. vulgare* L.	Ligustrum fruit Privet	Fruit, root	Infusion, leaf, young stem and berry used as dye	Tincture, decoction, and capsules from leafing branches
Linum usitatissimum L.	Flax	Whole plant	For oil, fiber, infusion of flowers	Fresh flowering stem or plants and seed for essential oil and used in perfumes and colognes
Liquidambar styraciflua L. *L. orientalis* L.	Gum tree, storax Oriental sweet gum	Balsam	Syrups	Tincture, or used in commercial flavoring of food and tobacco
Lobelia inflata L. *L. siphilitica* L. *L. pulmonaria* L.	Lobelia Blue lobelia Lungwort	Whole plant	Decoction, infusion and liquid extract	Tincture, an important ingredient of antismoking tobacco (imitating effects of nicotine) and cough mixture
Lomatium dissectum (Nutt.) Math. & Const.	Lomatium	Root, flower, leaf, seed	Decoction, infusion	Tincture from fresh plant, flower essence, oil from seed
Lonicera caerulea L. *L. caprifolium* L.	Mountain fly honeysuckle Dutch honeysuckle	Flower, leaf, stem	Infusion, decoction	Fresh and dried flowering stems for tincture, powder, poultices

Scientific name	Common name	Source	Uses	Value-added products
Lycium barbarum L. *L. pallidum* L. *L. chinesis* Mill.	Lycium Wolfberry Chinese wolfberry	Root bark, fruit	Decoction from bark and fruit	Tincture from crushed dried roots or flowering stems collected before flowering (side effect on large dose, especially pregnant women)
Lycopodium clavatum L.	Clubmoss	Whole plant	Infusion, decoction	Tincture from powder of spores
Lysimachia vulgaris L.	Loosestrife	Whole plant	Infusion	Tincture
Lythrum salicaria L.	Purple loosestrife	Whole plant	Decoction and infusion	Tincture, powder
Macropiper excelsum G. Forst	Kava-Kava	See *Piper methysticum*		
Mahonia aquifolium (Pursh) Nutt.	Oregon grape	Root, fruit	Used as a dye, decoction, liquid extract	Oil from seeds can be used as medicine, manufacture of paints, soap, and printer's ink
Marrubium vulgare L.	Horehound	Whole plant	Infusion, liquid extract, syrups	Fresh and dried flowering plants and seeds for tincture, powder
Matricaria chamomilla L. *M. recutita* L.	Chamomile German chamomile	Root, flower stem	Liquid extracts	Fresh or dry root for tincture, used in cosmetics as an antiallergenic agent, conditioner and lightener

Scientific name	Common name	Source	Uses	Value-added products
Matteuccia struthiopteris (L) Tod	Ostrich fern	Root, spores, shoots	Decoction	Powdered root, tincture, extract
Medicago sativa L.	Alfalfa	Seed, leaf	Young leaves used fresh, seeds used for sprouts, yellow dye from seeds	Tincture from fresh herb, a commercial source of chlorophyll, carotene, and vitamin K
Melaleuca alternifolia (Maid. & Bet.) Cheel	Melaleuca, medicinal tea tree	Leaf, twig	Oil from leaf and twig	Oil used to flavor candy, in perfumery, detergents, soap, and insect repellants, mouthwashes
Melilonus officinali (L.) Lamk. *M. arvensis* L.	Sweet clover Melilot	Flower, leaf, seed, root	Infusion, liquid extract, syrup	Tincture, essential oil, flower essence
Melissa officinalis L.	Lemon balm, balm	Whole plant	Infusion, liquid extract, ointments	Tincture
Mentha x piperita L. *M. spicata* L. *M. pulegium* L.	Peppermint Spearment Pennyroyal	Whole plant	Infusion, liquid extract	Tincture and capsules from root, flavor. Drug products "Menthol," "peppermint oil," "spearmint" are derived from these plants
Monarda odoratissima Bench. *M. punctata* T. J. Howell	Horsebalm Horsemint	Whole plant	Infusion	Fresh or dried flowering stems for tincture
Myosotis scorpioides L	Lily of the valley	See *Convallaria majalis*		

Scientific name	Common name	Source	Uses	Value-added products
Myrica penxylvanica Lois.	Bayberry	Whole plant	Decoction, infusion, liquid extract	Powder, fruit for wax extraction
Myristica fragrans Houtt.	Nutmeg	Seed	Oil, ground nutmeg for bakery products, decoction	Fatty oil (nutmeg butter) is used in the pharmaceutical industry, perfume, soap, and candle manufacturing, flavor
Myrtus communis L.	Myrtle	Leaf, fruit	Infusion	Tincture
Narcissus pseudonarcissus L.	Daffodil	Leaf, bulb	Infusion	Tincture
Nasturtium officinale L.	Watercress	Leaf, root	Fresh vegetable	Powdered root, tincture, extract, flower essences
Nepeta cataria L.	Catnip, catmint	Whole plant	Infusion, culinary purposes	Dried catnip is used to stuff cat toys
Nymphaea alba L.	White water lily	Root, flower, fruit, seed	Root for decoction, liquid extraction, infusion from flowering herb	Dried leaves for oil with lemon flavor
Ocimum basilicum L.	Basil	Whole plant	Juice, infusion, tonic from seed	Oil from fresh or partially dried herb for medicinal use and flavoring in toothpastes, mouthwashes, and a refreshing tea substitute, insect repellants
Ocotea bullata (Birth) E. May	Stinkwood	Leaf	Infusion	Tincture

Scientific name	Common name	Source	Uses	Value-added products
Oenothera biennis L.	Evening primrose	Flower, seed, leaf	Decoction, food	Fresh or dry plant for tincture, essential oil from seed is added to skin prep
Olea europaea L.	Olive	Leaf, fruit	Infusion, liquid extract	Tincture, oil is added to liniments, ointments, skin and hair preparations, and soap
Oplopanax horridus (Sm.) Miq.	Devil's club	Root, bark	Infusion, decoction	Tincture, flower essence, essential oil
Opuntia fragilis (Nutt.) Haw.	Prickly pear	Pad	Infusion	Juice, used in food, feed, fermentation such as jam, alcohol, food pigment, single-cell protein and microbial oil
Origanum majorana L. *Origanum vulgare* subsp. *hirtum* (Link) Ietswaart.	Sweet marjoram Marjoram, Oregano	Flower, leaf	Infusion and distill for oil	Oil is used in food flavoring, liqueurs, perfumery, soaps, and hair products
Paeonia lactiflora Pall. *P. officinalis* L. *P. suffruiticosa* Andr.	White peony (Bia shao) European peony Tree peony (Mu Dan Pi)	Whole plant, root	Boiled and dried root for decoctions, infusion from flowering stem	Dried flowering stems without the woody parts for tea mixture and tincture, externally in ointments for hemorrhoids
Panax ginseng C.A. Meyer. *P. notoginseng* (Burk.) F. H. Chen *P. quinquefolium* L.	Asian ginseng Tian Qi American ginseng	Root, flower, seed	Decoction, liquid extracts	Crushed, sliced, or powdered dried roots, seed oil, tincture, capsule

Scientific name	Common name	Source	Uses	Value-added products
Papaver braceatum L. *P. rhoens* L.	Iranian poppy Poppy, corn poppy	Flower, seed	Infusion, syrups	Oil from fresh herb and flowering stems used medicinally and in perfumery and food industry. Opium extract (paregoric), codeine, morphine, papaverine are derived from this plant
Passiflora incarnata L.	Passion flower	Whole plant	Infusion, liquid extract from flowering herb, fruit for culinary use	Tincture, powder
Pelargonium capitatum (L.) L'Her *P. graveolens* L'Her. ex. Aiton *P. crispum* (L.) L'Her. ex. Aiton	Wild rose geranium Lemon geranium Rose geranium	Leaf	Culinary purpose, infusion, tea mix	Distilled for oil immediately after harvesting, perfumery
Peltigera canina L.	Ground liverwort (Dog lichen)	Whole plant	Infusion, fresh leaf and stem used as dye	Tincture from fresh thallus, dye
Petroselinum crispum (Mill.) Nym.	Parsley	Leaf, root, seed	Infusion of fresh or dried herb	Tincture, oil distilled from leaf and seed
Phataris canariensis L.	Canary creeper	Leaf	Decoction	Tincture
Phellodendron chinensis Schneid.	Chinese corktree (Huang Bai)	Bark	Decoction	Dried flowering stems for herbal tea mixture, a flavor for cooking

Scientific name	Common name	Source	Uses	Value-added products
Phoradendron leucarpum (Raf.) Rev. & M. C. Johnst *P. flavescens* (Push.) Nutt.	Mistletoe American mistletoe	Whole plant	Fresh or dry root for infusion	Tincture and capsules
Phyllanthus emblica L.	Myrobalan, emblic	Root	Decoction	Tincture
Physalis alkekengi L.	Chinese lantern	Flower	Dried flowers, roots and seeds used in herbal mixtures, (never used on its own), dried root for decoction	Fruits used as garnish
Phytolacca americana L.	Pokeweed	Seed, root	Decoction, liquid extract, fresh or dried flowers for tea and tonic	Dried whole root, slice or powder for capsule, tablet, and tincture
Picea mariana (Mill.) Black *P. glauca* (Moench) Voss. *P. ables* (L.) Karst	Black spruce White spruce Norway spruce	Inner bark, buds, needles	Herbal tea, decoction	Tincture
Pimpinella anisum L.	Anise	Leaf, seed	Decoction, distill for oil	Oil from seeds add a bitter flavor to liqueurs and pharmaceutical products

Scientific name	Common name	Source	Uses	Value-added products
Pinus mugo Torra var. Pumilio *P. palustris* Mill. *P. strobus* L. *P. albicaulis* Engelm. *P. contorta* Dougl. ex. Loud	Dwarf mountain pine Southern pitch pine White pine Lodgepole pine	Flower, needle, park, pollen	Infusion and decoction from buds	Essential oil, pine nuts, pinewood oil for use in detergents, soaps and household cleaners, solvent
Piper methysticum G. Forst	Kava-Kava	Root	Decoction, liquid extract	Tincture, powder from roots, unripe fruit for oleoresin and oil
P. nigrum L.	English pepper	Fruit	Dried pepper, infusion	Volatile oil, spice
Plantago asiatica L. *P. psyllium* L. *P. lanceolata* L *P. major* L.	Psyllium Plantain	Whole plant	Dry or fresh herb for liquid extract, and infusion, decoction	Powder, tincture, a drug product "Psyllium husks" is derived from this plant.
Podophyllum peltatum L.	Mayapple	Root, unripe fruit, seed, leaf	Decoction, juice from fresh roots externally heals wounds and reduces swellings	Fresh or dried root for tincture and tea, oil from ripe fruits, root is used for extraction of resin
Polygala senega L.	Seneca snakeroot	Root	Decoction, infusion, liquid extract	Tincture and powder, a drug "Senaca fluid extract" is derived from this plant

Scientific name	Common name	Source	Uses	Value-added products
Polygonum bistorta L. *P. multiflorum* Thunb.	Bistort root Flowery knotweed Fu-ti	Root, stem	Decoction, infusion, liquid extracts	Tincture, powder
Populus balsamifera L. *P. candicans* L. *P. tremuloides* Michx.	Poplar Balm of gilead American aspen	Leaf, bud, root, bark	Infusion, liquid extract	Tincture, powder from bark
Primula vulgaris Huds. *P. veris* L.	Primrose Cowslip	Whole plant	Infusion, ointment	Tincture
Prunella vulgaris L.	Heal all (xu ku gao)	Whole plant	Infusion, ointment, decoction	Fresh berries in jam or dried berries made into a tea, tincture
Prunus africana L.	African prune	Flower, bark whole plant	Decoction, infusion	Extracts from bark
P. dulcis (Mill) D.A. Webb.	Almond	Seed	Decoction	Fruits are used in the manufacture of various medicines, seed oil used to flavor liqueurs, perfumery
P. mume Siebold & Zucc.	Japanese apricot	Unripe fruit	Decoction	Oil extracted from leaves
P. serotina J. F. Ehrb.	Black cherry, wild cherry	Bark	Infusion, liquid extracts	Powders, syrups, tinctures
P. virginiana L.	Chokecherry	Bark	Decoction, infusion	Tincture

Scientific name	Common name	Source	Uses	Value-added products
Pueraria lobata (Wild) Ohwi	Kudzu	Root, flower	Juice, decoction	Seed oil including oleic, linoleic, and linolenic acids
Pygeum africanum L.	Pygeum	Leaf	Infusion	Fresh or frozen juice from herb, seed oil (linoleic, oleic, and palmitic fatty acids), capsule
Pyrethrum cinerarifolium L.	Pyrethrom	Leaf	Infusion	Tincture
Quercus alba L.	White oak	Bark, root	Decoction and liquid extraction, bark and galls are used in tanning and in dyeing	Tincture from root
Ranunculus occidentalis Nutt. *R. ficaria* L.	Buttercup Pilewort	Whole plant, root	Ointments, infusion, liquid extracts	Tincture from fresh herb, dry root, tablets
Raphanus sativus L.	Radish	Root, seed, leaf	Decoction	Juice from root and leaf, syrup from root
Rhamnus catharticus L. *R. frangula* L. *R. purshiaanus* L.	Buckthorn Alder buckthorn Cascara sagrade	Seed, bark, fruit	Decoction, liquid extract, fruit for syrup	Tincture, powder from bark; extracts to flavor liqueurs, soft drinks, ice cream, and baked goods
Rheum palmatum L. *R. officinal* Baill. *R. tanguticum* L.	Chinese rhubarb Rhubarb Rhubarb	Leaf, rhizome	Decoction	Tincture from fresh or dry early spring leaf buds, root, extract with bitterness removed is used in food flavoring

Scientific name	Common name	Source	Uses	Value-added products
Rhodiola rosea (L) Scop.	Roseroot	See *Sedum rosea*		
Rhus radicans L. *R. toxicodendron* L.	Poison ivy Roseroot	Bark, flower	Decoction and liquid extract	Tincture and powder from bark and flower
Ribes nigrum L. *R. lacustre* (Pres.) Poir.	Black current Black gooseberry	Bark, root, flower bud	Decoction, infusion	Tincture of fresh flowering stems, externally as ointment, juice
Robertium macrorrhizum Pic.	Geranium	See *Geranium macrorrhizum*		
Rosa canina L. *R. damascena* Mill.	Dog rose Damask rose	Seed, seed husk	Rose water from petal for syrup, husk for decoction	Petals and seeds for oil, bark extract esters, perfumery, flavoring
Rosmarinus officinalis L.	Rosemary	Leaf, seed	Infusion, decoction	Extract oil from flower top and dried leaves, used in cosmetic and toilet preparations, liqueur industries, fatty oil from seeds
Rubus chamaemorus L. *R. fruiticosus* L.	Cloudberry Blackberry	Root	Young shoot of blackberry used as dye	Kernel for oil
R. idaeus L.	Raspberry	Root, fruit	Fruit is made into syrups, cordials, jam, wine and purees, flavor vinegar	Tincture, fruit essence for shampoos and bath preparation and flavoring agent for medicines

Scientific name	Common name	Source	Uses	Value-added products
Rumex acetosella L. *R. obtusifolia* L.	Sorrel Dock	Leaf	Whole plant and root used as dye, decoction, liquid extract	Tincture
Ruscus aculeatus L.	Butcher's broom	Leaf bud	Decoction, ointments, culinary, fruiting branches are used in Christmas decorations	Bark extracts, capsule from dry bark
Ruta graveolens L.	Rue, herb of grace	Leaf, root	Infusion, liquid extraction	Decoction and tincture from roots
Salix alba L. *S. discolour* Muhlenb *S. caprea* L.	White willow Pussy willow	Inner bark, leaf, catkin	Infusion, sap, decoction	Tincture, oil, powder from bark, cough syrup
Salvia officinalis L. *S. sclarea* L.	Sage Clary sage	Kernel, flower top, leaf	Distill flowers, infusion from fruits, syrup	Tincture, oil
Sambucus nigra L. *S. canadensis* L. *S. racemosa* L.	Elderberry Elder	Kernel, bark	Leaf and berry used as dye	Dried bark from young twigs for tincture, capsules
Sanguinaria canadensis L.	Bloodroot	Rhizome	Used as a dye	Dried rhizomes of 5- to 7-yr-old plants for tincture, it is also a component of herbal tea mixtures and digestive powders

Scientific name	Common name	Source	Uses	Value-added products
Sanicula marilandica L.	Black snakeroot	Root, flower	Decoction	Tincture, oil, flower essence
Santalum album L.	Sandalwood	Heart wood, root	Decoction, liquid extract, fresh fruits are made into syrup, jam, cordials and wines, or eaten raw or stewed	Powder and tincture, perfumery
Saponaria officinalis L.	Soapwort	Root, leaf	Decoction of the root, infusion of aerial part	Tincture from root, dried herb is used as a soap substitute for delicate materials, shampoo
Sarothamnus scoparius L.	Broom	Whole plant	Decoction, infusion, liquid extract	Tincture
Sassafras albidum (Nutt.) Nees	Sassafras	Bark, leaf	Tea made by macerating the crushed hips, decoction, infusion, fresh hips can be made into jam, syrup and medicinal wine	Tincture

Scientific name	Common name	Source	Uses	Value-added products
Satureja hortensis L.	Savory	Leaf, stem, flower	Infusion and oil extraction	Leaves for tincture and capsule, rosemary oil extracted from fresh leaves in toilet preparations, disinfectants and extensively in perfumery
Schisandra chinensis (Turcz.) Baill.	Schizandra (wu wei zi)	Fruit	Dried fruit for use in decoction	Tincture and powder from dried fruit
Scutellaria baicalensis Georgi. *S. lateriflora* L. *S. galericulata* L.	Baical skullcap Virginia skullcap Skullcap	Root (5-7 years old) Flower, leaf	Infusion, decoction, liquid extract	Tincture, powdered capsules
Sedum acre L. *S. rosea* (L.) Scop. subsp. *integrifolium*	Stonecrop, small houseleek Roseroot	Leaf, root, stem	Fresh plant juice, infusion	Tincture, essential oil, flower essence
Sempervivum tectorum L.	Houseleek, hens and chicks	Leaf	Fresh leaf for infusions, poultices	Tincture from leaves
Senecio vulgris L.	Groundsel	Whole plant, rhizome	Infusion and liquid extract from dried plant	Powder from dried root
Senna alexandrina L.	Alexandrian senna	Leaf, pod	See *Cassia angustifolia*	
Serenoa repens (Bartr.) Small.	Saw palmetto	Leaf, fruit	Dried fruit for elixirs, infusion, liquid extract	Dried leaves and fruit used in tincture

Scientific name	Common name	Source	Uses	Value-added products
Sesamum indicum L. *S. orientale* L.	Sesame	Leaf, seed	Leaves for infusion, seed for decoction, pressed for oil or ground into paste, residue is used in livestock feeds	Fresh or dry roots for tincture, capsules; seeds used in bakery products, oil from seeds for manufacture of margarine, lubricants, soaps, and pharmaceutical drugs
Shepherdia canadensis (L.) Nutt.	Buffalo berry	Leaf, young shoot	Infusion and decoction from dried leaves	Tincture
Silybum marianum (L.) Gaertn.	Milk thistle	Whole plant, seed	Whole plant is dried for use in infusion	Tincture from fresh plant, oil from young shoots with the basal leaves, extraction of silymarin
Simmondsia chinensis (Link) C. Schneid	Jojoba	Seed, bark	Seed for oil, dried bark from 2- to 3-yr-old twigs for decoction	Oil from seeds added to shampoos, moisturizers and sun screens, also used as a lubricant, detergent, and wetting agent
Solidago virgaurea L.	Goldenrod	Leaf	Dried leaf for infusion, flowering herb for decoction	Dried leaves for oil, tincture

Scientific name	Common name	Source	Uses	Value-added products
Sorbus aucuparia L.	Mountain ash	Whole plant	Dried flowers and fruits used in infusion, tea mixture (the leaves are purgative and should not be used)	Seed oil (palmitic, stearic, oleic, linoleic, and linolenic acids)
Spirea ulmaria L.	Meadowsweet	See *Filipendula ulmaria*		
Stachys officinalis (L.) Trev.	Betony, woundwort	Whole plant	Flowering plant for infusion and liquid extracts	Oil from leaf, ointment from flower, tincture from plant, dried leaves are included in herbal tobacco and snuff
Stellaria media (L.) Vill.	Chickweed	Whole plant	Fresh plant for juice or poultices, infusion, liquid extract	Fresh or dried plant for infusion, medicated oils, ointments, and tincture
Stevia rebaudana (Bertoni) Bertoni	Stevia Sweet herb of Paraguay	Leaf	Decoction, powdered leaf	Leaf extracts
Symhytum officinale L.	Comfrey	Root, leaf	Fresh plant used as dye, dried leaf for infusion, liquid extract and poultices, root for decoction	Oil from root for ointments, powders from dried root, a drug product “Allentonin” is derived from this plant

Scientific name	Common name	Source	Uses	Value-added products
Syringa vulgaris L.	Lilac	Leaf, bud, stem flower, bark	Infusion and decoction	Oil from plant parts including flower
Syzygium aromaticum (L.) Merr. & Perry	Clove	Flower buds	Infusion, powders of sun-dried buds	Oil extracts from flower buds, stem, and leaves
Tagetes lucida Cav.	Mexican mint marigold	Leaf	Plants for oil, infusion, ointment	Tincture
Tanacetum parthenium (L.) Schultz.	Feverfew	See *Chrysanthermum parthenium*		
T. cinerarifolium L.	Pyrethum	See *Chrysanthemum cinerarifolium*		
T. vulgare L.	Tansy	Leaf, stem, flower, fruit	Flowering top used as dye, infusion, liquid extract, dry berries alone or make decoction	Oil from leaves, tansy is used in preserving meat and to repel ants
Taraxacum officinale G. H. Weber ex. Wigg.	Dandelion	Root, leaf	Decoction from root, leaves used as vegetable, juice, infusion	Roots for tincture, leaves and roots flavor herbal beers and soft drinks, roasted ground root as caffeine-free coffee substitute, used in medicinal preparations
Taxus x media Rehd. *T. brevifolia* Nutt.	Yew	Bark, root	Heartwood chips used as dye	Fresh and dry berries for tincture, capsule

Scientific name	Common name	Source	Uses	Value-added products
Thymus citriodorus (Pers.) Schreb. *Thymus vulgaris* L	Lemon thyme Garden thyme	Whole plant	Distilled for oil, elixirs, liquid extract, infusion from plant	Oil from seeds for perfumery, cooking, and salads, it is an important ingredient of toothpastes and mouthwash, a drug product "Thymol" is derived from *T. vulgaris*
Tillia cordata Mill. *T. europaea* L.	Linden, small-leaved lime	Berry, flower	Dried flower for infusion, liquid extract	Ripe fruits for tinctures, fluid extract, capsules and tablets, tincture of seeds is used in homeopathy
Tribulus terrestris L.	Puncture vine	Flower, fruit	Infusion, decoction	Tincture
Trifolium incarnatum L. *T. pratense* L.	Clover (crimson) Clover (red)	Flower, seed	Flower heads with upper leaves for infusion, liquid extracts, ointment	Oil from seeds, tincture from flower heads
Trigonella foenum-gracecum L.	Fenugreek	Leaf, seed	Fresh or dried leaf for infusion, seed for decoction, pastes and powder	Seeds are used for extracts that are used in synthetic maple syrup and flavoring for food industry
Tropaeolum majus L.	Canary creeper, Nasturtium	Whole plant	Infusion, juice	Tincture

Scientific name	Common name	Source	Uses	Value-added products
Tussilago farfara L.	Colisfoot	Berry, leaf, flower	Whole plant used as dye, flowers for decoction, liquid extract, syrup, pressed juice from fresh fruits	Dried fruits are the raw material for the manufacture of sorbose (a sweetening agent for diabetics), dried leaves are an ingredient of herbal tobaccos and are used in curing pipe tobaccos
Ulmus rubra Muhl. *U. procera* L.	Slippery elm, sweet elm English elm	Inner bark	Decoction, liquid extract, ointments, poultices	Tincture from fresh or dried flowering stems with the basal leaves and inner bark, powder from dried inner bark
Umbelluslaria california Nutt.	California laurel Bay laurel	Leaf	Infusion	Tincture, powder
Uncaria tomentosa (Willd.) DC. *U. guianensis* (Aubl.) Gmel.	Cat's claw	Root, bark	Decoction	Tincture, powder
Urtica dioica L. *U. urens* L.	Stinging nettle	Root, herb	Infusion, liquid extracts, ointments	Fresh and dry herb for tincture and powder; plants are processed commercially for extraction of chlorophyll, which is used as coloring agent in food and medicine
Vaccinium macrocarpon L.	Cranberry	Berry, leaf	Leaves for decoction, fruits for decoction and liquid extract	Fruits are added to wine, extracts are used to flavor liqueurs, fruit juice with medicinal value

Scientific name	Common name	Source	Uses	Value-added products
Vaccinium vitis-idaea L.	Mountain cranberry, cowberry, foxberry	Berry	Fruits for decoction and liquid extract	Tincture
V. mrytilloides Michx. *V. myrtillus* L. *V. oreophilum* Rydb.	Blueberry Bilberry	Root, berry, flower, leaf	Liquid extract, decoction of berry, fresh juice, jam	Tincture, flower essence
Valeriana officinalis L.	Valerian	Rhizome	Decoction, infusion, liquid extract	Tincture of flowering herb and root, powder and oil from root
Verbascum thapsus L.	Mullein	Whole plant	Infusion, liquid extract	Tincture, fresh or frozen flowers for oil and syrup
Veronica officinalis L.	Speedwell	Whole plant	Infusion	Fluid extract and tincture from roots, flowering stems, leaves, also for wine, pressed juice from stalks or leaves (for warts), dried herb may be added to tea blends
Vetiveria zizaniodes L. (Nash.)	Vetiver	Root	Dried roots are woven into scented mats, screens, for drinks	Extract essential oil; oil is used in soaps, cosmetics and as a fragrance fixative, food flavoring, soap, perfumes
Viburnum opulus L.	Highbush cranberry	Bark	Decoction, liquid extract	Tincture from fresh needles, extract from needle and bark for pharmaceutical use

Scientific name	Common name	Source	Uses	Value-added products
Vinca minor L.	Periwinkle	Leaf, root	Infusion, liquid extract	Distillation from flowering stems is used in the manufacture of gargles, toothpastes, mouthwashes, tincture
Viola tricolor L.	Pansy	Whole plant	Decoction, infusion and liquid extract	Flowers to flavor and color candy and breath fresheners, plant is dried and powdered for skin cream
Viscum album L.	Mistletoe	Leaf, stem	Infusion, dried flowers as linden tea	Extracts, tincture
Vitex agnus-castus L. *V. negundo* L.	Chaste tree Chinese chaste	Whole plant	Infusion, poultices, decoction, dried flower heads as tea mixture	Extract from fruit, tincture, fresh leaves are burned with grass as a fumigant against mosquitoes
Vitis labrusca L. *V. vinifera* L.	Grape	Leaf, flower, root	Decoction, liquid extract, infusion of flower heads, tea mixture	Tincture, liquid from stem is used directly as an eyewash and diuretic, essential oil from seeds
Withania somnifera Dunal.	Ashwagandha, Indian ginseng	Bark, root	Pastes	Oil, powder
Yucca aloifolia L.	Yucca	Leaf	Infusion, tea mixture, a decoction is used externally in compresses and bath	Tincture of flower head and young leaves
Y. glauca L.	Soapweed	Leaf	Decoction as tea	Soap, shampoo

Scientific name	Common name	Source	Uses	Value-added products
Zanthoxylum americanum Mill.	Prickly ash, toothache tree	Whole plant	Bark and fruit for decoction, liquid extracts	Tincture from bark
Zea mays L.	Corn	Flower	Corn silk for decoction, infusion, liquid extract and syrup	Ears for vegetable, or processing as cereals, flour, oil, and syrup
Zingiber officinale Roscoe	Ginger	Flower, rhizome	Ground rhizome for infusion, decoction, oil from unpeeled dried rhizomes	Water-alcohol extract, flowering stems and the leaves for tea, beverage, beer also used as a hair tonic and face wash

References

1. Abram, V. and M. Donko. 1999. Tentative identification of polyphenols in *Sempervivum tectorum* and assessment of the antimicrobial activity of *Sempervivum* L. J. Agric. Food Chem. 47: 485-489.

2. Ahlemeyer, B. and J. Krieglstein. 1998. Neuroprotective effects of *Ginkgo biloba* extract. In: Lawson, L. D. and R. Bauer (eds.). Phytomedicines of Europe, Chemistry and Biological Activity. Am. Chem. Soc., Washington, DC. p. 210-220.

3. Ahluwalia, S. S. 1997. Goldenseal, American Gold. Walden House, Bronx, NY. 22 p.

4. Annoymous. 1952. Insects, The Yearbook of Agriculture 1952. USDA, Washington, DC. 780 p.

5. Arctander, S. 1960. Perfume and Flavor Materials of Natural Origin. Published by the author, Elizabeth, NJ. 736 p.

6. Arctander, S. 1969. Perfume and Flavor Chemicals. Vol. 1 and 2. Published by the author, Elizabeth, NJ.

7. Bandra, K. A., N. P. Peries, I. D. R. Kumar and V. Karunaratne. 1990. Insecticidal activity of *Acarus calamus* and *Glycosmis mauritata* Tanaka against *Aphis craccivora*. Tropical Agri. 67: 223-228.

8. Baranyk, P., V. Zeleny, H. Zukalova and P. Horejs. 1995. Oil content of some species of alternative oil plants. Rostlinna-Vyroba. 41: 433-438.

9. Baresova, H. 1988. *Centaurium erythraea* Rafn: micropropagation and the production of secoiridoid glucosides. In: Bajaj, Y. P. S. (ed.). Biotechnology in Agriculture and Forestry 4. Medicinal and Aromatic Plants I. Springer-Verlag Co., New York. p. 350-366.

10. Bauer, R. 1998. *Echinacea:* biological effects and active principles. In: Lawson, L. D. and R. Bauer (eds.). Phytomedicines of Europe, Chemistry and Biological Activity. Am. Chem. Soc. Washinton, DC. 324 p.

11. Beveridge, T., T. S. C. Li, B. D. Oomah and A. Smith. 1999. Sea buckthorn products: manufacture and composition. J. Agric. Food Chem. 47: 3480-3488..

12. Bisset, N. G. 1994. Herbal Drugs and Phytopharmaceuticals. CRC Press, London. 566 p.

13. Bottoni, D. 1998. Tea tree oil. Int. J. Pharmaceutical Compounding 2: 284-285.

14. Bown, D. 1987. *Acorus calamus* L: a species with a history. Aroideana (International Aroid

Society, South Miami, FL.) 10: 11-14.

15. Bown, D. 1995. Encyclopedia of Herbs & Their Uses. Reader's Digest Assoc. (Canada) Ltd., Montreal, PQ. 424 p.

16. Bremness L. 1994. The Complete Book of Herbs. Dorling Kindersley Ltd., London. 304 p.

17. Bunney, S. 1992. The Illustrated Encyclopedia of Herbs, Their Medicinal and Culinary Uses. Chancellor Press, London. 320 p.

18. Cavallini, A., L. Natali and I. Castorena Sanchez. 1991. *Aloe barbadensis* Mill. (=*A. vera* L.). In: Bajaj, Y. P. S. (ed.). Biotechnology in Agriculture and Forestry 15. Medicinal and Aromatic Plants III. Springer-Verlag Co., New York. p. 95-106.

19. Chapple, C. C. S. and B. E. Ellis. 1991. *Syringa vulgaris* L. (Common lilac): in vitro culture and the occurrence and biosynthesis of phenylpropanoid glycosides. In: Bajaj, Y. P. S. (ed.). Biotechnology in Agriculture and Forestry 15. Medicinal and Aromatic Plants III. Springer-Verlag Co., New York. p. 478-497.

20. Charlwood, B. V. and K. A. Charlwood. 1991. *Pelargonium* spp. (Geranium): in vitro culture and the production of aromatic compounds. In: Bajaj, Y. P. S. (ed.). Biotechnology in Agriculture and Forestry 15. Medicinal and Aromatic Plants III. Springer-Verlag Co., New York. p. 339-352.

21. Chen, H. M., T. T. Fang. 1993. Studies on the preparation of mei juice VI. Distribution pattern of volatile compounds of mei fruit juice. Memoirs College Agric. National Taiwan Univ. 33: 163-179.

22. Chevallier, A. 1996. The Encyclopedia of Medicinal Plants. Dorling Kindersley Ltd., London. 336 p.

23. Collin, H. A. and S. Isaac. 1991. *Apium graveolens* L. (Celery): in vitro culture and the production of flavors. In: Bajaj, Y. P. S. (ed.). Biotechnology in Agriculture and Forestry 15. Medicinal and Aromatic Plants III. Springer-Verlag Co., New York. p. 73-94.

24. Cordeiro, M. C., M. S. Pats and P. E. Brodelius. 1998. *Cynara cardunculus* subsp. *flavescens* (Cardoon): in vitro culture, and the production of cyprosins—milk-clotting enzymes. In: Bajaj, Y. P. S. (ed.). Biotechnology in Agriculture and Forestry 41. Medicinal and Aromatic Plants X. Springer-Verlag Co., New York. p. 132-153.

25. Corto-Cositet, M. F., L. Chapuis and J. P. Delbecque. 1998. *Chenopodium album* L. (Fat hen): in vitro cell culture, and production of secondary metabolites (phytosterols and ecdysteroids). In: Bajaj, Y. P. S. (ed.). Biotechnology in Agriculture and Forestry 41. Medicinal and Aromatic Plants X. Springer-Verlag Co., New York. p. 97-112.

26. Crellin, J. K. 1990. Home Medicine, the Newfoundland Experience. J. K. Crellin Publication. 280 p.

27. Cunio, L. 1994. *Vaccinium myrtillus*. Australian J. Medical Herbalism 5: 81-85.

28. Dube, S., P. D.Upadhyay and S. C. Tripathi. 1991. Fungitoxic and insect repellent efficacy of some spices. Indian Phytopathology 44: 1, 101-105.

29. Dobelism, I. N. 1990. Magic and Medicine of Plants (5th edition). The Reader's Digest Assoc., Inc., New York. 464 p.

30. Duke, J. A. 1985. CRC Handbook of Medicinal Herbs. CRC Press, Inc., Boca Raton, FL. 677 p.

31. Duke, J. A. and J. L. duCellier. 1993. CRC Handbook of Alternative Cash Crops. CRC Press, London. 536 p.

32. Ernst, D. 1989. *Pimpinella anisum* L. (Anise): cell culture, somatic embryogenesis, and the production of anise oil. In: Bajaj, Y. P. S. (ed.). Biotechnology in Agriculture and Forestry 7. Medicinal and Aromatic Plants II. Springer-Verlag Co., New York. p. 381-397.

33. Foster, S. 1993 Herbal Renaissance. Gibbs Smith Publisher, Utah. 234 p.

34. Fujii, Y. 1991. *Podophyllum* spp.: in vitro regeneration and the production of podophyllotoxins. In: Bajaj, Y. P. S. (ed.). Biotechnology in Agriculture and Forestry 15. Medicinal and Aromatic Plants III. Springer-Verlag Co., New York. p. 362-375.

35. Furmanowa, M. and J. Guzewska. 1989. *Dioscorea:* in vitro culture and the micropropagation of diosgenin-containing species. In: Bajaj, Y. P. S. (ed.). Biotechnology in Agriculture and Forestry 7. Medicinal and Aromatic Plants II. Springer-Verlag Co., New York. p. 162-184.

36. Furmanowa, M., D. Sowinska and A. Pietrosiuk. 1991. *Carum carvi* L. (Caraway): in vitro culture, embryogenesis, and the production of aromatic compounds. In: Bajaj, Y. P. S. (ed.). Biotechnology in Agriculture and Forestry 15. Medicinal and Aromatic Plants III. Springer-Verlag Co., New York. p. 177-192.

37. Furuya, T. and T. Yoshikawa. 1991. *Carthamus tinctorius* L. (safflower): production of vitamin E in cell culture. In: Bajaj, Y. P. S. (ed.). Biotechnology in Agriculture and Forestry 15. Medicinal and Aromatic Plants III. Springer-Verlag Co., New York. p. 142-155.

38. Genest, K. and W. Hughes. 1969. Natural products in Canadian pharmaceuticals IV. Can. J. Pharm. Sci. 4: 41-45.

39. Guenther, E. and D. Althausen (eds.). 1949. The Essential Oils. Vol. 1, 2, and 3. D. Van

Nostrand Co. Inc., New York. 427 p., 852 p., and 777 p.

40. Halstead, B. W. and L. L. Hood. 1942. *Eleutherococcus senticosus,* Siberian Ginseng: An Introduction to the Concept of Adaptogenic Medicine. Oriental Healing Arts Inst., Long Beach, CA. 94 p.

41. Hamana, K., S. Matsuzaki, M. Niitsu and K. Samejima. 1994. Distribution of unusual polyamines in aquatic plants and gramineous seeds. Can. J. Bot. 72: 1114-1120.

42. Hamdi, M. 1997. Prickly pear cladodes and fruits as a potential raw material for the bioindustries. Bioprocess Engineering 17: 387-391.

43. Hanczakowski, P., B. Szymczyk and T. Wolski. 1993. The nutritive value of the residues remaining after oil extraction from seeds of evening primrose (*Oenothera biennis* L.). J. Sci. Food Agric. 63: 375-376.

44. Haunold, A. 1993. Agronomic and quality characteristics of native North American hops. Am. Soc. Brew. Chem. J. 51: 133-137.

45. Heale, J. B., T. Legg and S. Connell. 1989. *Humulus lupulus* L. in vitro culture: attempted production of bittering components and novel disease resistance. In: Bajaj, Y. P. S. (ed.). Biotechnology in Agriculture and Forestry 7. Medicinal and Aromatic Plants II. Springer-Verlag Co., New York. p. 264-285.

46. Henkel, A. 1904. Weeds used in medicine. U. S. Dept. Agric. Farmers' Bull. No. 188.

47. Henkel, A. 1907. American root drugs. U. S. Dept. Agric. Bur. Pl. Indus. Bull. No. 107.

48. Henkel, A. and G. F. Klugh. 1908. The cultivation and handling of goldenseal. U. S. Dept. Agric. Bur. Plant Ind. Circ. No. 6. 19 p.

49. Henning, J. 1995. Aromatic *Agastache.* Am. Nurseryman 181: 59-67.

50. Henry, M. 1989. *Saponaria officinalis* L.: in vitro culture and the production of triterpenoidal saponins. In: Bajaj, Y. P. S. (ed.). Biotechnology in Agriculture and Forestry 7. Medicinal and Aromatic Plants II. Springer-Verlag Co., New York. p. 431-442.

51. Henry, M., A. M. Edy, P. Desmarest and J. Du Manuir. 1991. *Glycyrrhiza glabra* L. (Licorice): cell culture, regeneration, and the production of glycyrrhizin. In: Bajaj, Y. P. S. (ed.). Biotechnology in Agriculture and Forestry 15. Medicinal and Aromatic Plants III. Springer-Verlag Co., New York. p. 270-282.

52. Hornok, L. 1992. Cultivation and Processing of Medicinal Plants. John Wiley & Sons, New York. 338 p.

53. Huizing, H. J. and J. H. Sietsma. 1991. *Symphytum officinale* (Comfrey): in vitro culture, regeneration, and biogenesis of pyrrolizidine alkaloids. In: Bajaj, Y. P. S. (ed.). Biotechnology in Agriculture and Forestry 15. Medicinal and Aromatic Plants III. Springer-Verlag Co., New York. p. 464-477.

54. Hutchens, A. R. 1991. Indian Herbalogy of North America. Shambhala Publication Inc., London. 382 p.

55. Ishikura, N. 1989. *Crytomeria japonica* Don (Japanese cedar): in vitro production of volatile oils. In: Bajaj, Y. P. S. (ed.). Biotechnology in Agriculture and Forestry 7. Medicinal and Aromatic Plants II. Springer-Verlag Co., New York. p. 129-134.

56. Ishimaru, K., N. Tanaka, T. Kamiya, T. Sato and K. Shimomura. 1998. *Cornus kousa* (Dogwood): in vitro culture, and the production of tannins and other phenolic compounds. In: Bajaj, Y. P. S. (ed.). Biotechnology in Agriculture and Forestry 41. Medicinal and Aromatic Plants X. Springer-Verlag Co., New York. p. 113-131.

57. Jayatilake, G. S. and A. J. MacLeod. 1987. Volatile constituents of *Centella asiatica*. In: Martens, M., G. A. Dalen and H. Russwurm Jr. (eds.). Flavour Science and Technology. John Wiley & Sons Ltd., New York. p. 79-82.

58. Jonard, R. 1989. *Jasminum* spp. (Jasmine): micropropagation and the production of essential oils. In: Bajaj, Y. P. S. (ed.). Biotechnology in Agriculture and Forestry 7. Medicinal and Aromatic Plants II. Springer-Verlag Co., New York. p. 315-331.

59. Kato, M. 1989. *Camellia sinensis* L. (Tea): in vitro regneration. In Bajaj, Y. P. S. (ed.). Biotechnology in Agriculture and Forestry 7. Medicinal and Aromatic Plants II. Springer-erlag Co., New York. p. 82-98.

60. Katsarska, P. and D. Desev. 1976. Study on the essential oil (concrete and absolute) of *Mahonia aquifolia* with a view to its utilization in the perfumery and cosmetic industry. Perfumer & Flavorist 2: 62.

61. Kelen, A., G. Y. Hangay, Z. S. Kernoczi and J. D. Voonar. 1992. Sports cosmetics containing medicinal herb extracts. Acta Hort. 306: 281-289.

62. Kubo, A., C. S. Lunde and I. Kubo. 1995. Antimicrobial activity of the olive oil flavor compounds. J. Agric. Food Chem. 43: 1629-1633.

63. Kuusi, T., K. Hardh and H. Kanon. 1984. Experiments on the cultivation of dandelion for salads use. II. The nutritive value and intrinsic quality of dandelion leaves. J. Agric. Sci. Finland 56: 23-31.

64. Leung, A. Y. 1980. Encyclopedia of Common Natural Ingredients Used in Food, Drugs, and Cosmetics. John Wiley & Son, New York. 409 p.

65. Li, T. S. C. 1995. Asian and American ginseng, a review. HortTechnol. 5: 27-34.

66. Li, T. S. C. 1998. *Echinacea:* cultivation and medicinal value. HortTechnol. 8: 122-129.

67. Li, T. S. C. and W. R. Schroeder. 1996. Sea buckthorn (*Hippophae rhamnoides* L.): a multipurpose plant. HortTechnol. 6: 370-380.

68. Li, T. S.C. and L. C. H. Wang. 1998. Physiological components and health effects of ginseng, echinacea and sea buckhtorn. In: Mazza, G. (ed.). Functional Foods, Biochemical and Processing Aspects. Technomic Publishing Co., Inc., Lancaster, PA. 460 p.

69. McGimsey J. 1993. Sage, *Salvia officianalis* WWW. Crop.cri.nz/broadshe/sage.htm.

70. McKeown, A. W. and J. Potter. 1994. Native wild grasses and flowers: new possibilities for nematodes. Agri-Food Res. Ontario (Dec., 1994): 20-25.

71. Mansell, R. L. and C. A. McIntosh. 1991. *Citrus* spp.: in vitro culture and the production of naringin and limonin. In: Bajaj, Y. P. S. (ed.). Biotechnology in Agriculture and Forestry 15. Medicinal and Aromatic Plants III. Springer-Verlag Co., New York. p. 193-210.

72. Mazza, G. and T. Cottrell. 1999. Volatile components of roots, stems, leaves, and flowers of *Echinacea* species. J. Agric. Food Chem. 47: 3081-3085.

73. Mazza, G. and F. A. Kiehn. 1992. Essential oil of *Agastache foeniculum,* a potential source of methyl chavicol. J. Essential Oil Res. 4: 295-299.

74. Minker, E. and K. Szendrei. 1986. About the green wave. Gyogyszereszet 30: 323-325.

75. Miyazawa, M. and H. Kameoka 1987. Volatile flavor components of *Astragali radix* (*Astragalus membranaceus* Bunge.). Agric. Biol. Chem. 51: 3153-3154.

76. Miyazawa, M. and H. Kameoka. 1988. Volatile flavor components of *Puerariae radix* (*Pueraria lobata* Ohwi). Agric. Biol. Chem. 52: 1053-1055.

77. Murray, M. T. 1995. The Healing Power of Herbs. Prima Publishing, Rocklin, CA. 410 p.

78. Natali, A. C. and I. C. Sanchez. 1991. *Aloe barbadensis* Mill. (= *A. vera* L.). In: Bajaj, Y. P. S. (ed.). Biotechnology in Agriculture and Forestry 15. Medicinal and Aromatic Plants III. Springer-Verlag Co., New York. p. 95-106.

79. O'Dowd, N. A., P. G. McCauley, G. Wilson, J. A. N. Parnell, T. A. K. Kavanagh and D. J. McConnell. 1998. *Ephedra* species: in vitro culture, micropropagation, and the production ephedrine and other alkaloids. In: Bajaj, Y. P. S. (ed.). Biotechnology in Agriculture and Forestry 41. Medicinal and Aromatic Plants X. Springer-Verlag Co., New York. p. 154-193.

80. Ofek, I., J. Goldhar and N. Sharon. 1996. Anti-*Escherichia coli* adhesin activity of cranberry and blueberry juices. Adv. Exp. Med. Biol. 408: 179-183.

81. Ogasawara, T., K. Chiba and M. Tada. 1998. *Sesamum indicum* L. (Sesame): in vitro culture, and the production of naphthoquinone and other secondary metabolites. In: Bajaj, Y. P. S. (ed.). Biotechnology in Agriculture and Forestry 41. Medicinal and Aromatic Plants X. Springer-Verlag Co., New York. p. 366-393.

82. Ortega-Calvo, J. J., C. Mazuelos, B. Hermosin and C. Saiz-jimenez. 1993. Chemical composition of Spirulina and eukaryotic algae food products marketed in Spain. J. Appl. Phycol. 5: 425-435.

83. Parry, E. J. 1922. The Chemistry of Essential Oils and Artificial Perfumes Vol. I. Scott, Greenwood and Son, London. 365 p.

84. Parry, E. J. 1922. The Chemistry of Essential Oils and Artificial Perfumes Vol. II. Scott, Greenwood and Son, London. 549 p.

85. Petit-Paly, G., K. G. Ramawat, J. C. Chenieux and M. Rideau. 1989. *Ruta graveolens:* in vitro production of alkaloids and medicinal compounds. In: Bajaj, Y. P. S. (ed.). Biotechnology in Agriculture and Forestry 7. Medicinal and Aromatic Plants II. Springer-Verlag Co., New York. p. 488-505.

86. Pisano, R. C. 1986. The future of natural essential oils in the flavour and fragrances industry. Perfumer and Flavorist 11: 35-42.

87. Puri, D., N. Baral and B. P. Upadhyaya. 1997. Indigenous plant remedies in Nepal used in heart diseases. J. Nepal Med. Assoc. 36: 123, 334-337.

88. Purseglove, J. W., E. G. Brown, C. L. Green and S. R. J. Robbins. 1981. Spices. Vol. 1. Longman, London.

89. Ritch-Krc, E.M., N. J. Turner and G. H. N. Towers. 1996. Carrier herbal medicine: an evaluation of the antimicrobial and anticancer activity in some frequently used remedies. J. Ethnopharmacol. 52: 151-156.

90. Rizvi, M. A., G. R. Sarwar and A. Mansoor. 1998. Poisonous plants of medicinal use growing around Madinat al-Hikmah. Hamdard Medicus 41: 88-95.

91. Rogers, R. D. 1997. Sundew, Moonwort, Medicinal Plants of the Prairies. Vol. 1 and 2. Edmonton, AB. 282 p.

92. Rogers, R. and L. Szott-Rogers. 1997. Prairie Deva, Flower Essences. Edmonton, AB. 40 p.

93. Russell, G. A. 1921. Drying crude drugs. U. S. Dept. Agric. Farmers' Bull. No. 1231.

94. Sakamura, F. and T. Suga. 1989. *Zingiber officinale* Roscoe (Ginger): in vitro propagation and the production of volatile constituents. In: Bajaj, Y. P. S. (ed.). Biotechnology in Agriculture and Forestry 7. Medicinal and Aromatic Plants II. Springer-Verlag Co., New York. p. 524-538.

95. Schreiner, O. 1900. Saw palmetto oil. Pharmaceutical Rev. 1900. p. 217-222.

96. Segura, J. and M. C. Calvo. 1991. *Lavandula* spp. (Lavender): in vitro culture, regeneration of plants and the formation of essential oils and pigments. In Bajaj, Y. P. S. (ed.). Biotechnology in Agriculture and Forestry 15. Medicinal and Aromatic Plants III. Springer-Verlag Co., New York. p. 283-310.

97. Shellard, E. J. 1987. Medicines from plants with special reference to herbal products in Great Britain. Planta Medica 53: 121-123.

98. Short, K. C. and A. V. Roberts. 1991. *Rosa* spp. (Roses): in vitro culture, micropropagation, and the production of secondary products. In: Bajaj, Y. P. S. (ed.). Biotechnology in Agriculture and Forestry 15. Medicinal and Aromatic Plants III. Springer-Verlag Co., New York. p. 376-397.

99. Shoyama, Y., I. Nishioka and K. Hatano. 1991. *Aconitum* spp. (Monkshood): somatic embryogenesis, plant regeneration, and the production of aconitine and other alkaloids. In: Bajaj, Y. P. S. (ed.). Biotechnology in Agriculture and Forestry 15. Medicinal and Aromatic Plants III. Springer-Verlag Co., New York. p. 58-72.

100. Singh, I. S. and R. K. Pathak. 1987. Evaluation of aonla (*Emblica officinalis* Gaertn.) varieties for processing. Acta Hort. 208: 173-177.

101. Skrzypczak, L., B. Thiem and M. Wesolowska. 1998. *Oenothera* species (Evening primrose): in vitro regeneration, production of flavonoids, fatty acids and other secondary metabolites. In: Bajaj, Y. P. S. (ed.). Biotechnology in Agriculture and Forestry 41. Medicinal and Aromatic Plants X. Springer-Verlag Co., New York. p. 286-304..

102. Small, E. (ed.). 1997. Culinary Herbs. NRC Research Press, Ottawa. 710 p.

103. Small, E. and P. M. Catling (eds.). 1999. Canadian Medicinal Crops. NRC Research Press, Ottawa. 240 p.

104. Solet, J. M., A. Simon-Ramiasa, L. Cosson and J. L. Guignard. 1998. *Centella asiatica* (L.) Urban. (Pennywort): cell culture, production of terpenoids, and biotransformation capacity. In: Bajaj, Y. P. S. (ed.). Biotechnology in Agriculture and Forestry 41. Medicinal and Aromatic Plants X. Springer-Verlag Co., New York. p. 81-96.

105. Strenath, H. L. and K. S. Jagadishchandra. 1991. *Cymbopogon* Spreng. (Aromatic grasses): in vitro culture, regeneration, and the production of essential oils. In: Bajaj, Y. P. S. (ed.). Biotechnology in Agriculture and Forestry 15. Medicinal and Aromatic Plants III. Springer-Verlag Co., New York. p. 211-236.

106. Tawfik, A. A., P. E. Read and S. L. Cuppett. 1998. *Rosmarinus officinalis* L. (Rosemary): in vitro culture, regeneration of plants, and the level of essential oil and monoterpenoid constituents. In: Bajaj, Y. P. S. (ed.). Biotechnology in Agriculture and Forestry 41. Medicinal and Aromatic Plants X. Springer-Verlag Co., New York. p. 349-365..

107. Teas, J. 1973. The dietary intake of *Laminaria,* a brown seaweed, and breast cancer prevention. Nutr. Cancer 4: 217-222.

108. Terreaux, C., M. Maillard, K. Hostettmann, G. Lodi, E. Hakizamungu. 1994. Analysis of the fungicidal constituents from the bark of *Ocotea usambarensis* Engl. (Lauraceae). Phytochemical Analysis 5: 233-238

109. Toleva, P., L. Tolev, P. Katsarska and D. Desev. 1977. Study on the essential oil (concrete and absolute) of *Mahonia aquifolia* with a view to its utilization in the perfumery and cosmetic industry. Higher Inst. Food Industry, Higher Inst. Medicine 26, Lenin Boul, 4000 Plovdiv, Bulgaria [from Perfumer and Flavorist. Vol. 2 (Oct. 1977), # 52].

110. Tyler, V. E. 1986. Plant drugs in the twenty-first century. Economic Bot. 40: 279-280.

111. Tyler, V. E. 1993. The Honest Herbal. A Sensible Guide to the Use of Herbs and Related Remedies. Pharmaceutical Products Press, New York. 375 p.

112. Van den Berg, A. J. J. and R. P. Labadie. 1988. Rhamnus spp.: in vitro production of anthraquinones, anthrones, and dianthrones. In: Bajaj, Y. P. S. (ed.). Biotechnology in Agriculture and Forestry 4. Medicinal and Aromatic Plants I. Springer-Verlag Co., New York. p. 513-528.

113. Vanwagenen, B. C. and J. H. H. Cardellina. 1986. Native American food and medicinal plants 7. Antimicrobial tetronic acids from *Lomatium dissectum.* Tetrahedron 42: 1117-1122.

114. Vasander, H. and T. Lindholm. 1987. Use of mires for agricultural, berry and medical plant production in Soviet Karelia. Hort. Abst. 1988. 735.

115. Walkiw, O. and D. E. Douglas. 1975. Health food supplements prepared from kelp—a source of elevated urinary arsenic. Clin. Toxicol. 8: 325-331.

116. Wichtl, M., and H. D. V. Prendergast 1998. Quality control and efficacy evaluation of phytopharmaceuticals. Proc. Soc. Economic Botany. London, UK. 1-6 July, 1996. p. 309-316.

117. Wickremesinhe, E. R. M. and R. N. Arteca. 1998. *Taxus* species (Yew): in vitro culture, and the production of taxol and other secondary metabolites. In: Bajaj, Y. P. S. (ed.). Biotechnology in Agriculture and Forestry 41. Medicinal and Aromatic Plants X. Springer-Verlag Co., New York. p. 415-442.

118. Worwood, V. A. 1991. The Complete Book of Essential Oils and Aromatherapy. New World Library, San Rafael, CA. 423 p.

119. Yamamoto, H. 1988. *Paeonia* spp.: in vitro culture and the production of paeoniflorin. In Bajaj, Y. P. S. (ed.). Biotechnology in Agriculture and Forestry 4. Medicinal and Aromatic Plants I. Springer-Verlag Co., New York. p. 464-483..

120. Yamamoto, H. 1991. *Scutellaria baicalensis* Georgi: in vitro culture and the production of flavonoids. In: Bajaj, Y. P. S. (ed.). Biotechnology in Agriculture and Forestry 15. Medicinal and Aromatic Plants III. Springer-Verlag Co., New York. p. 398-418.

121. Yokoyama, M. and S. Inomata. 1998. *Catharanthus roseus* (Periwinkle): in vitro culture, and high-level production of arbutin by biotransformation. In: Bajaj, Y. P. S. (ed.). Biotechnology in Agriculture and Forestry 41. Medicinal and Aromatic Plants X. Springer-Verlag Co., New York. p. 67-80.

122. Yukawa, C., H. Iwabuchi and K. H. C. Baser. 1995. Volatile constituents of Japanese apricot (mume, *Prunus mume* sieb. Et Zucc.). Proc. 13th International Cong. Flavours, fragrances and essential oils. Istanbul, Turkey. Vol. 2: Plenary Lectures, Analytical Composition, Trade-Industry Agriculture Botany. p. 276-282.

123. Zhang, S. and K. Cheng. 1989. *Angelica sinensis* (Oliv.) Diels.: in vitro culture, regeneration, and the production of medicinal compounds. In: Bajaj, Y. P. S. (ed.). Biotechnology in Agriculture and Forestry 7. Medicinal and Aromatic Plants II. Springer-Verlag Co., New York. p. 1-22.

CHAPTER 5

Cultivation and Harvesting

In the past, medicinal plants were collected from the wild in sufficient quantities to meet the market demand. Over the years, some popular species became scarce and endangered. To meet future increasing demand, medicinal plants must be cultivated. In addition to ensuring a sufficient supply, cultivation of medicinal plants permits production of uniform quality raw materials of known identity and properties.

Growing medicinal plants commercially requires consideration of soil and environmental requirements, reliability of seed sources, horticultural techniques, harvesting, and costs of production Most of the active ingredients in medicinal herbs are secondary metabolites such as alkaloids, and their biosynthesis is genetically controlled. However, environmental conditions play an important role in plant growth and in the formation and quantity of the secondary metabolites. Because most alkaloids are formed in young, actively growing tissues, factors that influence plant growth can also influence the production of the secondary metabolites. Environmental factors such as temperature, light intensity and photoperiod, soil fertility, and water supply determine the suitable growing area and life form of medicinal plants, and also have a substantial effect on production. Cultural practices, such as plant propagation, sowing dates, irrigation and fertilization, improving drainage, herbicides and pesticides, and postharvest treatments, also have a great impact on the final biomass and phytochemical yields of medicinal plants.

The cultivation of medicinal plants is a recent and relatively new horticultural practice. The requirements for growing them vary widely. Some medicinal plants have specific environmental requirements, such as shade, winter temperature, and chilling requirements. The best method of understanding how to cultivate a plant is to grow the plant in soil and conditions resembling as closely as possible their natural habitat. It is not easy to determine the environmental factors responsible for changes in the level of active ingredients; however, it is important to determine which part or parts of the plant contain the active ingredient and how a certain environmental factor may influence these parts during production. Medicinal plants are more heterogeneous genetically than vegetable plants and fruit trees in their botanical and physiological properties. Similar to ornamental plants, medicinal plants include annual, biennial, and perennial species with herbaceous or woody stalks, shrub or tree forms with or without fruit.

One of the most important procedures in the cultivation of medicinal plants is harvesting. The complexity of the operation depends on which part or parts (because the kind and level of active ingredients are different in various parts) of the plant are going to be harvested, such as flowers, leaves, bark, stems, rhizomes, roots, or seeds. Thus, the mechanization of harvesting is difficult. Furthermore, medicinal plants usually must be harvested over a very short optimal time period, because losses of active ingredients will result over a longer period, with consequent losses in value. Therefore, most medicinal plants can only be harvested with a high amount of labor.

The time of collection of medicinal plants is most important, because the composition of the plant and the level of active ingredient vary at different times of the year. Roots and rhizomes of biennial and perennial plants should be dug in the autumn of the first year, and in the autumn of the second year or after, respectively. Bark should be collected in spring when the sap begins to flow, but the process may be performed at any time during the winter. Healthy green leaves are collected in the early spring when the plant is in bloom. Where the whole green plant or herb is used, only the younger branches, flowers, and leaves should be harvested. Flowers are collected just after they open and before they begin to wither. Fruits and seeds are picked when mature with few exceptions.

Table 5. Origin of medicinal plants and their life cycle, plant form, propagation and growing site and planting space.

Scientific name	Common name	Origin[a]	Life cycle[b]	Plant form[c]	Culti-vated	Growing site[d]	Propagation	Planting space
Abies balsamea (L.) Mill.	Balsam fir	Te	P	W (T)	No	S or Sh	Seeds	15 m
Achillea millefolium L.	Yarrow	Te	P	H	Yes	FS	Division, seeds	30 cm
Aconitum napellus L.	Monkshood	Te	P	H	No	Sh	Division, seeds	40 cm
Acorus calamus L. var. Americanus Wolff. *A. tatarinowii* L.	Calamus, sweet flag Shi Chang Pu	Tr	P	H	Yes	S	Division	30 cm
Actaea alba L. *A. rubra* (Ait.) Willd.	Baneberry Red baneberry	Te	P	W (S)	No	S	Seeds, division	5-10 m
Aesculus hippocastanum L.	Horse chestnut	Te	P	W (T)	Yes	S or PSh	Seeds, grafting	5-8 m
Agastache foeniculum L.	Aniseed	Te	P	W (S)	Yes	S	Seeds, cutting, division	50 cm
Agrimonia eupatoria L.	Agrimony	Te	P	H	No	S	Division, seeds	30 cm
Agropyron repens (L.) Beauvois	Couch grass	Te	P	H	No	S or PSh	Seeds, rhizome cuttings	3-5 cm
Alchemilla vulgaris L. *A. xanthochlora* Rothm.	Lady's mantle	Te	A	H	No	S or SSh	Seeds, division	60 cm
Allium cepa L.	Onion	Te	A	H	Yes	FS	Seeds	20 cm

Scientific name	Common name	Origin[a]	Life cycle[b]	Plant form[c]	Culti-vated	Growing site[d]	Propagation	Planting space
A. sativum L.	Garlic	Te	A	H	Yes	FS	Cloves	15 cm
A. schoenoprasum L.	Chives	Te	A	H	Yes	FS	Seeds, bulb	15 cm
Alnus crispus (Ait.) Pursh. *A. incana* (L.) Moench. subsp. *tenufolia* *A. glutinosa* (L.) Gaertn.	Alder	Te	P	W (T)	No	S or PSh	Seeds	10 m
Aloe vera (L.) Burm. *A. barbadensis* L.	Aloe	Tr	P	H	Yes	S	Division	24 cm
Althaea officinalis L.	Marshmallow	Te	P	H	No	S	Seeds, division	30 cm

Amelanchier alnifolia Nutt.	Saskatoon	Te	P	W (S)	Yes	S	Seeds, suckers	3-5 m
Ananas comosus (L.) Merr.	Pineapple	Tr	P	H	Yes	FS	Seeds, offsets	1 m
Anaphalis margaritacea (L.) Bench & Hook	Everlastings	Te	P	H	No	S	Seeds	2-4 m
Anethum graveolens L.	Dill	Te	A	H	Yes	S	Seeds	20 cm
Angelica archangelica L. *A. sinensis* (Oliv.) Diels	Angelica Dong quai	Te	P	H	Yes	S or PSh	Seeds, division	1 m
Antennaria magellanica Schultz	Everlastings	See *Anaphalis margaritaceas*						
Anthemis nobilis L.	Chamomile	See *Chamaemelum nobile*						

Scientific name	Common name	Origin[a]	Life cycle[b]	Plant form[c]	Culti-vated	Growing site[d]	Propagation	Planting space
Apium graveolens L.	Celery	Te	A	H	Yes	S or Psh	Seeds	30 cm
Aralia racemosa L. *A. nudicaulis* L.	Spikenard	Te	P	W (S)	No	PSh	Seeds	30 cm
Arctium lappa L.	Burdock	Te	B	W (S)	No	S or LSh	Seeds, division	1 m
Arctostaphylos uva-ursi (L.) Spreng	Bearberry, Uva-ursi	Te	P	W (S)	Yes	S or PSh	Seeds, layering, cutting, grafting	60 cm
Armoracia rusticana Gaerin, Mey & Scherb.	Horseradish	Te	P	H	Yes	S or PSh	Seeds, division	30 cm
Arnica latifolia Bong. *A. montana* L. *A. chamissonis* L. subsp. *foliosa* *A. condifolia* Hook *A. fulgens* Pursh *A. sororia* Greene	Arnica	Te	P	W (S)	Yes	S	Seeds, division	15 cm
Artemisia absinthium L.	Wormwood	Te	P	H	No	S	Division, seeds	60 cm
A. annua L.	Qing Hao	Te	A	H	No	S	Division, seeds	60 cm
A. dracunculus L.	Tarragon	Te	P	H	Yes	S	Seeds	45 cm
A. tridentata Nutt.	Sagebrush	Te	P	W (S)	No	FS	Division, seeds	1-3 m
A. vulgaris L.	Mugwort	Te	P	H	No	S	Division	1 m

Scientific name	Common name	Origin[a]	Life cycle[b]	Plant form[c]	Culti-vated	Growing site[d]	Propagation	Planting space
Asarum canadensis L.	Wild ginger	Tr	P	H	No	Psh	Division	1 m
Astragalus americana Bunge. *A. membranaceus* Bunge. *A. sinensis* L.	Astragalus Huang Qi	Te	A P	H	No	S	Seeds, division	40 cm
Avena sativa L.	Oat	Te	A	H	Yes	S	Seeds	2-3 cm
Baptisia tinctora (L.) R.Br. Ex Ait.f.	Wild indigo	Te	P	H	No	S	Seeds, division	50 cm
Bellis perennis L.	Daisy	Te	A	H	No	S or PSh	Seeds	10 cm
Berberis aquifolium L.	Oregon grape	See *Mahonia aquifolium*						
Berberis vulgaris L.	Barberry	Te	P	W (S)	No	S	Seeds	1-2 m
Beta vulgaris L.	Beet root	Te	P	H	Yes	S	Seeds	15 cm
Betula lenta L.	Birch	Te	P	W (T)	No	S or Sh	Seeds	5-10 m
Borago officinalis L.	Borage	Te	A	H	No	FS	Seeds, cutting, division	60 cm
Boswellia sacra Roxb. ex. Colebr. *B. carteri* Birdwood	Frankincense	Te	P	H	No	FS	Seeds	1-3 m
Brassica nigra (L.) Kock.	Black mustard	Te	A	H	Yes	FS	Seed	30cm

Scientific name	Common name	Origin[a]	Life cycle[b]	Plant form[c]	Culti-vated	Growing site[d]	Propagation	Planting space
Buxus sempervirens L.	Boxwood	Te	P	W (S)	No	S or Sh	Cutting, suckers, seeds division, layering,	1-3 m
Calamintha nepeta (L.) Savi *C. ascendens* L. *C. officinalis* L.	Basil thyme Calamint Mountain mint	Te	P	H	Yes	S	Seeds	40 cm
Calendula officinalis L.	Calendula, Marigold	Te	A	H	Yes	FS	Seeds	50 cm
Camellia sinensis (L.) Kunize	Green tea	Tr	P	W (S)	Yes	S or PSh	Cutting, grafting	1-2 m
Cananga odorata (Lam.) J. D. Hook & T. Thompson	Ylang-Ylang	Tr	P	W (S)	No	S	Seeds	5-10 m
Cannabis sativa L.	Hemp	Te	A	H	Yes	S	Seeds	40 cm
Capsella bursa-pastoris (L.) Medik.	Shepherd's purse	Tr	A	H	No	S or PSh	Seeds	10 cm
Capsicum annuum L. var. Annuum. *C. annum* L. var. Grossum. *C. annum* L. var. Acuminatum.	Cayenne pepper Sweet pepper Chilli pepper	Tr	A	H	Yes	S	Seeds	30 cm

Scientific name	Common name	Origin[a]	Life cycle[b]	Plant form[c]	Cultivated	Growing site[d]	Propagation	Planting space
Carbenia benedicta L.	Blessed thistle	See *Cnicus benedictus*						
Carduus benedita L.	Blessed thistle	See *Cnicus benedictus*						
Carica papaya L.	Papaya	Tr	P	W (T)	Yes	S	Seeds, cutting, grafting	3-5 m
Carthamus tinctorius L.	Safflower	Tr	A	H	No	FS	Seeds	50 cm
Carum caroi L.	Caraway	Te	B	H	Yes	FS	Seeds	40 cm
Cassia senns L. *C. angustifolia* Vahl.	Alexandria senna Tinnevelly senna	Tr	P	H	Yes	S	Seeds, division	30 cm
Castanea sativa Mill	Chestnut	Te	P	W (T)	Yes	S	Budding, grafting	10-15 m
Catharanthus roseus (L.) G. Don	Periwinkle	See *Vinca minor*						
Caulophyllum thalietroides (L.) Michx. *C. giganteum* (F.) Loconte & W. H. Blackwell	Blue cohosh	Te	P	W (S)	No	Sh	Seeds	1-2 m
Cedrus libani A. Rich subsp. *atlantica*	Cedarwood	Te	P	W (T)	No	S	Seeds	10 cm
Centaurea calcitrapa L.	Star thistle	Te	P	H	Yes	S	Seeds	25 cm

Scientific name	Common name	Origin[a]	Life cycle[b]	Plant form[c]	Culti-vated	Growing site[d]	Propagation	Planting space
Centella asiatica (L.) Urb.	Gotu Kola	Tr	P	H	No	S, Sh	Seeds	1-2 m
Chamaelirium luteum (L.) A. Gray	False unicorn root	Te	P	H	No	PSh	Seeds, division	50 cm
Chamaemelum nobile (L.) All.	Roman chamomile	Te	P	W (S)	Yes	FS	Seeds, division	30 cm
Chamaenerion angustifolium (L.) Scop.	Firewood	See *Epilobium angustifolium*						
Chenopodium album L.	Lamb's quarter	Tr.	A	H	Yes	S or Sh	Seeds	30 cm
C. ambrosioides L.	Wormseed	Tr	A	H	No	S or Sh	Seeds	30 cm
Chrysanthemum cinerarifolium (Trevir) Vis.	Pyrethrum	Te	P	H	No	S	Seeds, division	40 cm
C. parthenium (L.) Berhn.	Feverfew	Te	P	H	Yes	S	Seeds, division	30 cm
C. vulgare L.	Tansy	See *Tanacetum vulgare*						
Cichorium intybus L.	Chicory	Te	P	H	Yes	S	Seeds	45 cm
Cimicifuga racemosa (L.) Nutt.	Black cohosh	Te	P	H	Yes	PSh	Seeds	60 cm
Cinnamomum verum J. Pres. *C. cassia* (Nees) Nees & Eber	Cinnamon Camphor	Tr	P	W (S)	Yes	S or Psh	Seeds, cutting, root cutting, suckers	2-3 m

Scientific name	Common name	Origin[a]	Life cycle[b]	Plant form[c]	Cultivated	Growing site[d]	Propagation	Planting space
Citrus limon (L.) Burm. *C. bergamia* Risso & Poit *C. reticulata* Blanco *C. aurantifolia* Swingle *C. aurantium* L. *C. paradisi* Macfad	Lemon Bergamot orange Mandarin Lime Bitter orange Grapefuit	Tr	P	W (T)	Yes	S	Grafting	2-5 m
Cnicus benedictus L.	Blessed thistle	Te	A	H	No	S	Seeds, division	50 cm
Cochlearia amorucia L.	Horseradish	See *Armoracia rusticana*						
Codonopsis pilosula (Franch) Nannfeldt. *C. tangshen* Oliver	Codonopsis, dang shen Bellflower	Te	P	H (V)	No	SSh	Seeds	1-2 m
Coffea arabica L.	Coffee	Tr	P	W (S)	Yes	SSh	Cutting, seeds	1-2 m
Colchicum autumnale L.	Crocus, Meadow saffron	Tr	P	H	No	S or SSh	Division, seeds	30 cm
Collinsonia canadensis L.	Pilewort	Te	P	H	No	PSh	Seeds, division	50 cm
Commiphoe molmol Engl. ex Tschirch *C. myrrha* (Nees) Engl.	Myrrh	Te	P	W (T)	No	S	Cuttings, seeds	1 m
Convallaria majalis L.	Lily of the valley	Te	P	H	No	S	Division, seeds	15 cm
Convolvulus arvensis L. *C. sepium* L.	Bindweed	Te	A	H	No	S	Seeds, cutting	1 m

Scientific name	Common name	Origin[a]	Life cycle[b]	Plant form[c]	Culti-vated	Growing site[d]	Propagation	Planting space
Coriandrum sativum L.	Coriander, Cilantro	Te	A	H	Yes	S	Seeds	30 cm
Cornus canadensis L.	Bunchberry	Te	P	W (S)	No	S or SSh	Cutting, layering, budding, grafting	8-10 m
C. florida L.	Dogwood	Te	P	W (T)	No	S	Seeds, cutting	8-10 m
Corylus cornuta Marsh. *C. rostrata* Marsh. *C. avellana* L.	Hazelnut	Te	P	W (T)	Yes	S	Seeds	15 cm
Crataegus monogyna Jacq. *C. pinnatifida* Bge.	English hawthorn Chinese hawthorn	Te	P	W (T)	No	S or PSh	Seeds, budding, grafting on *C. oxyacanthoides* or other species	5 m
Crocus sativus L.	Saffron	Tr	P	H	No	FS	Seeds	10 cm
Cryptotaenia japonica Hassk.	Japanese parsley, Mitsuba	See *Petroselinum crispum*						
Cucurbita pepo L.	Pumpkin	Te	A	H	Yes	S	Seeds	3-4 m
Cuminum cyminum L.	Cumin	Tr	A	H	No	FS	Seeds	15 cm
Curcuma domestica L. *C. xanthorrhiza* L. *C. longa* L.	Curcuma, Jiang Huang	Tr	P	H	Yes	S	Division	2-6 m

Scientific name	Common name	Origin[a]	Life cycle[b]	Plant form[c]	Culti-vated	Growing site[d]	Propagation	Planting space
Cuscuta epithymum Murr.	Dodder	Te	P	H	No	S	Seeds, cutting	2-4 m
Cymbopogon citratus (DC. ex Nees) Stapf. *C. nardus* L. *C. winterlanus* Jowitt *C. martinii* (Roxb) Wats	Lemon grass Citronella Palmarosa, gingergrass	Tr	P	W (S)	Yes	S	Seeds, division	1 m
Cynara scolymus L. *C. cardunculus* L.	Artichoke Wild artichoke	Te	A	H	Yes	S	Seeds	1 m
Cypripedium calceolus L.	Lady's slipper	Te	A	H	No	Sh	Seeds	40 cm
Cytisus scoparius (L.) Link.	Broom	Te	P	W (S)	No	S	Seeds, division	75 cm
Daucus carota L.	Carrot	Te	A	H	Yes	S or PSh	Seeds	15 cm
Digitalis purpurea L.	Foxglove	Te	B	H	No	PSh	Seeds	50 cm
Dioscorea oppositae Thunb. *D. villosa* L.	Chinese yam, ShenYao Wild or Mexican yam	Tr	P	H	Yes	S or PSh	Seeds, cuttings	2 m
Echinacea angustifolia DC. *E. pallida* (Nutt.) Nutt. *E. pururea* (L.) Moench.	Narrowleaf echinacea Pale-flower echinacea Purple coneflower	Te	P	H	Yes	S	Seeds, division	30 cm

Scientific name	Common name	Origin[a]	Life cycle[b]	Plant form[c]	Culti-vated	Growing site[d]	Propagation	Planting space
Echinopanax horridus (Sm.) Decne. & Planch. ex. H. A. T. Harms	Devil's club	See *Oplopanax horridus*						
Elaeagnus angustifolia L.	Russian olive	Te	P	W (S)	No	S	Seeds	2-4 m
Eleutherococcus senticosus (Rupr. ex Maxim) Maxim.	Siberian ginseng	Te	P	W (S)	Yes	S or PSh	Seeds, cuttings	25 cm
Elymus repens L.	Couch grass	See *Agropyron repens*						
Ephedra distachya L. *E. nevadensis* Wats. *E. sinica* Stapf.	Ephedra Mormon tea Ma Huang	Te	P	H	No	S	Seeds, division, suckers, layering	75 cm
Epilobium angustifolium L.	Fireweed	Te	P	H	Yes	S	Seeds, division	50 cm
Epilobium parviflorum Schreb.	Small-flowered willow herb	Te	P	H	Yes	S	Seeds, division	50 cm
Equisetum arvense L. *E. telmateia* Ehrh.	Horsetail	Te	P	H	No	S or PSh	Division, seeds	1 m
Erythroxylum coca Lam.	Coca	Tr.	P	W (S)	Yes	S	Seeds	2 m
Eschscholtzia californica Cham.	California poppy	Te	A	H	Yes	FS	Seeds	30 cm
Eucalyptus citriodora Hool. *E. globulus* Labill.	Gum tree Eucalyptus	Tr	P	W (T)	No	S	Seeds	1.5 m

Scientific name	Common name	Origin[a]	Life cycle[b]	Plant form[c]	Cultivated	Growing site[d]	Propagation	Planting space
Eugenia caryophyllata L.	Clove	See *Syzygium aromaticum*						
Eupatorium perfoliatum L.	Boneset	Te	P	W (S)	No	S or PSh	Seeds, sucker	90 cm
Euphrasia officinalis L.	Eyebright	Te	P	H	No	S	Seeds	20 cm
Fagopyrum tataricum L.	Buckwheat	Te	A	H	Yes	S	Seeds	4-8 cm
Fagus grandfolia Ehrh. *F. sylvatica* L.	Beech European beech	Te	P	W (T)	No	S or PSh	Seeds	5-10 m
Filipendula ulmaria (L.) Maxim.	Meadow sweet	Te	P	H	No	S or PSh	Seeds, division	30 cm
Foeniculum vulgare Mill.	Fennel	Te	B	H	Yes	FS	Seeds	50 cm
Forsythia suspensa (Thunb.) Vahl.	Forsythia, Lian qiao	Te	P	W (S)	No	S or PSh	Cuttings	2-4 m
Fragaria vesca L.	Strawberry	Te	A	H	Yes	S or PSh	Seeds, division	30 cm
Fraxinus americana L. *F. excelsior* L.	White ash	Te	P	W(T)	No	S	Seeds	3-5 m
Fucus vesiculosus L.	Kelp	Te	P	H	Yes	S	Seeds, cuttings	30 cm
Galium aparine L.	Cleavers	Te	P	H	Yes	Sh	Seeds, division	20 cm
Gaultheria procumbens L.	Wintergreen, teaberry	Te	P	W(S)	No	PSh	Seeds, layering, sucker, division, cutting	10 cm

Scientific name	Common name	Origin[a]	Life cycle[b]	Plant form[c]	Culti-vated	Growing site[d]	Propagation	Planting space
Geranium macrorhizum L.	Geranium, Am. cranesbill	Te	A	H	Yes	S or Psh	Seeds, cuttings	40 cm
Ginkgo biloba L.	Ginkgo	Te	P	W(T)	Yes	S	Seeds, cutting	2-4 m
Glechoma hederacea L.	Ground ivy	Te	P	H	No	S or Sh	Seeds, division	50 cm
Glycine max (L.) Merrill	Soybean	Tr	A	H	Yes	FS	Seeds	15 cm
Glycyrrhiza glabra L. *G. uralensis* Fisch ex DC	Licorice Chinese licorice	Te	P	W(S)	No	S	Seeds, division	40 cm
Gossypium hirsutum L.	Cotton	Tr	A	H	Yes	S	Seeds	1 m
Grindelia robusta Nutt. *G. squarrosa* (Pursh) Donal.	Gumweed	Te	P	H	No	S	Seeds, division	50 cm
Hamamelis virginiana L.	Witch hazel	Te	P	W(S)	Yes	S or PSh	Seeds	2-4 m
Harpagophytum procumbens DC.	Devil's claw	Tr	P	W(T)	Yes	S	Seeds	2-4 m
Hedeoma pulegiodes (L) Pers	Pennyroyal	Te	A	H	Yes	S or PSh	Seeds	10 cm
Hedera helix L.	English ivy	Te	P	W(V)	No	S or PSh	Cuttings	3-4 m
Helianthus annuus L.	Sunflower	Te	A	H	Yes	S	Seeds	40 cm
Helichrysum angustifolium (Lam.) DC.	Curry plant	Tr	A	H	Yes	S	Seeds	30 cm

Scientific name	Common name	Origin[a]	Life cycle[b]	Plant form[c]	Culti-vated	Growing site[d]	Propagation	Planting space
Helonias dioica L.	False unicorn root	See *Chamaelirum luteum*						
Heracleum maximum Bartr. *H. lanatum* Michx. *H. sphondylium* L.	Cow parsley Cow parsnip Cow parsnip	Te	P	H	No	S or Psh	Seeds	60 cm
Hibiscus sabdariffa L. *H. rosa-sinensis* L.	Hibiscus	Tr	P	W(S)	Yes	S	Cuttings, grafting	1-2 m
Hierochloe odorata (L.) Beauv.	Sweetgrass	Te	P	H	No	S or PSh	Seeds, division	2 m
Hippophae rhamnoides L	Sea buckthorn	Te	P	W(S)	Yes	S	Seeds, sucker, cuttings	1 m
Humulus lupulus L.	Hops	Te	P	W(V)	Yes	S or PSh	Seeds	90 cm
Hydrangea arborescens L.	Hydrangea	Te	P	W(S)	No	S or PSh	Seeds, cuttings, sucker	1-2 m
Hydrastis canadensis L.	Golenseal	Te	P	H	Yes	Sh	Seeds, division	15 cm
Hypericum perforatum L.	St. John's wort	Te	P	H	Yes	S	Seeds	20 cm
Hyssopus officinalis L.	Hyssop	Te	P	H	Yes	S	Seeds	60 cm
Ilex aquifolium L.	English holly	Te	B	W(S)	Yes	S or Sh	Cuttings	2-4 m
Inula helenium L.	Elecampane	Te	P	W(S)	No	S	Seeds	90 cm

Scientific name	Common name	Origin[a]	Life cycle[b]	Plant form[c]	Culti-vated	Growing site[d]	Propagation	Planting space
Iris versicolor L.	Blue flag	Te	P	H	Yes	S	Division, offsets, seeds	70 cm
Jasminum grandiflorum L. *J. auriculatum* Vahl.	Royal jasmine	Tr	P	W(S)	Yes	S	Cuttings, layering, seeds	1.5 m
Juglans regia L. *J. nigra* L.	Walnut Black walnut	Te	P	W(T)	Yes	S	Seeds, grafting	10-15 m
Juniperus communis L.	Juniper	Te	P	W(S)	Yes	S or PSh	Seeds, cuttings	10-15 m
Laminaria digitata (Hudss.) Lamk *L. saccharina* (L.) Lamk	Kelp	See *Fucus vesiculosus*						
Lamium album L.	Nettle	Te	P	H	No	S or PSh	Seed, division	60 cm
Larrea tridentata (Sesse. & Moc. ex DC.) Coville.	Chapatral	Te	P	W(S)	No	S	Seeds	75 cm
Laurus nobilis L.	Bay laurel	Te	P	W(T)	No	S or PSh	Cuttings, seeds	5-10 m
Lavandula angustifolia Mill.	Lavender	Te	P	H	Yes	S	Seeds, cuttings	15 cm
Ledum latifolium Jacq.	Labrador tea	Te	P	W(S)	No	S or PSh	Seeds, layering, division	1 m
Lemna minor L.	Duckweed	Te	P	H	No	S	Seeds, division	1 m
Leonurus cardiaca L.	Motherwort	Te	P	H	No	S or PSh	Seeds, division	50 cm

Scientific name	Common name	Origin[a]	Life cycle[b]	Plant form[c]	Culti-vated	Growing site[d]	Propagation	Planting space
Levisticum officinale W. Koch.	Lovage	Te	P	H	No	S or PSh	Seeds, division	60 cm
Ligustrum lucidium W. T. Aiton. *L. vulgare* L.	Ligustrum fruit Privet	Te	P	W(S)	No	S	Cuttings, division, grafting, seeds	2-3 m
Linum usitatissimum L.	Flax	Te	A	H	Yes	S	Seeds	10 cm
Liquidambar styraciflua L. *L. orientalis* L.	Gum tree, Storax Oriental sweet gum	Te	P	W(T)	No	S or PSh	Seeds	20 cm
Lobelia inflata L. *L. siphilitica* L.	Lobelia Blue lobelia	Tr	P	H	No	S or PSh	Cuttings, division	1 m
L. pulmonaria L.	Lungwort	Te	P	H	No	S or PSh	Cuttings, division	1 m
Lomatium dissectum (Nutt.) Math. & Const.	Lomatium	Te	P	H	No	S	Seeds, division	1 m
Lonicera caerulea L. *L. caprifolium* L.	Mountain fly honeysuckle Dutch honeysuckle	Te	P	W(S)	No	S or PSh	Seeds, cuttings, division	10 m
Lycium barbarum L. *L. pallidum* L. *L. chinesis* Mill.	Lycium Wolfberry Chinese wolfberry	Tr	P	W(S)	No	S	Cuttings, suckers, layers, seeds	2-3 m
Lycopodium clavatum L.	Clubmoss	Te	P	H	No	PSh	Cuttings	80 cm

Scientific name	Common name	Origin[a]	Life cycle[b]	Plant form[c]	Culti-vated	Growing site[d]	Propagation	Planting space
Lysimachia vulgaris L.	Loosestrife	Te	P	H	No	FS	Division	20 cm
Lythrum salicaria L.	Purple loosestrife	Te	P	H	o	S or PSh	Division, seeds	60 cm
Macropiper excelsum G. Forst	Kava-Kava	See *Piper methysticum*						
Mahonia aquifolium (Pursh) Nutt.	Oregon grape	Te	P	W(S)	No	S or Psh	Seeds, suckers, layers, cuttings	1.5 m
Marrubium vulgare L.	Horsehound	Te	P	H	No	S	Seeds, division	30 cm
Matricaria chamomilla L. *M. recutita* L.	Chamomile German chamomile	Te	A	H	Yes	S	Seeds	15 cm
Matteucia struthiopteris (L.) Tod	Ostrich fern	Te	P	H	No	Sh	Division	20 cm
Medicago sativa L.	Alfalfa	Te	P	H	Yes	S	Seeds	5 cm
Melaleuca alternifolia (Maid. & Bet.) Cheel	Melaleuca, Medicinal tea tree	Tr	P	W(S)	No	S	Seeds, cuttings	1-2 m
Melilotus officinalis (L.) Lamk. *M. arvensis* L.	Sweet clover Melilot	Te	B	H	No	S	Seeds	45 cm
Melissa officinalis L.	Lemon balm, Balm	Te	P	H	No	S or PSh	Seeds, cuttings division	30 cm

Scientific name	Common name	Origin[a]	Life cycle[b]	Plant form[c]	Culti-vated	Growing site[d]	Propagation	Planting space
Mentha piperita L. *M. spicata* L. *M. pulegium* L.	Peppermint Spearment Pennyroyal	Te	P	H	Yes	S or PSh	Seeds, division from stolons (7-10 cm)	60 cm
Monarda odoratissima Benth. *M. punctata* T. J. Howell	Horsebalm Horsemint	Te	A	H	No	S	Division	45 cm
Myosotis scorpioides L.	Lily of the valley	See *Convallaria majalis*						
Myrica penxylvanica Lois	Bayberry	Te	P	W(S)	No	S or PSh	Seeds, layering, suckers	2-3 m
Myristica fragrans Houtt.	Nutmeg	Tr	P	W(T)	No	S	Seeds, grafting	5 m
Myrtus communis L.	Myrtle	Te	P	W(S)	No	S	Cuttings, seeds	2 m
Narcissus pseudonarcissus L.	Daffodil	Te	A	H	Yes	PSh or S	Bulbs, seeds	15 cm
Nasturtium officinale L.	Watercress	Te	P	H	No	S	Seeds, division, cuttings	1-2 m
Nepeta cataria L.	Catnip, catmint	Te	P	H	No	S	Division	30 cm
Nymphaea alba L.	White water lily	Te	P	H	Yes	FS	Division	1 m
Ocimum basilicum L.	Basil	Tr	A	H	Yes	S	Seeds	40 cm
Ocotea bullata (Birch) E. May	Stinkwood	Te	P	W(S)	No	S	Suckers, seeds	1-2 m
Oenothera biennis L.	Evening primrose	Te	B	H	Yes	S	Seeds	30 cm

Scientific name	Common name	Origin[a]	Life cycle[b]	Plant form[c]	Culti-vated	Growing site[d]	Propagation	Planting space
Olea europaca L.	Olive	Te	P	W(T)	Yes	S	Seeds, grafting	10 m
Oplopanax horridus L.	Devil's club	Te	P	H	Yes	S or Psh	Seeds, suckers, root cuttings	75 cm
Opuntid fragilis (Nut.) Haw.	Prickly pear	Te	P	H	Yes	FS	Seeds, cuttings	60 cm
Origanum majorana L. *Origanum vulgare* L. subsp. *hirtum* (Link) Ietswaart.	Sweet marjoram Marjoram, Oregano	Te	P	W(S)	No	S	Seeds	45 cm
Paeonia lactiflora Pall. *P. officinalis* L. *P. suffruiticosa* Andr.	White peony, Bia shao European peony Tree peony, Mu Dan Pi	Te	P	H	Yes	S or PSh	Seeds, division	80 cm
Panax ginseng C.A. Meyer. *P. notoginseng* (Burk.) F. H. Chen *P. quinquefolium* L.	Asian ginseng Tian Qi American ginseng	Te	P	H	Yes	SSh	Seeds	10 cm
Papaver bracteatum L. *P. rhoens* L.	Iranian poppy Poppy, corn poppy	Te .	P	H	Yes	S	Seeds	40 cm
Passiflora incarnata L.	Passion flower	Tr	P	W(V)	No	S	Seeds, cutting	5-6 m

Scientific name	Common name	Origin[a]	Life cycle[b]	Plant form[c]	Culti-vated	Growing site[d]	Propagation	Planting space
Pelargonium capitatum (L.) L'Her *P. crispum* L'Her. ex. Aiton *P. graveolens* (L.) L'Her ex. Aiton	Wild rose geranium Lemon geranium Rose geranium	Te	A	H	Yes	S	Seeds, cutting	60 cm
Peltigera canina L.	Ground liverwort, Dog lichen	Te	P	H	No	Psh	Seeds, division	1 m
Petroselinum crispum (Mill.) Nym.	Parsley	Te	A	H	Yes	S or PSh	Seeds	25 cm
Phataris canariensis L.	Canary creeper	Te	P	H	No	S	Seeds, division	1 m
Phellodendron chinensis Schneid.	Chinese corktree, Huang Bai	Te	P	W(T)	No	S	Division	4 m
Phoradendron leucarpum (Raf.) Rev. & M. C. Johnst. *P. flavescens* (Push.) Nutt.	Mistletoe American mistletoe	Tr	P	W(S)	No	S of PSh	Seeds, cuttings	4 m
Phyllanthus emblica L.	Myrobalan, emblic	Tr	P	W(T)	No	S	Seeds	2-4 m
Physalis alkekengi L.	Chinese lantern	Te	P	W(S)	Yes	S	Seeds	1 m
Phytolacca americana L.	Pokeweed	Tr	P	H	No	S or PSh	Seeds	1 m
Picea mariana (Mill) Black *P. glauca* (Moench) Voss. *P. ables* (L.) Karst.	Black spruce White spruce Norway spruce	Te	P	W(T)	No	S	Seeds	5-8 m

Scientific name	Common name	Origin[a]	Life cycle[b]	Plant form[c]	Culti-vated	Growing site[d]	Propagation	Planting space
Pimpinella anisum L.	Anise	Te	A	H	Yes	S	Seeds	20 cm
Pinus mugo var. Pumilio *P. palastris* Mill	Dwarf mountain pine	Te	P	W(S)	No	S	Seeds	1-2 m
P. strobus L. *P. albicaulis* Engelm. *P. contorta* Dougl. ex. Loud.	Southern pitch pine, white pine Lodgepole pine	Te	P	W(T)	No	S	Seeds	2-4 m
Piper methysticum G. Forst	Kava-Kava	Tr	P	W(S)	No	LSh	Cuttings, seeds	5-10 m
P. nigrum L.	English pepper	Te	A	H	Yes	S	Seeds	30 cm
Plantago asiatica L *P. psyllium* L.	Psyllium	Tr	P	H	Yes	S	Seeds	20 cm
P. lanceolata L. *P. major* L.	Plantains	Tr	P	H	Yes	S or PSh	Seeds	30 cm
Podophyllum peltatum L.	Mayapple	Te	P	H	No	S or PSh	Division, seeds	30 cm
Polygala senega L.	Seneca snakeroot	Te	P	H	Yes	S	Seeds	50 cm
Polygonum bistorta L.	Bistort root	Tr	P	H	No	S or PSh	Seeds	45 cm
P. multiflorum Thunb.	Flowery knotweed, fu-ti	Tr	P	H	No	S or PSh	Seeds, division	40 cm

Scientific name	Common name	Origin[a]	Life cycle[b]	Plant form[c]	Culti-vated	Growing site[d]	Propagation	Planting space
Populus balsamifera L. *P. candicans* L. *P. tremuloides* Michx.	Poplar, Balm of gilead American aspen	Te	P	W(T)	Yes	S	Seeds	5-8 m
Primula vulgaris Huds. *P. veris* L.	Primrose Cowslip	Te	P	H	No	S or PSh	Division, seeds	20 cm
Prunella vulgaris L.	Heal all, selfheal	Te	P	H	No	S of Lsh	Seeds	1 m
Prunus africana L.	African prune	Tr	P	W(T)	No	S	Grafting	3-4 m
P. dulcis L.	Almond	Te	P	W(T)	Yes	S	Seeds, grafting	4-6 m
P. mume Siebold & Zucc.	Japanese apricot	Te	P	W(T)	Yes	S	Grafting	3-4 m
P. serotina J. F. Ehrb.	Black cherry, wild cherry	Te	P	W(T)	Yes	S	Grafting	3-4 m
P. virginiana L.	Chokecherry	Te	P	W(T)	Yes	S	Seeds, suckers	3-4 m
Pueraria lobata (Willd) Ohwi	Kudzu	Tr	P	H	No	S	Division, seeds, cuttings	15 cm
Pygeum africanum L.	Pygeum	Tr	P	H	No	S	Seeds, division	1 m
Pyrethrum cinerarifolium L.	Pyrethrum	See *Chrysanthemum cinerarifolium*						
Quercus alba L.	White oak	Te	P	W(T)	No	S	Seeds	10-15 m
Ranunculus occidentalis Nutt.	Buttercup	Te	P	H	No	S or Sh	Seeds	30 cm

Scientific name	Common name	Origin[a]	Life cycle[b]	Plant form[c]	Culti-vated	Growing site[d]	Propagation	Planting space
R. ficaria L.	Pilewort	Te	P	H	No	S or Sh	Seeds, division	30 cm
Raphanus sativus L.	Radish	Tr	A	H	Yes	S	Seeds	10 cm
Rhamnus catharticus L. *R. frangula* L. *R. purshianus* L.	Buckthorn Alder buckthorn Cascara sagrada	Te	P	W(S)	No	S	Seeds, cutting, grafting	2-4 m
Rheum palmatum L. *R. officinal* Baill. *R. tanguticum* L.	Chinese rhubarb Rhubarb Rhubarb	Te	P	H	Yes	S	Seeds, division	70 cm
Rhodiola rosea (L.) Scop.	Roseroot	See *Sedum rosea*						
Rhus radicans L.	Poison ivy	Te	P	H	No	S	Seeds, division	1-2 m
R. toxicodendron L.	Roseroot	See *Sedum rosea* subsp. *integrifolium*						
Ribes nigrum L. *R. lacustre* (Pers.) Poir.	Black current Black gooseberry	Te	P	W(S)	Yes	S or PSh	Seeds, grafting	1-2 m
Robertium macrorrhizum Pic.	Geranium	See *Geranium macrorhizum*						
Rosa canina L. *R. damascena* L.	Dog rose Damask rose	Te	P	W(S)	Yes	S	Grafting, cuttings	1-2 m
Rosmarinus officinalis L.	Rosemary	Te	P	H	Yes	S	Seeds, cutting	50 cm
Rubus chamaemorus L.	Cloudberry	Te	P	H	No	S or PSh	Seeds, suckers, cuttings	1-2 m

Scientific name	Common name	Origin[a]	Life cycle[b]	Plant form[c]	Culti-vated	Growing site[d]	Propagation	Planting space
R. fruiticosus L.	Blackberry	Te	P	W(S)	No	S	Seeds, suckers, cuttings	3-4 m
R. idaeus L.	Raspberry	Te	P	W(S)	Yes	S	Seeds	1-2 m
Rumex acetosella L. *R. obtusifolia* L.	Sorrel Dock	Te	P	H	No	S or PSh	Seeds, cuttings	30 cm
Ruscus aculeatus L.	Butcher's broom	Te	P	W(S)	No	S of Sh	Seeds, divisions	1 m
Ruta graveolens L.	Rue, herb of grace	Te	P	W(S)	No	S	Seeds, divisions	40 cm
Salix alba L. *S. discolour* Muhlenb. *S. caprea* L.	White willow Pussy willow	Te	P	W(T)	No	S	Seeds	5-8 m
Salvia officinalis L. *S. sclarea* L.	Sage Clary sage	Te	P	W(S)	No	S	Seeds divisions	50 cm
Sambucus nigra L. *S. canadensis* L.	Elderberry	Te	P	W(S)	No	S of PSh	Seeds	3-4 m
S. racemosa L.	Elder	Te	P	W(S)	No	S	Seeds	3-4 m
Sanguinaria canadensis L.	Bloodroot	Te	P	H	Yes	Sh	Seeds, division	45 cm
Santalum album L.	Sandalwood	Tr	P	W(S)	No	PSh	Seeds, root cutting	2.5 m
Saponaria officinalis L.	Soapwort	Te	P	H	No	S or PSh	Seeds, division	60 cm

Scientific name	Common name	Origin[a]	Life cycle[b]	Plant form[c]	Culti-vated	Growing site[d]	Propagation	Planting space
Sarothamnus scoparius L.	Broom	See *Cytisus scoparius*						
Sassafras albidum (Nutt.) Nees.	Sassafras	Te	P	W(T)	No	S of Sh	Seeds, suckers, cuttings	20 m
Satureja hortensis L. *S. montana* L.	Savory	Te	P	W(S)	Yes	S	Seeds	45 cm
Schisandra chinensis (Turcz.) Baill.	Schizandra (wu wei zi)	Te	P	W(V)	No	S or Psh	Seeds	6-8 m
Scutellaria baicalensis Georgi. *S. lateriflora* L. *S. galericulata* L.	Baical skullcap Virginia skullcap Skullcap	Te	P	H	No	S or PSh	Seeds, divisions	40 cm
Sedum acre L.	Stonecrop, small houseleek	Te	P	W(S)	No	S	Seeds, divisions	1 m
S. rosea (L.) Scop. subsp. *integrifolium*	Roseroot	Te	P	H	Yes	S	Seeds	80 cm
Sempervivum tectorum L.	Houseleek, hens and chicks	Te	P	H	No	S	Divisions	25 cm
Senecio vulgris L.	Groundsel	Te	P	H	No	S or PSh	Seeds, divisions	30 cm
Senna alexandrina L.	Alexandrina	Tr	P	H	No	S	Seeds, cutting	1 m
Serenoa repens (Bartr.) Small.	Saw palmetto	Tr	P	W(S)	No	S	Seeds, suckers	1-3 m

Scientific name	Common name	Origin[a]	Life cycle[b]	Plant form[c]	Culti- vated	Growing site[d]	Propagation	Planting space
Sesamum indicum L. *S. orientale* L.	Sesame	Tr	P	W(S)	Yes	S	Seeds	60 cm
Shepherdia canadensis L.	Buffalo berry	Te	P	W(S)	No	S	Seeds, suckers	3-4 m
Silybum marianum (L) Gaert.	Milk thistle	Te	P	H	Yes	S	Seeds	40 cm
Simmondsia chinensis (Link) C. Schneid	Jojoba	Tr	P	W(T)	No	S	Seeds	1-2 m
Solidago virgaurea L.	Goldenrod	Te	P	W(S)	Yes	S or PSh	Seeds, division	50 cm
Sorbus aucuparia L.	Mountain ash	Te	P	W(T)	No	S	Seeds	3-4 m
Spirea ulmaria L.	Meadowsweet	See *Filipendula ulmaria*						
Stachys officinalis (L.) Trev.	Betony, woundwort	Te	P	H	No	S or PSh	Seeds	45 cm
Stellaria media (L.) Vill.	Chickweed	Te	A	H	No	S or PSh	Seeds	15 cm
Stevia rebaudiana (Bertoni) Bertoni	Stevia, Sweet herb of Paraguay	Tr	P	H	Yes	S	Seeds	75 cm
Symhytum officinale L.	Comfrey	Te	P	H	Yes	S or PSh	Seeds	60 cm
Syringa vulgaris L.	Lilac	Te	P	W(S)	No	S	Seeds, cuttings, suckers	1-3 m
Syzygium aromaticum (L.) Merrr. & Perry	Clove	Tr	P	W(T)	Yes	S	Seeds	7-8 m

Scientific name	Common name	Origin[a]	Life cycle[b]	Plant form[c]	Culti-vated	Growing site[d]	Propagation	Planting space
Tagetes lucida Cav.	Mexican mint marigold	Tr	P	H	Yes	S	Seeds	30 cm
Tanacetum parthenium (L.) Schultz	Feverfew	See *Chrysanthemum parthenium*						
T. cinerarifolium L.	Pyrethum	See *Chrysanthemum cinerarifolium*						
T. vulgare L.	Tansy	Te	P	H	Yes	S	Seeds	75 cm
Taraxacum officinale G. H. Weber ex Wigg.	Dandelion	Te	P	H	Yes	S	Seeds	10 cm
Taxus x media Rehd. *T. brevifolia* Nutt.	Yew	Te	P	W(T)	Yes	S or Sh	Seeds, cuttings	1 m
Thymus citriodorus (Pers.) Schreb. *Thymus vulgaris* L	Lemon thyme Thyme	Te	P	W(S)	Yes	S	Seeds, division	40 cm
Tillia cordata Mill. *T. europaea* L.	Linden, small-leaved lime	Te	P	W(T)	No	S or PSh	Seeds, division	20 m
Tribulus terrestris L.	Puncture vine	Tr	P	H	No	S	Seeds, cuttings	1 m
Trifolium incarnatum L. *T. pratense* L.	Clover (crimson) Clover (red)	Te	P	H	Yes	S	Seeds	60 cm
Trigonella foenum-gracecum L.	Fenugreek	Te	A	H	No	S	Seeds	10 cm

Scientific name	Common name	Origin[a]	Life cycle[b]	Plant form[c]	Culti-vated	Growing site[d]	Propagation	Planting space
Tropaeolum majus L.	Canary creeper, Nasturtium	Tr	A	H	No	S	Seeds	20 cm
Tussilago farfara L.	Colisfoot	Te	P	H	No	S or PSh	Seeds	1 m
Ulmus rubra Muhl. *U. procora* L.	Slippery elm, sweet elm English elm	Te	P	W(T)	No	S	Seeds	15 m
Umbelluslaria california Nutt.	California laurel Bay laurel	Te	P	W(S)	No	S	Seeds	10 m
Uncaria tomentosa (Willd.) DC. *U. guianensis* (Aubl.) Gmel.	Cat's claw	Tr.	P	W(S)	Yes	S	Seeds, suckers	2-4 m
Urtica dioica L. *U. urens* L.	Stinging nettle	Te Te	P A	H	No	S or DSh	Seeds, cuttings, suckers	2 m
Vaccinium macrocarpon L. *V. vitis-idaea* L.	Cranberry Mountain cranberry, cowberry, foxberry	Te	P	W(S)	Yes	S or PSh	Seeds	75 cm
V. mrytilloides Michx. *V. myrtillus* L. *V. oreophilum* Rydb.	Blueberry Bilberry	Te	P	W(S)	Yes	LSh	Seeds	1 m
Valeriana officinalis L.	Valerian	Te	P	H	Yes	S or Sh	Seeds	60 cm

Scientific name	Common name	Origin[a]	Life cycle[b]	Plant form[c]	Culti-vated	Growing site[d]	Propagation	Planting space
Verbascum thapsus L.	Mullein	Te	B	H	No	S	Seeds	60 cm
Veronica officinalis L.	Speedwell	Te	P	H	No	S or PSh	Seeds	25 cm
Vetiveria zizanioides L. (Nash)	Vetiver	Tr	P	H	No	S	Seeds, root division	80 cm
Viburnum opulus L.	Highbush cranberry	Te	P	W(S)	No	S or PSh	Seeds, suckers, cuttings	3-4 m
Vinca minor L.	Periwinkle	Te	P	H	Yes	S or Psh	Seeds	50 cm
Viola tricolor L.	Pansy	Te	A	H	Yes	S of PSh	Seeds	15 cm
Viscum album L.	Mistletoe	See *Phoradendron leucarpum*						
Vitex agnus-castus L. *V. negundo* L.	Chaste tree Chinese chaste tree	Te	P	W(S)	No	S	Seeds	4-5 m
Vitis labrusca L. *V. vinifera* L.	Grape	Te	P	W(V)	Yes	S	Seeds, grafting	2-3 m
Withania somnifera Dunal.	Ashwagandha, Indian ginseng	Tr	P	H	No	S	Seeds	1-2 m
Yucca aloifolia L. *Y. glauca* L.	Yucca Soapweed	Te	P	H	Yes	S	Division	1 m

Scientific name	Common name	Origin[a]	Life cycle[b]	Plant form[c]	Culti-vated	Growing site[d]	Propagation	Planting space
Zanthoxylum americanum Mill	Prickly ash, toothache tree	Te	P	W(T)	No	S or Sh	Seeds, root cuttings	5-8 m
Zea mays L.	Corn	Te	A	H	Yes	S	Seeds	30 cm
Zingiber officinale Roscoe	Ginger	Tr	P	H	Yes	S or PSh	Root division	20 cm

[a] Origin: Tr - tropical region, Te - temperate region
[b] Life cycle: A - annual, B - biennial, P - perennial
[c] Plant form: W - woody, H - herbaceous, S - shrub, T - tree, V - vine
[d] Growing site: S - sunny, FS - full sun, Sh - shady, PSh - partial shade, SSh - semi shade, LSh - light shade, Dsh - dappled shade

Table 6. Soil conditions and requirements and the harvesting of medicinal plants.

Scientific name	Common name	Soil conditions and requirements	Harvesting
Abies balsamea (L.) Mill.	Balsam fir	Deep, moist, well-drained, and slightly acid soil	Leaves and young shoots are collected in spring.
Achillea millefolium L.	Yarrow	Well-drained average or poor soil	Leaves and flowers in late summer.
Aconitum napellus L.	Monkshood	Deep, moisture-retentive soil	Plants are dug in the fall.
Acorus calamus L. var. Americanus Wolff. *A. tatarinowii* L.	Calamus, sweet flag Shi Chang Pu	Wet and damp soil	Plants can be dig at any time after flowering.
Actaea alba L. *A. rubra* (Ait.) Willd.	White baneberry Red baneberry	Well-drained soil	Berries are picked ripe, roots are dug in autumn.
Aesculus hippocastanum L.	Horse chestnut	Rich, well-drained soil	Bark and seeds are collected in the fall.
Agastache foeniculum L.	Aniseed	Well-drained, humus-laden soil with low fertility	Leaves are collected in the spring and summer, flowers during blooming.
Agrimonia euparoria L.	Agrimony	Well-drained soil with wide range of pH	Plants are collected when flowering.
Agropyron repens (L.) Beauvois	Couch grass	Any kind of soil	Rhizomes are dug in spring.
Alchemilla vulgaris L.	Lady's mantle	Moist and well-drained soil	Leaves during flower period.

Scientific name	Common name	Soil conditions and requirements	Harvesting
Allium cepa L.	Onion	Rich and well-drained soil	Harvested in the summer and early autumn.
A. sativum L.	Garlic	Rich and well-drained soil	Dig bulbs in summer.
A. schoenoprasum L.	Chives	Rich and well-drained soil, tolerates damper conditions, heavier soil	Cut the leaves and leaving 5 cm for regrowth, pick flowers when they open.
Alnus crispus (Ait.) Pursh *A. incana* (L.) Moench. subsp. *tenufolia* *A. glutinosa* (L.) Gaertn.	Alder	Rich and moist soil	Peel off fresh bark of young twigs or two year old branches.
Aloe vera (L.) Burm. *A. barbadensis* L.	Aloe	Well-drained soil, minimum temperature for cultivation is 5°C	Cut leaves from at least 2-yr-old plants.
Althaea officinalis L.	Marshmallow	Constant moist soil is required	Collect seeds when ripe, dig up roots in autumn.
Amelanchier alnifolia Nutt.	Saskatoon	Well-drained fertile sandy loam soil, it may survive in marginal land	Berries are picked ripe, roots are dug in autumn, bark is stripped in early spring.
Ananas comosus (L.) Merr.	Pineapple	Tropical plant, planted in well-drained, rich, and sandy loam soil	Harvest fruit when it is mature, collect leaf anytime for leaf fiber.
Anaphalis margaritacea (L.) Bench & Hook	Everlastings	Any type of soil	Leaves are picked at the beginning of flowering, flowers are picked during bloom.

Scientific name	Common name	Soil conditions and requirements	Harvesting
Anethum graveolens L.	Dill	Well-drained, sandy loam, neutral to slightly acid soil	Gather leaves when young, pick flowering tops just as fruits begin to form, collect seeds after flowering head turns brown.
Angelica archangelica L. *A. sinensis* (Oliv.) Diels	Angelica Dong quai	Rich, moist soil	Cut stems before midsummer for crystallizing, harvest leaves before flowering, collect ripe seeds in late summer, dig up root in autumn of first year.
Antennaria magellanica Schultz	Everlastings	Well-drained soil	Whole plants or flower heads are harvested before the flowers are fully open.
Anthemis nobilis L.	Chamomile	See *Chamaemelum nobile*	
Apium graveolens L.	Celery	Rich and damp soil	Pick leaves in late summer.
Aralia racemosa L. *A. nudicaulis* L.	Spikenard	Rich, moist soil	Roots are dug in the fall.
Arctium lappa L.	Burdock	Moist, neutral to alkaline soil	Young leaf stalks are cut in spring, roots are dug in autumn.
Arctostaphylos uva-ursi (L.) Spreng	Bearberry, Uva-ursi	Moist, sandy soil, tolerate dry habitats	Leaves are picked at any time during the growing season, fruit pick ripe.
Armoracia rusticana Gaertn, Mey & Scherb.	Horseradish	Well-drained rich soil	Dig up root in autumn, pick young leaves.

Scientific name	Common name	Soil conditions and requirements	Harvesting
Arnica latifolia Bong. *A. montana* L. *A. chamissonis* L. subsp. *foliosa* *A. condifolia* Hook *A. fulgens* Pursh *A. sororia* Greene	Arnica	Well-drained humus-rich acid soil	Flowers are picked when fully open, root dig in the fall.
Artemisia absinthium L. *A. annua* L.	Wormwood Quing Hao	Well-drained, neutral to slightly alkaline soil	Pick flowering tops and leaves in mid- to late summer.
A. dracunculus L.	Tarragon	Well-drained, neutral to slightly alkaline soil	Harvest leaves in late summer and leave 1/3 for regrowth.
A. tridentata Nutt. *A. vulgaris* L.	Sagebrush Mugwort	Sandy soil with slightly alkaline soil	Pick leaves during the growing season.
Asarum canadensis L.	Wild ginger	Well-drained, moist soil, high in organic matters	Whole or part of rhizome are harvested in the fall.
Astragalus americana Bunge. *A. membranaceus* Bunge. *A. sinensis* L.	Astragalus Huang Qi	Well-drained soil	Gum is collected from second-year or older plants. Roots are dug in the fall.
Avena sativa L.	Oat	Well-drained fertile soil	Mature seeds are collected and de-husked for liquid extracts and tinctures.
Baptisia tinctora (L.) R.Br. Ex Ait.f.	Wild indigo	Deep, rich soil	Roots are lifted in autumn.

Scientific name	Common name	Soil conditions and requirements	Harvesting
Bellis perennis L.	Daisy	Well-drained soil	Leaves are picked in spring and summer.
Berberis aquifolium L.	Oregon grape	See *Mahonia aquifolium*	
Berberis vulgaris L.	Barberry	Neutral or slightly acid soil	Fruits, stems and roots are harvested in fall.
Beta vulgaris L.	Beet root	Well-drained fertile soil	Whole plant are cut in summer, roots are dug in autumn.
Betula lenta L.	Birch	Well-drained sandy acid soil	Leaf buds and young leaves are picked in spring.
Borago officinalis L.	Borage	Well-drained moist soil	Pick flowers and leaves at any time.
Boswellia sacra Roxb. ex Colebr. *B. carteri* Birdwood	Frankincense	Well-drained, slightly dry soil	Gum resin is collected all year.
Brassica nigra (L.) Kock.	Black mustard	Rich, well-drained soil	Gather flowers as they open, pick seed pods before they open in late summer, cut young leaves for salad 8-10 days after sowing.
Buxus sempervirens L.	Boxwood	Well-drained neutral to alkaline soil	Leaves are picked in early spring before flowering. Bark is stripped in the fall.

Scientific name	Common name	Soil conditions and requirements	Harvesting
Calamintha nepeta (L.) Savi *C. ascendens* L. *C. officinalis* L.	Basil thyme Calamint Mountain mint	Well-drained slightly dry soil neutral to alkaline	Flowering plants are cut in the summer.
Calendula officinalis L.	Marigold, calendola	Will tolerate poor but well-drained soil	Pick flowers when open, leaves when young.
Camellia sinensis (L.) Kunize	Green tea	Rich, moist soil	Pick leaves when young.
Cananga odorata (Lam.) J. D. Hook & T. Thompson	Ylang-Ylang	Well-drained, moist soil under high humidity environment, minimum temperature 15°C	Flowers are picked at night.
Cannabis sativa L.	Hemp	Well-drained sandy loam soil	Harvest stem and seeds in the late summer.
Capsella bursa-pastoris (L.) Medik.	Shepherd's purse	Well-drained soil	Cut the whole plant during growing season.
Capsicum annuum L. var. Annuum *C. annum* L. var. Grossum *C. frutescens* L. *C. annum* L. var. Acuminatum	Cayenne pepper Sweet pepper Chilli pepper	Well-drained soil, minimum 18°C for full growth	Ripe fruits are picked in summer, unripe fruits are picked at anytime.
Carbenia benedicta L.	Blessed thistle	See *Cnicus benedictus*	

Scientific name	Common name	Soil conditions and requirements	Harvesting
Carduus benedita L.	Blessed thistle	See *Cnicus benedictus*	
Carica papaya L.	Papaya	Rich, moist soil with high humidity, minimum temperature 13°C	Leaves are picked as required, papaia is extracted from unripe fruit.
Carthamus tinctorius L.	Safflower	Well-drained soil	Flower hears are picked in summer.
Carum caroi L.	Caraway	Well-drained sandy loam soil rich in organic matter are preferred	Gather leaves when young, pick seed heads when seeds are brown, dip up roots in second year.
Cassia senns L. *C. angustifolia* Vahl.	Alexandria senna Tinnevelly senna	Well-drained soil	Leaves are picked before and during flowering, pods in autumn.
Castanea satuva Mill	Chestnut	Well-drained soil	Leaves are picked in summer.
Catharanthus roseus (L.) G. Don	Periwinkle	See *Vinca minor*	
Caulophyllum thalietroides (L.) Michx. *C. giganteum* (F.) Loconte & W. H. Blackwell	Blue cohosh	Rich, moist soil	Rhizomes and roots are dug in autumn.
Cedrus libani A. Rich subsp. *atlantica*	Cedarwood	Well-drained soil	Branches collected anytime for oil.
Centaurea calcitrapa L.	Star thistle	Well-drained soil	Flowers collected during blooming.

Scientific name	Common name	Soil conditions and requirements	Harvesting
Centella asiatica (L.) Urb.	Gotu Kola	Well-damped soil, swampy area	Leaves are picked at any time.
Chamaelirium luteum (L.) A. Gray	False unicorn root	Moist, well-drained humus rich soil	Dig up roots in autumn.
Chamaemelum nobile (L.) All.	Roman chamomile	Moist, moderately heavy to light loam soil, well-drained and rich in humus	Pick leaves at any time, flowers when fully open.
Chamaenerion angustifolium (L.) Scop.	Firewood	See *Epilobium angustifolium*	
Chenopodium album L. *C. ambrosioides* L.	Lamb's quarters Wormseed	Rich, well-drained soil	Pick leaves at anytime after the plants are fully developed, pick flowering spikes as they begin to open.
Chrysanthemum cinerarifolium (Trev.) Vis *C. parthenium* (L.) Berhn.	Pyrethrum Feverfew	Well-drained stony soil	Gather open flowers, whole plants are cut when flowering.
C. vulgare L.	Tansy	See *Tanacetum vulgare*	
Cichorium intybus L.	Chicory	Rich, well-drained, neutral to alkaline soil	Gather leaves when young, dig up root in first autumn and chitons in winter.
Cimicifuga racemosa (L.) Nutt.	Black cohosh	Moist, humus-rich soil	Rhizomes and roots are dug in the fall after the fruits have matured.

Scientific name	Common name	Soil conditions and requirements	Harvesting
Cinnamomum verum J. Presl. *C. camphora* (L.) Nees & Eberm.	Cinnamon Camphor	Moist, well-drained, deep fertile sandy loam soil, best grown above 15°C	Leaves and bark are collected at anytime from older trees. Collect leaves as required.
Citrus limon (L.) Burm. *C. bergamia* Risso & Poit *C. reticulata* Blanco *C. aurantifolia* Swingle *C. aurantium* L. *C. paradisi* Macfad	Lemon Bergamot orange Mandarin Lime Bitter orange Grapefruit	Rich, well-drained soil with ample moisture during the growing season, minimum temperature around 7°C	Leaves are picked as required, fruits picked ripe.
Cnicus benedictus L.	Blessed thistle	Well-drained soil	Whole plants are cut during blooming.
Cochlearia amorucia L.	Horseradish	See *Armoracia rusticana*	
Codonopsis pilosula (Franch) Nannfeldt. *C. tangshen* Oliver	Codonopsis, Dang Shen, Bellflower	Well-drained sandy soil	Roots of at least 3-yr-old plants are dug in the fall.
Coffea arabica L.	Coffee	Well-drained moisture retentive soil, minimum of 10°C	Fruits are picked when ripe.
Colchicum autumnale L.	Crocus, meadow saffron	Moist, well-drained soil	Corms are dug in summer.
Collinsonia canadensis L.	Pilewort	Moist, well-drained soil	Rhizomes are dug in autumn.

Scientific name	Common name	Soil conditions and requirements	Harvesting
Commiphora molmol Engl. ex Tschirch *C. myrrha* (Nees) Engl.	Myrrh	Well-drained soil, minimum of 10°C	Resin is collected from cut branches.
Convallaria majalis L.	Lily of the valley	Rich, moist soil	Leaves and flowers are picked in spring.
Convolvulus arvensis L. *C. sepium* L.	Bindweed	Well-drained humus-rich soil	Roots are dug in autumn.
Coriandrum sativum L.	Coriander, Cilantro	Well-drained soil in a cool damp spring and followed by a hot dry summer	Pick young leaves anytime, collect seeds when brown before drop, dig up roots in autumn.
Cornus canadensis L. *C. florida* L.	Bunchberry Dogwood	Well-drained soil	Fruits are picked when ripe.
Corylus cornuta Marsh. *C. rostrata* Marsh. *C. avellana* L.	Hazelnut	Well-drained woodland soil	Fruits are collected ripe, leaves are picked young.
Crataegus monogyna Jacq. *C. pinnatifida* Bge.	English hawthorn Chinese hawthorn	Almost any kind of soil with wide range of pH.	Fruits are picked when ripe.
Crocus sativus L.	Saffron	Well-drained soil	Flowers are picked when open.
Cryptotaenia japonica Hassk.	Japanese parsley, Mitsuba	Rich, moist soil	Young leaves are picked at anytime.
Cucurbita pepo L.	Pumpkin	Rich, well-drained soil	Seeds are collected from ripe fruit.

Scientific name	Common name	Soil conditions and requirements	Harvesting
Cuminum cyminum L.	Cumin	Well-drained medium to heavy loam, rich soil	Seeds are collected right after when ripe to avoid shedding.
Curcuma domestica L. *C. xanthorrhiza* L. *C. longa* L.	Curcuma, Jiang Huang	Well-drained soil	Roots are dug during dormant period.
Cuscuta epithymum Murr. *C. chinensis* Lam.	Dodder	Survives on host plants only	Seeds are collected when ripe.
Cymbopogon citratus (DC. ex Nees) Stapf. *C. nardus* L. *C. winterianus* Jowitt *C. martinii* (Roxb) Wats.	Lemon grass Citronella Palmarosa, gingergrass	Well-drained light loam soil, rich in organic matter, pH 5.8-8, minimum temperature around 7°C	Harvest 4-6 months after planting, stems are cut at 15 cm above the ground.
Cynara scolymus L. *C. cardunculus* L.	Artichoke Wild artichoke	Deep, rich, well-drained soil	Leaves are picked before flowering.
Cypripedium calceolus L.	Ladyslipper	Humus-rich soil	Rhizomes are dug in autumn.
Cytisus scoparius (L.) Link.	Broom	Well-drained soil	Young shoots are cut as flowering begins.
Daucus carota L.	Carrot	Well-drained, fertile, alkaline soil	Whole plants are cut in summer, roots dug as needed.
Digitalis purpurea L.	Foxglove	Well-drained, neutral to acid soil	Leaves are picked before flowering.

Scientific name	Common name	Soil conditions and requirements	Harvesting
Dioscorea oppositae Thunb. *D. villosa* L.	Chinese yam (Shen yao) Wild or Mexican yam	Rich, well-drained soil	Tubers are lifted in autumn.
Echinacea angustifolia DC. *E. pallida* (Nutt.) Nutt. *E. pururea* (L.) Moench.	Narrow leaf echinacea Pale-flower echinacea Purple coneflower	Rich, well-drained loam and sandy loam soils, pH 6-7, irrigation needed during the growing season	Roots are dug in autumn after 3 years of cultivation, flowers collected during blooming.
Echinopanax horridus (Sm.) Decne. & Planch. ex. H. A. T. Harms.	Devil's club	See *Oplopanax horridus*	
Elaeagnus angustifolia L.	Russian olive	Marginal land, woodland soil	Leaves are picked during flowering, fruits are picked ripe.
Eleutherococcus senticosus (Rupr. ex Maxim) Maxim.	Siberian ginseng	Well-drained rich moist soil	Roots are dug in autumn.
Elymus repens L.	Couch grass	See *Agropyron repens*	
Ephedra distachya L. *E. nevadensis* Wats. *E. sinica* Stapf.	Ephedra, horsetail Mormon tea Ma Huang	Well-drained dry soil	Stems are collected at any time.

Scientific name	Common name	Soil conditions and requirements	Harvesting
Epilobium angustifolium L. *Epilobium parviflorum* Schreb.	Fireweed Small-flowered willow herb	Well-drained woodland or embankment soil	Leaves are picked before flowering.
Equisetum arvense L. *E. telmatecia* Ehrb.	Horsetail	Moist soil	Stems are cut at anytime during the growing season.
Erythroxylum coca Lam.	Coca	Tropical high-rainfall regions with 1500 m altitudes	Leaves are picked when they begin to curl.
Eschscholzia california Cham.	California poppy	Well-drained marginal soil	Whole plants are cut when flowering.
Eucalyptus citriodora Hool. *E. globulus* Labill.	Gum tree Eucalyptus	Fertile, well-drained medium to light soil, neutral to acid, minimum of 5°C for *E. citriodora*	Leaves are harvested as required. If planted for essential oil purpose, harvest leaf after 6 months when tree reached 1 m, subsequent harvests every 4 months.
Eugenia caryophyllata L.	Clove	See *Syzygium aromaticum*	
Eupatorium perfoliatum L.	Boneset	Moist soil	Whole plants are cut when in bud.
Euphrasia officinalis L.	Eyebright	Grows in natural grassland near host plants	Plants are cut when flowering.
Fagopyrum tataricum L.	Buckwheat	Well-drained sandy soil	Leaves and flowers are collected as flowering begins.

Scientific name	Common name	Soil conditions and requirements	Harvesting
Fagus grandifolia Ehrb. *F. sylvatica* L.	Beech European beech	Well-drained moist to dry soil	Branches are cut at anytime.
Filipendula ulmaria (L.) Maxim.	Meadow sweet	Rich, moisture retentive to wet soil	Gather young leaves before flowering, pick flowers when young.
Foeniculum vulgare Mill.	Fennel	Well-drained sandy loam soil	Pick young stems and leaves at any time, dig up bulb in autumn, fruits are harvested just before turning gray.
Forsythia suspensa (Thunb.) Vahl.	Forsythia, Lian qiao	Moist soil	Fruits are collected when ripe.
Fragaria vesca L.	Strawberry	Humus-rich soil	Pick fruit as ripe, dig up roots in autumn.
Fraxinus americana L. *F. excelsior* L.	White ash	Well-drained neutral to alkaline soil	Sap is collected from cutting branches.
Fucus vesiculosus L.	Kelp	Coastline, poly-culture	Plants are collected in summer.
Galium aparine L.	Cleavers	Moist, well-drained neutral to alkaline soil	Pick leaves and flowering stem at any time.
Gaultheria procumbens L.	Wintergreen, teaberry	Acid soil	Leaves are gathered from spring throughout the growing season.
Geranium macrorrhizum L.	Geranium, Amer. cranesbill	Moist to wet soil	Plants are cut as flowering begins, roots dug in the fall.

Scientific name	Common name	Soil conditions and requirements	Harvesting
Ginkgo biloba L.	Ginkgo	Rich, well-drained soil	Leaves are picked as they change color in autumn.
Glechoma hederacea L.	Ground ivy	Moist, well-drained soil	Plants are collected during blooming.
Glycine max (L.) Merrill.	Soybean	Fertile, well-drained sandy loam soil, neutral or slightly acid in pH	Seeds collected when mature.
Glycyrrhiza glabra L. *G. uralensis* Fisch ex DC	Licorice Chinese licorice	Deep, rich, sandy soil	Three or four year old stolons are dug in early autumn.
Gossypium hirsutum L.	Cotton	Rich, well-drained soil	Leaves are picked during the growing season.
Grindelia robusta Nutt. *G. squarrosa* (Pursh) Donal.	Gumweed	Well-drained soil	Plants are cut in full bloom.
Hamamelis virginiana L.	Witch hazel	Moist, humus-rich neutral to acid soil	Leaves are picked in summer, twigs collected autumn, winter and early spring.
Harpagophytum procumbens DC. ex Me.	Devil's claw	Sandy soil	Tubers are dug when dormant.
Hedeoma pulegiodes (L.) Pers.	Pennyroyal	Rich sandy soil	Plants are cut when flowering.
Hedera helix L.	English ivy	Any type of soil under any environment	Leaves are picked at anytime.

Scientific name	Common name	Soil conditions and requirements	Harvesting
Helianthus annuus L.	Sunflower	Well-drained soil	Cut flower heads when they droop and hang until seeds fall, gather stem in autumn.
Helichrysum angustifolium (Lam.) DC.	Curry plant	Well-drained sandy soil	Pick leaves anytime, gather flowers as they open.
Helonias dioica L.	False unicorn root	See *Chamaelirium luteum*	
Heracleum maximum Bartr. *H. lanatum* Michx. *H. sphondylium* L.	Cow parsley Cow parsnip Cow parsnip	Moist or wet soil	Leaves are cut just before flowering, fruit harvested when ripe, roots are dug in the fall.
Hibiscus sabdariffa L. *H. rosa-sinensis* L.	Hibiscus	Well-drained soil, minimum temperature at 7°C	Leaves are picked when young, stems are cut at any time.
Hierochloe odorata (L.) Beauv.	Sweet grass	Well-drained moisture fertile soil enriched with humus	Leaves are collected in summer.
Hippophae rhamnoides L	Sea buckthorn	No specific requirement for soil, as long as well-drained	Leaves are collected before autumn, fruits harvested when ripe.
Humulus lupulus L.	Hops	Moist, well-drained soil	Pick young shoots in spring, pick ripe flowers in early autumn, collect stems in late autumn.
Hydrangea arborescens L.	Hydrangea	Moist, well-drained, humus-rich soil	Roots are dug in autumn.

Scientific name	Common name	Soil conditions and requirements	Harvesting
Hydrastis canadensis L.	Goldenseal	Rich, moist, well-drained soil	Rhizomes are dug in autumn after foliage has died down.
Hypericum perforatum L.	St. John's wort	Well-drained to dry soil	Plants are cut as flowering begins.
Hyssopus officinalis L.	Hyssop	Well-drained to dry neutral to alkaline soil	Pick flowers and young flowering tops as flowering begins.
Ilex aquifolium L.	English holly	Moist, well-drained soil	Leaves are picked in early summer.
Inula helenium L.	Elecampane	Moist, well-drained soil	Dig up second or third year roots in autumn.
Iris versicolor L.	Blue flag	Well-drained neutral to alkaline soil	Rhizomes are dug in late summer and early autumn.
Jasminum grandiflorum L. *J. auriculatum* Vahl.	Royal jasmine	Rich, well-drained sandy loam soil, water log is not suitable	Flowers are picked early in the morning.
Juglans regia L. *J. nigra* L.	Walnut Black walnut	Well-drained, sandy loam, deep rich soil.	Leaves are picked during the growing season, fruits are picked unripe or when ripe in autumn.
Juniperus communis L.	Juniper	Tolerate acid and alkaline conditions, dry and wet soil	Fruits are collected ripe.
Laminaria digitata (Hudss.) Lamk. *L. saccharina* (L.) Lamk.	Kelp	Seashore	Pick at anytime.

Scientific name	Common name	Soil conditions and requirements	Harvesting
Lamium album L.	Nettle	Moist, well-drained soil	Whole plants are cut during flowering.
Larrea tridentata (Sesse. & Moc. ex DC.) Coville.	Chapatral	Dry sandy soil	Pick leaves young, roots are dug in autumn.
Laurus nobilis L.	Bay laurel	Well-drained soil, needs protection from frost and cold in areas with hard winter	Leaves are collected in summer.
Lavandula angustifolia Mill.	Lavender	Well-drained light loam, neutral to alkaline (7-8.4) soil	Gather flower with stems (12 cm) in full bloom during bright sunny day.
Ledum latifolium Jacq.	Labrador tea	Moist to wet, acid soil	Leaves and shoots are collected in late summer.
Lemna minor L.	Duckweed	Well-drained woodland soil	Whole plants are cut in the summer, flowers are picked during blooming.
Leonurus cardiaca L.	Motherwort	Well-drained, moist soil	Plants are cut when flowering before seeds are set.
Levisticum officinale W. Koch.	Lovage	Deep, rich, moist soil	Pick leaves (except young central leaves) and gather young branched stem in spring, dig second and third season roots before flowers open each year.
Ligustrum lucidium W. T. Aiton. *L. vulgare* L.	Ligustrum fruit Privet	Well-drained soil	Fruits are collected when ripe.

Scientific name	Common name	Soil conditions and requirements	Harvesting
Linum usitatissimum L.	Flax	Well-drained, dry sandy soil	Plants are cut when mature, seeds harvested when ripe.
Liquidambar styraciflua L. *L. orientalis* L.	Gum tree, storax Oriental sweet gum	Deep, rich, moist, neutral to slightly acid soil	Balsam is collected as a natural exudate or from cuts in the bark.
Lobelia inflata L. *L. siphilitica* L. *L. pulmonaria* L.	Lobelia Blue lobelia Lungwort	Rich, moist soil	*L. inflata:* whole plants are cut when lower fruits are ripe. Other species: whole plants are cut when flowering.
Lomatium dissectum (Nutt.) Math. & Const.	Lomatium	Well-drained soil	Roots are dug in autumn, leaves collected before flowering, flowers are picked during full bloom. Seeds collected from ripe fruit.
Lonicera caerulea L. *L. caprifolium* L.	Mountain fly honeysuckle Dutch honeysuckle	Well-drained soil	Stems are cut in autumn and winter.
Lycium barbarum L. *L. chinense* Mill. *L. pallidum* L.	Lycium Chinesewolfberry Wolfberry	Moist, well-drained, alkaline sandy soil	Bark is stripped from roots in winter.
Lycopodium clavatum L.	Clubmoss	Well-drained, moist, sandy soil	Whole fungi are collected in summer.
Lysimachia vulgaris L.	Loosestrife	Any type of soil with good drainage	Flowering stems are cut during blooming.

Scientific name	Common name	Soil conditions and requirements	Harvesting
Lythrum salicaria L.	Purple loosestrife	Moist to wet, neutral to silt soil	Plants are cut during flowering.
Macropiper excelsum G. Forst	Kava-Kava	See *Piper methysticum*	
Mahonia aquifolium (Pursh) Nutt.	Oregon grape	Well-drained, humus-rich soil	Roots and root bark are collected in late autumn.
Marrubium vulgare L.	Horsehound	Well-drained, neutral to alkaline soil	Pick leaves and flowering tops at flowering time.
Matricaria chamomilla L. *M. recutita* L.	Chamomile German chamomile	Well-drained, moist to dry, neutral to slightly acid soil	Flowers are collected as soon as fully opened.
Matteuccia strothiopteris (L.) Tod	Ostrich fern	Shady woodland soil	Roots are dug in autumn, shoots are cut in early summer.
Medicago sativa L.	Alfalfa	Light, well-drained, neutral to alkaline dry soil	Plants are cut before flowering.
Melaleuca alternifolia (Maid. & Bet.) Cheel	Melaleuca, medicinal tea tree	Moisture-retentive, neutral to acid slightly wet soil	Leaves and twigs are collected at anytime for oil.
Melilotus officinalis (L.) Lamk. *M. arvensis* L.	Sweet clover Melilot	Well-drained, neutral to alkaline dry soil	Collect at anytime.
Melissa officinalis L.	Lemon balm, balm	Moist soil	Pick leaves when flowers begin to open.

Scientific name	Common name	Soil conditions and requirements	Harvesting
Mentha piperita L. *M. spicata* L. *M. pulegium* L.	Peppermint Spearment Pennyroyal	Damp, sandy, acid soil	Pick leaves during full bloom. In sub-tropical areas, especially with short days, mint flowers very late or does not flower at all, harvest the leaves after 120-130 days after planting.
Monarda odoratissima Benth. *M. punctata* T. J. Howell	Horsebalm Horsemint	Rich, moist soil	Collect leaves when flowers formed, pick flowers when open.
Myosotis scorpiodes L.	Lily of the valley	See *Convallaria majalis*	
Myrica penxylvanica Lois.	Bayberry	Well-drained, damp sandy acid soil	Leaves are picked during the growing season, bark and root in late autumn or early spring.
Myristica fragrans Houtt.	Nutmeg	Well-drained, humus-rich sandy soil, needed minimum of 15°C to grow	Seeds are collected from ripe fruit.
Myrtus communis L.	Myrtle	Well-drained, neutral to alkaline soil	Pick buds, flowers and ripe berries as available, pick leaves when in flower.
Narcissus pseudonarcissus L.	Daffodil	Well-drained woodland soil	Collect flowers when fully open, bulbs are dug in late summer.
Nasturtium officinale L.	Watercress	Shallow, flowing, slightly alkaline water, minimum of 10°C	Leaves are picked at anytime.

Scientific name	Common name	Soil conditions and requirements	Harvesting
Nepeta cataria L.	Catnip, catmint	Moist, well-drained soil	Young leaves and flowering tops at anytime.
Nymphaea alba L.	White water lily	Rich soil in still water up to 45 cm deep, minimum of 21°C	Flowers are cut when open, fruits and seeds are harvested when ripe.
Ocimum basilicum L.	Basil	Rich, medium to light, well-drained dry soil, pH 4.3-8.3, required temperature not lower than 9°C	Pick leaves as flowers in full bloom, when the lower leaves start turning yellow.
Ocotea bullata (Birch) E. May	Stinkwood	Marginal woodland soil	Leaves are picked in spring.
Oenothera biennis L.	Evening primrose	Variety of soil types with good drainage in a range of pH	Gather leaves and stem bark when flowering stems have grown, dig up roots in second year.
Olea europaea L.	Olive	Well-drained soil	Leaves are collected as required.
Oplopanax horridus (Sm.) Miq.	Devil's club	Well-drained to poorly drained soil with pH 3.8-6.0, usually shaded or sunny side	Roots are dug in autumn, bark is stripped in early spring.
Opuntia fragilis (Nutt.) Haw.	Prickly pear	Very well-drained sandy soil or sand	Pads are cut in spring.
Origanum majorana L. *Origanum vulgare* L. subsp. *hirtum* (Link) Ietswaart.	Sweet marjoram Marjoram, Oregano	Well-drained, neutral to alkaline dry soil	Pick young leaves anytime. If leaves are to be used for preserving, gather just before flowers open.

Scientific name	Common name	Soil conditions and requirements	Harvesting
Paeonia lactiflora Pall. *P. officinalis* L. *P. suffruiticosa* Andr.	White peony, Bia shao European peony Tree peony, Mu Dan Pi	Rich, well-drained soil	Roots are dug from 4- to 5-yr-old plants in late summer.
Panax ginseng C.A. Meyer. *P. notoginseng* (Burk.) F. H. Chen *P. quinquefolium* L.	Asian ginseng Tian Qi American ginseng	Moist to dry, well-drained sandy loam soil with pH 6-7.5, it thrives under 75% shade	Roots are dug after 4- and 6-yr-old plants for *P. quinquefolium* and *P. ginseng*, respectively.
Papaver braceatum L. *P. rhoens* L.	Iranian poppy Poppy, corn poppy	Well-drained soil	Collect seeds when capsule is ripe.
Passiflora incarnata L.	Passion flower	Well-drained, slightly acid sandy soil	Plants are cut when fruiting.
Pelargonium capitatum (L.) L'Her *P. crispum* L'Her. ex. Aiton. *P. graveolens* (L.) L'Her ex. Aiton	Wild rose geranium Lemon geranium Rose geranium	Well-drained deep light soil, rich in organic matters, neutral to alkaline (pH 5.8-6.6) soil	Pick leaves just before flowers open. First harvest is obtained 4-6 months after planting, green leafy shoots are harvested.
Peltigera canina L.	Ground liverwort, Dog lichen	Marginal woodland soil	Whole plants are cut in spring.

Scientific name	Common name	Soil conditions and requirements	Harvesting
Petroselinum crispum (Mill.) Nym.	Parsley	Rich, well-drained, neutral to alkaline soil	Pick leaves during first year, dig roots in autumn of second year.
Phataris canariensis L.	Canary creeper	Woodland soil	Leaves are picked young.
Phellodendron chinensis Schneid.	Chinese corktree, Huang Bai	Rich, well-drained neutral to alkaline soil	Bark is stripped in winter.
Phoradendron leucarpum (Raf.) Rev. & M. C. Johnst. *P. flavescens* (Push.) Nutt.	Mistletoe American mistletoe	Survive on host plants	Leaves are picked before flowering.
Phyllanthus emblica L.	Myrobalan, emblic	Well-drained soil	Roots are dug in autumn or early spring, flowers are collected during blooming.
Physalis alkekengi L.	Chinese lantern	Well-drained soil	Fruits are harvested when ripe.
Phytolacca americana L.	Pokeweed	Rich, moist, well-drained soil	Roots and fruits are collected in autumn.
Picea mariana (Mill) Black. *P. glauca* (Moench) Voss. *P. ables* (L.) Karst	Black spruce White spruce Norway spruce	Woodland soil, slightly alkaline soil	Buds are collected before open, needles are picked young, bark is stripped from older branches or trunk in autumn.

Scientific name	Common name	Soil conditions and requirements	Harvesting
Pimpinella anisum L.	Anise	Aniseed can be cultivated in a wide range of soil with pH range of 5-8. Moderate to heavy loam soil, well-drained and with good water holding capacity, is ideal	Pick lower leaves as required. Collect flowers as they open, for seed, cut plant at ground level when fruit begins to turn gray-green at the tips. Gather stems and dig up roots in autumn.
Pinus mugo Torra var. Pumilio *P. palustris* Mill. *P. strobus* L. *P. albicaulis* Engeim. *P. contorta* Dougl. ex. Loud	Dwarf mountain pine Southern pitch pine White pine Lodgepole pine	Well-drained, neutral to acid soil	Leaves and young shoots are picked during the growing season.
Piper methysticum G. Forst	Kava-Kava	Well-drained, stony soil with ample water, minimum temperature of 15°C	Roots are dug as required.
P. nigrum L.	English peper	Rich, well-drained soil	Fruits are picked ripe.
Plantago asiatica L. *P. psyllium* L. *P. lanceolata* L.	Psyllium Plantain	Well-drained soil	Plants are cut during the growing season.
Podophyllum peltatum L.	Mayapple	Humus-rich, moist soil with low to neutral pH, shady location	Rhizomes are dug in autumn.
Polygala senega L.	Seneca snakeroot	Well-drained, moist soil	Roots (4-yr-old) are dug in autumn.

Scientific name	Common name	Soil conditions and requirements	Harvesting
Polygonum bistorta L. *Polygonum multiflorum* Thunb.	Bistort root Flowery knotweed	Rich, moist soil; *P. multiflorum* needs extra organic matters for better growth and protection from severe winter	Rhizomes are dug in autumn. *P. multiflorum* is harvested from 3- to 4-yr-old plant.
Populus balsamifera L. *P. candicans* L. *P. tremuloides* Michx.	Poplar, Balm of gilead American aspen	Deep, moist, well-drained soil	Buds are collected in spring before opening. Bark is stripped from side branches in autumn or early spring.
Primula veris L.	Cowslip	Dry, neutral to alkaline soil	Roots are dug in spring.
Primula vulgaris Huds.	Primrose	Moist, well-drained soil	Pick leaves and flowers as they open, harvest roots in autumn of second year.
Prunella vulgaris L.	Heal all, selfheal	Moist, well-drained soil	Plants are cut in summer when flowering.
Prunus africana L.	African prune	Well-drained, neutral to alkaline soil	Seeds from ripe fruits.
P. dulcis (Mill.) D. A. Webb.	Almond	Well-drained, neutral to alkaline soil	Seeds from ripe fruits.
P. mume Siebold & Zucc.	Japanese apricot	Well-drained, neutral to alkaline soil	Unripe fruit harvested in summer.
P. serotina J. F. Ehrb.	Black cherry, wild cherry	Well-drained, neutral to alkaline soil	Bark is stripped in autumn.
P. virginiana L.	Chokecherry	Well-drained, neutral to alkaline soil	Bark is stripped in autumn.
Pueraria lobata (Wild) Ohwi.	Kudzu	Well-drained soil	Roots are dug from autumn to spring.

Scientific name	Common name	Soil conditions and requirements	Harvesting
Pygeum africanum L.	Pygeum	Well-drained sandy soil	Leaves are picked young, bark is stripped in autumn.
Pyrethrum cinerarifolium L.	Pyrethrum	See *Chrysanthemum cinerarifolium*	
Quercus alba L.	White oak	Deep, well-drained soil	Bark is stripped in spring from older trees.
Ranunculus occidentalis Nutt.	Buttercup	Moist, neutral to alkaline soil	Plants are cut after flowering.
R. ficaria L.	Pilewort	Moist, neutral to alkaline soil	Plants including roots are cut after flowering.
Raphanus sativus L.	Radish	Rich, moist, well-drained soil	Leaves are picked when young.
Rhamnus catharticus L.	Buckthorn	Well-drained alkaline soil	Bark is stripped from young plant in spring to early summer, fruit collected when ripe.
R. frangula L.	Alder buckthorn	Well-drained neutral to acid soil	Bark is stripped from 2-yr-old plants.
R. purshianus L.	Cascara sagrade	Well-drained soil	Bark is stripped from 2-yr-old plants.
Rheum palmatum L. *R. officinal* Baill. *R. tanguticum* L.	Chinese rhubarb Rhubarb Rhubarb	Well-drained, moist soil with high organic matter	Rhizomes are dug in autumn from 3-yr-old plants.
Rhodiola rosea (L.) Scop.	Roseroot	See *Sedum rosea*	

Scientific name	Common name	Soil conditions and requirements	Harvesting
Rhus radicans L.	Poison ivy	Well-drained soil	Roots are dug as required.
Ribes nigrum L. *R. lacustre* (Pers.) Poir.	Black current Black gooseberry	Moist, clay soil, needs protection from cold winds and late frost	Leaves are picked during the growing season.
Robertium macrorrhizum Pic.	Geranium	See *Geranium macrorrhizum*	
Rosa canina L. *R. damascena* Mill.	Dog rose Damask rose	Well-drained, moist, fertile, neutral to slightly acid soil	Pick buds when formed, petals when first open, hips when ripe and after the first frost when softened.
Rosmarinus officinalis L.	Rosemary	Well-drained, light, neutral to alkaline soil	Gather leaves when flowering has started, 30-50 cm of the top along with leaves and flowers.
Rubus chamaemorus L.	Cloudberry	Moist, well-drained soil	Fruits harvested when ripe.
R. fruiticosus L.	Blackberry	Moist, well-drained soil	Leaves are picked before flowering, roots are dug in summer, fruits when ripe.
R. idaeus L.	Raspberry	Moist, well-drained soil	Leaves are picked before flowering, fruits when ripe.
Rumex acetosella L. *R. obtusifolia* L.	Sorrel Dock	Moist soil	Gather leaves when young before flowering.
Ruscus aculeatus L.	Butcher's broom	Well-drained, dry soil	Plants are cut in late spring, root in autumn.

Scientific name	Common name	Soil conditions and requirements	Harvesting
Ruta graveolens L.	Rue, herb of grace	Well-drained, neutral to alkaline soil	Pick young leaves just before flowers open.
Salix alba L. *S. discolour* Muhlenb. *S. caprea* L.	White willow Pussy willow	Moist to wet, heavy soil	Leaves are collected during the growing season.
Salvia officinalis L. *S. sclarea* L.	Sage Clary sage	Well-drained to dry, acidic (pH 4.5), clay loam soil	Flower tops harvested in full bloom, when the seeds of 2-3 flowers at the main raceme become brown in color.
Sambucus nigra L. *S. canadensis* L. *S. racemosa* L.	Elderberry Elder	Rich, moist, neutral to alkaline soil	Leaves are picked in summer.
Sanguinaria canadensis L.	Bloodroot	Well-drained soil with high organic matter, grows better in shade	Rhizomes (2-yr-old or more) are dug in autumn or immediately after flowering.
Sanicula marilandica L.	Black snakeroot	Well-drained sandy soil	Roots are dug in the fall, flowers collected at the beginning of opening.
Santalum album L.	Sandalwood	Moist, fertile clay to sandy loam soil	Heartwood from older tree (30-yr-old).
Saponaria officinalis L.	Soapwort	Well-drained, moist, neutral to alkaline soil	Pick flowers, leaves, stems, and roots in autumn.
Sarothamnus scoparius L.	Broom	See *Cytisus scoparius*	

Scientific name	Common name	Soil conditions and requirements	Harvesting
Sassafras albidum (Nutt.) Nees	Sassafras	Deep, rich, neutral to acid soil	Leaves are picked in spring.
Satureja hortensis L. *S. montana* L.	Savory	Well-drained, neutral to alkaline dry soil	Pick leaves just as flower buds form, collect flowering tops in late summer.
Schisandra chinensis (Turcz.) Baill.	Schizandra, wu wei zi	Rich, well-drained moist soil	Fruits are collected after the first frost.
Scutellaria baicalensis Georgi.	Baical skullcap	Very well-drained light soil, may tolerate drought	Roots are dug in autumn or spring.
S. lateriflora L. *S. galericulata* L.	Virginia skullcap Skullcap	Light, well-drained damp soil	Plants are cut when flowering.
Sedum acre L.	Stonecrop (small houseleek)	Well-drained soil	Whole plants are cut during flowering, stolons are dug in the fall.
S. rosea (L.) Scop. subsp. *integrifolium*	Roseroot	In crevices or among mats of moss and other vegetation near shores	Leaves are picked young, roots are dug in the fall.
Sempervivum tectorum L.	Houseleek, hens and chicks	Well-drained, gritty or stony soil	Harvest the thickest leaves anytime.
Senecio vulgris L.	Groundsel	Damp to wet soil	Plants are cut before flowering.
Senna alexandrina L.	Alexandrina	See *Cassia angustifolia*	
Serenoa repens (Bartr.) Small.	Saw palmetto	Well-drained moist soil, minimum temperature of 10°C	Fruits are harvested when ripe and used partly dried.

Scientific name	Common name	Soil conditions and requirements	Harvesting
Sesamum indicum L. *S. orientale* L.	Sesame	Well-drained sandy soil	Leaves are picked during the growing season, seeds are collected when ripe.
Shepherdia canadensis (L.) Nutt.	Buffalo berry	Well-drained marginal land	Berries are picked ripe, leaves and shoots are collected in the spring.
Silybum marianum (L.) Gaertn.	Milk thistle	Well-drained soil	Plants are cut when flowering, seeds when ripe.
Simmondsia chinensis (link) C. Schneid	Jojoba	Well-drained sandy or gravelly dry soil	Seeds are collected when ripe.
Solidago virgaurea L.	Goldenrod	Well-drained, moisture-retentive soil	Leaves and flowering tops are picked before flowers are fully opened.
Sorbus aucuparia L.	Mountain ash	Well-drained woodland soil	Berries are picked ripe, leaves picked in spring, bark is stripped in autumn.
Spirea ulmaria L.	Meadowsweet	Damp woodland alkaline soil	Flowers and leaves are picked during flowering, roots are dug in autumn.
Stachys officinalis (L.) Trev.	Betony, woundwort	Neutral to acid dry soil	Flowering plants are cut in summer.
Stellaria media (L.) Vill.	Chickweed	Moist soil	Plants are cut at any time.
Stevia rebaudiana (Bertoni) Bertoni	Stevia, Sweet herb of Paraguay	Well-drained, fertile soil	Leaves are picked before flowering.

Scientific name	Common name	Soil conditions and requirements	Harvesting
Symhytum officinale L.	Comfrey	Moist to wet soil	Pick leaves in summer, dig up roots in autumn or winter.
Syringa vulgaris L.	Lilac	Well-drained dry soil	Flowers are gathered full bloom.
Syzygium aromaticum (L.) Merr. & Perry	Clove	Well-drained, fertile sandy or medium loam soil, minimum temperature of 15°C	Flower buds are collected before opened from 4-5 years old.
Tagetes lucida Cav.	Mexican mint marigold	Well-drained, fertile soil	Harvest open flowers, plants during flowering.
Tanacetum parthenium (L.) Schultz-Bip.	Feverfew	See *Chrysanthemum parthenium*	
T. cinerarifolium L.	Pyrethum	See *Chrysanthemum cinerarifolium*	
T. vulgare L.	Tansy	Well-drained to dry stony soil	Pick leaves anytime, gather flowers when open.
Taraxacum officinale G. H. Weber ex Wigg.	Dandelion	Moist to dry, neutral to alkaline soil	Plants are cut in early summer, leaves are picked in spring.
Taxus x media Rehd. *T. brevifolia* Nutt.	Yew	It grows better on cool, moist flats with moist and well-drained soil	Leaves are picked in early autumn or spring, bark is stripped in autumn to spring.
Thymus citriodorus (Pers.) Schreb. *Thymus vulgaris* L	Lemon thyme Thyme	Well-drained soil, both are hardy to -15°C, avoid wet soil surface in the winter	Whole plants and flowering tops are collected in summer as flowering begins.

Scientific name	Common name	Soil conditions and requirements	Harvesting
Tillia cordata Mill. *T. europaea* L.	Linden, small-leaved lime	Moist, well-drained, neutral to alkaline soil	Flowers are picked in summer.
Tribulus terrestris L.	Puncture vine	Woodland slightly alkaline soil	Flowers are collected during full bloom, fruits are picked ripe.
Trifolium incarnatum L. *T. pratense* L.	Clover (crimson) Clover (red)	Moist, well-drained, neutral soil	Flower heads with upper leaves are picked in summer as they open.
Trigonella foenum-gracecum L.	Fenugreek	Well-drained, fertile soil	Pick young leaves as needed, cut whole plant in autumn.
Tropaeolum majus L.	Canary creeper Nasturtium	Well-drained, moist, average to poor soil	All parts are picked in summer.
Tussilago farfara L.	Colisfoot	Moist, neutral to alkaline soil, hardy to -29°C	Leaves are cut when full grown.
Ulmus rubra Muhl. *U. procera* L.	Slippery elm, sweet elm English elm	Moist, deep soil, hardy to -37°C	Inner bark is stripped from trunks or larger branches in spring.
Umbelluslaria california Nutt.	California laurel Bay laurel	Well-drained, moist soil	Leaves are picked as required.
Uncaria tomentosa (Willd.) DC *U. guianensis* (Aubl.) Gmel.	Cat's claw	Well-drained moist soil, minimum temperature of 10°C	Roots are dug in autumn, bark is stripped in early spring.

Scientific name	Common name	Soil conditions and requirements	Harvesting
Urtica dioica L. *U. urens* L.	Stinging nettle	Moist soil with high nitrogen	Whole plants are cut as flowering begins in summer.
Vaccinium macrocarpon L. *V. vitis-idaea* L.	Cranberry Mountain cranberry Cowberry, foxberry	Moist, acid muck soil in cool regions, avoid high water level	Leaves are picked in spring, fruits are collected in late summer.
V. mrytilloides Michx. *V. myrtillus* L. *V. oreophilum* Rydb.	Blueberry Bilberry	Moist, acid and fertile soil	Fruits are harvested in late summer.
Valeriana officinalis L.	Valerian	Moist soil	Roots are dug in second season in late summer.
Verbascum thapsus L.	Mullein	Well-drained dry soil	Collect flowers as they open, and leaves in their first season.
Veronica officinalis L.	Speedwell	Dry, slightly acid soil	Plants are cut during flowering.
Vetiveria zizanioides L. (Nash.)	Vetiver	Wet to dry light to medium loam soil, pH neutral, it can also grow in river beds, minimum temperature of 10-15°C	Roots are dug 15-18 months after planting.
Viburnum opulus L.	Highbush cranberry	Deep, moist soil	Bark is stripped before leaves change color in autumn, or before leaf buds open in spring.

Scientific name	Common name	Soil conditions and requirements	Harvesting
Vinca minor L.	Periwinkle	Moist soil	Plants are cut during flowering.
Viola tricolor L.	Pansy	Well-drained, moisture-retentive soil	Pick leaves in early spring, gather flowers when newly opened and roots in autumn.
Viscum album L.	Mistletoe	Grows on young branches of host tree such as oak	Leafy stem are cut in spring.
Vitex agnus-castus L. *V. negundo* L.	Chaste tree Chinese chaste	Rich, moist to dry soil	Leaves are picked in early summer.
Vitis labrusca L. *V. vinifera* L.	Grape	Deep, moist, humus-rich, neutral to alkaline soil	Leaves and stems are collected in early summer.
Withania somnifera Dunal.	Ashwagandha, Indian ginseng	Dry, stony soil	Roots are dug at anytime.
Yucca aloifolia L. *Y. glauca* L.	Yucca Soapweed	Fertile, well-drained soil	Leaves are cut early in the spring.
Zanthoxylum americanum Mill.	Prickly ash, toothache tree	Fertile, moist soil	Leaves are picked during the growing season.
Zea mays L.	Corn	Rich, well-drained soil	Corn silk is collected in summer before it withers, ears are cut when immature or ripe.
Zingiber officinale Roscoe	Ginger	Well-drained, humus-rich, light sandy loam, neutral to alkaline soil	Rhizomes are dug during the growing season after 8-9 months of planting.

References

1. Adams, J. 1927. Medicinal Plants and Their Cultivation in Canada. Dominion Expt. Farm Publication, Ottawa, Canada. 29 p.

2. Ahluwalia, S. S. 1997. Goldenseal, American Gold. Walden House, Bronx, NY. 22 p.

3. Baskin, C. C. and J. M. Baskin. 1993. Germination requirements of *Oenothera biennis* seeds during burial under natural seasonal temperature cycles. Can. J. Bot. 72: 779-782.

4. Bodkin, G. E. 1915. Report of the economic biologist. Rept. Dept. Sci. Agric. 1914-1915, 17th Sept. 1915, 11 p.

5. Bomme, U., R. Rinder and K. Voit. 1991. Influence of substrates and fertilization on raising transplants of *Arnica montana* L. Gartenbauwissenschaft 56: 106-113.

6. Bown, D. 1987. *Acorus calamus* L: a species with a history. Aroideana (International Aroid Society, South Miami, FL) 10: 11-14.

7. Bown, D. 1995. Encyclopedia of Herbs & Their Uses. Reader's Digest Association (Canada) Ltd. Montreal, PQ. 424 p.

8. Bremness, L. 1994. The Complete Book of Herbs. Dorling Kindersley Ltd., London. 304 p.

9. Brown, M. S. and R. J. Molyneux. 1996. Effects of water and mineral nutrient deficiencies on pyrrolizidine alkaloid content of *Senecio vulgaris* flowers. J. Sci. Food Agric. 70: 209-211.

10. Bunney, S. 1992. The Illustrated Encyclopedia of Herbs, Their Medicinal and Culinary Uses. Chancellor Press, London. 320 p.

11. Caballero, R., M. Haj-Ayed, J. F. Galvez, P. J. Hernaiz and M. H. Ayed. 1995. Yield components and chemical composition of some annual legumes and oat under continental Mediterranean conditions. Agricoltura Mediterranea 125: 222-230.

12. Chevallier, A. 1996. The Encyclopedia of Medicinal Plants. Dorling Kindersley Ltd., London. 336 p.

13. Dobelism, I. N. 1990. Magic and Medicine of Plants (5th edition). The Reader's Digest Association Inc., New York. 464 p.

14. Duke, J. A. 1985. CRC Handbook of Medicinal Herbs. CRC Press, Inc., Boca Raton, FL. 677 p.

15. Duke, J. A. and J. L. duCellier. 1993. CRC Handbook of Alternative Cash Crops. CRC Press, London. 536 p.

16. Eaton, G. W., A. Y. Shawa and P. A. Bowen. 1983. Productivity of individual cranberry uprights in Washington and British Columbia, Canada. *Vaccinium macrocarpon*, yield. Sci. Hortic. (Amsterdam) 20: 178-184.

17. Everett, T. H. 1960. New Illustrated Encyclopedia of Gardening. Vol. 14. Greystone Press, New York. 192 p.

18. Foster, S. 1991. Harvesting medicinals in the wild. HerbalGram 24: 10-16.

19. Foster, S. 1993. Herbal Renaissance. Gibbs Smith Publisher, Utah. 234 p.

20. Fuentes-Granados, R. G., M. P. Widrlechner and L. A. Wilson. 1998. An overview of *Agastache* research. J. Herbs, Spices & Medicinal Plants 6: 69-97.

21. Fyles, F. 1913. Golden seal. Dom. Canada Exp. Farms Rep. 1913. p. 495-496.

22. Golca, L., T. Czabajski, S. Kordana, B. Zorawska and M. Polkowski. 1975. Trials on the cultivation of peppermint, foxglove and opium poppy on deforested soils in the Pulawy region. Herba Polonica 21: 70-78.

23. Grigor'ev, A. G. 1985. Tree and shrubs for landscaping on the west coast of the Crimea. Hort. Abst. 1987. 57: 9702.

24. Hamana, K., S. Matsuzaki, M. Niitsu and K. Samejima. 1994. Distribution of unusual polyamines in aquatic plants and gramineous seeds. Can. J. Bot. 72: 1114-1120.

25. Haunold, A. 1993. Agronomic and quality characteristics of native North American hops. Am. Soc. Brew. Chem. J. 51: 133-137.

26. Hedberg, I. 1987. Research on medicinal and posionous plants of the tropics, past, present and future. In: Leeuwenberg, A. J. M. (ed.). Medicinal and Poisonous Plants of the Tropics. Pudoc. Wageningen, The Netherlands. p. 9-15.

27. Henkel, A. and G. F. Klugh. 1908. The cultivation and handling of goldenseal. U. S. Dept. Agric. Bur. Plant Ind. Circ. No. 6. 19 p.

28. Henning, J. 1995. Aromatic *Agastache*. American Nurseryman 181: 59-67.

29. Heptinstall, S. and D. V. C. Awang. 1998. Feverfew: a review of its histroy, its biological and medicinal properties, and the status of commercial preparations of the herb. In: Lawson, L. D. and R. Bauer (eds.). Phytomedicines of Europe, Chemistry and Biological Activity. Am. Chem. Soc., Washington, DC. p. 158-175.

30. Hornok, L. 1992. Cultivation and Processing of Medicinal Plants. John Wiley & Sons, New York. 338 p.

31. Hunkanen, E. and T. Pyysalo. 1976. The aroma of cloudberries (*Rubus chamaemorus* L.). Zeitschrift fur Lebensmittel Untersuchung und Forschung 160: 393-400.

32. Husain, A. 1994. Essential Oil Plants and Their Cultivation. Central Inst. Medicinal Aromatic Plants, Lucknow, India. 292 p.

33. Ikeda, H., Y. Yoshida, and T. Osawa. 1985. Effects of NO_3 NH_4 ratios and temperature of the nutrient solution on the growth of Japanese honewort, garland chrysanthemum and Welsh onion. J. Japanese Soc. Hort. Sci. 54: 58-65.

34. Janick, J. and J. E. Simon (eds.). 1991. New Crops. John Wiley and Sons, Inc., New York. 710 p.

35. Khodzhimatov, K., S. F. Fakhrutdinov, G. S. Aprasidi, N. P. Kuchni and K. Karimov. 1987. *Codonopsis clematidea* Schreuk, a valuable medicinal plant. Referativnyi Zhurnal 10: 790.

36. Kuusi, T., K. Hardh and H. Kanon. 1984. Experiments on the cultivation of dandelion for salads use. 1. Study of cultivation methods and their influence on yield and sensory quality. J. Agric. Sci. Finland 56: 9-22.

37. Kuusi, T., K. Hardh and H. Kanon. 1984. Experiments on the cultivation of dandelion for salads use. II. The nutritive value and intrinsic quality of dandelion leaves. J. Agric. Sci. Finland 56: 23-31.

38. Lee, S., S. Kim. G. Min, J. Cho, B. Choi, and S. Lee. 1996. Agronomic characteristics and aromatic compositions of Korea wild *Codonopsis lanceolata* collections cultivated in the field. Korean J. Crop Sci. 41: 188-199.

39. Li, T. S. C. 1995. Asian and American ginseng, a review. HortTechnology 5: 27-34.

40. Li, T. S. C. 1998. *Echinacea:* cultivation and medicinal value. HortTechnology 8: 122-129.

41. Li, T. S. C. and W. R. Schroeder. 1996. Sea buckthorn (*Hippophae rhamnoides* L.): a multipurpose plant. HortTechnology 6: 370-380.

42. Loconte, H. and W. H. Blackwell. 1981. A new species of blue cohosh (*Caulophyllum*, Berberidaceae) in eastern North America. Phytologia 49: 483.

43. McGimsey, J. 1993. Sage, *Salvia officianalis* WWW. Crop.cri.nz/broadshe/sage.htm.

44. Mazza, G. and F. A. Kiehn. 1992. Essential oil of *Agastache foeniculum*, a potential source of methyl chavicol. J. Essential Oil Res. 4: 295-299.

45. Meng, K., X. Yang, W. Pan, and L. Yu. 1997. Techniques for afforestation and amelioration of coastal saline-alkaline soils in Liaoning. Scintia Silvae Sinicae 33: 25-33.

46. Miller, F. A. 1914. Propagation of medicinal plants. Bull. Torrey Bot. Club. 41: 105-129.

47. Minker, E. and K. Szendrei. 1986. About the green wave. Gyogyszereszet 30: 323-325.

48. Mulligan, G. A. and D. B. Munro. 1990. Poisonous Plants of Canada. Canadian Government Publishing Centre, Ottawa, Ontario. 96 p.

49. Obregon Vilches, L. E. 1995. Cat's Claw (3rd edition). Institute de Fitoterapia Americano, Lima, Peru. 169 p.

50. Omer, E. A., A. M. Refaat, S. S. Ahmed, A. Kamel and F. M. Hammouda. 1993. Effect of spacing and fertilization on the yield and active constituents of milk thistle, *Silybum marianum.* J. Herbs Spices and Medicinal Plants 1: 17-23.

51. Palevitch, D. 1988. Agronomy applied to medicinal plant conservation. In: Akerele, O., V. Heywood and H. Synge (eds.). The Conservation of Medicinal Plants. Cambridge Univ. Press, New York. 362 p.

52. Purseglove, J. W., E. G. Brown, C. L. Green and S. R. J. Robbins. 1981. Spices. Vol. 1. Longman, London.

53. Rajan, S. and S. Ram. 1982. A new approach towards vegetative propagation of aonla through cuttings. Progressive Hort. 14: 190-191.

54. Rizvi, M. A., G. R. Sarwar, A. Mansoor. 1998. Poisonous plants of medicinal use growing around Madinat al-Hikmah. Hamdard Medicus 41: 88-95.

55. Rodov, V. S., P. S. Bugorskii, R. G. Butenko and A. S. Popov. 1988. The capacity of tissue cultures of higher plants for biotransformation of monoterpenes. Fiziologiya-Rastenli 35: 526-533.

56. Rogers, R. D. 1997. Sundew, Moonwort, Medicinal Plants of the Prairies. Vol. 1 & 2.

Edmonton, AB. 282 p.

57. Rogiers, S. Y. and N. R. Knowles. 1997. Physical and chemical changes during growth, maturation, and ripening of saskatoon. Can. J. Bot. 75: 1215-1225.

58. Schumacher, H. M. 1988. Biotechnology in the production and conservation of medicinal plants. In: Akerele, O., V. Heywood and H. Synge (eds.). The Conservation of Medicinal Plants. Cambridge Univ. Press, New York. 362 p.

59. Sharma, M. P., R. Prasad and R. K. Pathak. 1987. Effect of sodicity on growth and mineral composition of aonla (*Emblica officinalis* Gaern) seedlings. Indian J. Agri. Chemistry 20: 195-199.

60. Shellard, E. J. 1987. Medicines from plants with special reference to herbal products in Great Britain. Planta Medica 53: 121-123.

61. Singh, S., S. Kumar, S. Singh and S. Kumar. *Withania somnifera*: The Indian Ginseng Ashwagandha. Central Inst. Medicinal and Aromatic Plants, India. 293 p.

62. Small, E. (ed.). 1997. Culinary Herbs. NRC Research Press, Ottawa. 710 p.

63. Small, E. and P. M. Catling (eds.). 1999. Canadian Medicinal Crops. NRC Research Press, Ottawa. 240 p.

64. Sokolov, N. 1916. Progressive fruit-growing and market-gardening. Petrograd. xii, no. 51, 2nd Jan. 1916, p. 1347.

65. Stockberger, W. W. 1922. Drug plants under cultivation. U.S.D.A. Farmers' Bull No. 663.

66. Takeda, O., S. Azuma, H. Mizukami, T. Ikenaga and H. Ohashi. 1986. Cultivation of *Polygala senega* var. latifolia. II. Effect of soil moisture content on the growth and senegin content. Shoyakugaku Zasshi 40: 434-437.

67. True, R. H. 1903. Cultivation of Drug Plant in the United States. U.S.D.A. Year Book 1903. p. 337-346.

68. Van Fleet, W. 1914. Golden seal under cultivation. U.S.D.A. Farmers' Bull. No. 613.

69. Van Fleet, W. 1915. The cultivation of peppermint and spearmint. U.S.D.A. Farmers' Bull. No. 694.

70. Vasander, H. and T. Lindholm. 1987. Use of mires for agricultural, berry and medical plant production in Soviet Karelia. Hort. Abst. 1988. 735 p.

71. Wahab, J. 1998. Herb agronomy. Sask. Irrigation Development Centre Ann. Review 1997-1998. p. 167-173.

72. Wang, X. L., L. Peng and G. F. Jiang. 1991. On adequate harvesting time for *Lycium barbarum* fruit in Ningxia. Ningxia J. Agro-Forestry Sci. & Technology 6: 23-24.

73. Weiss, E. A. 1997. Essential Oil Crops. CAB International, New York. 590 p.

74. Yuknyavichene, G. K., A. V. Morkunas and N. A. Stankyavichene. 1973. Some biological characterisitics and essential oil content in *Pyrethrum*. Poleznye Rasterniya Pribaltiskikh Respubik i Belorussii. p. 299-303.

CHAPTER 6

Major Diseases and Insects Found in Medicinal Plants

Many crops now depend on scientifically bred and carefully grown seed stock, and they are protected by carefully timed and devised applications of crop protection chemicals. However, there has been very little research on disease and pest control for medicinal plants. The diseases and insects in Tables 7 and 8 are listed mainly for awareness of possible infestations. Identifying the disease or insect from host plants and understanding the life cycle of the pest are important steps in devising an effective control program.

Diseases and insects are the major factors affecting the success of medicinal plant cultivation. Most control measures are yet to be developed. For proper control of infestations, prevention and exclusion are the most important steps, followed by eradication. Exclusion measures are designed to keep the pathogen from entering the area in which the host is growing, or to reduce its presence to a minimum. These measures include quarantines and inspection of planting stocks. Eradication measures are concerned with the elimination of the pathogen after it has become established in an area. These measures include the eradication or removal of sources of primary inoculums, such as alternate or overwintering hosts, plant debris, field sterilization, and crop rotation.

Many cultural practices that tend to reduce the progress of disease and insect infestations have come into use. Avoidance of cultivation of row crops while dew is present on the foliage reduces the spread of waterborne fungal spores, insect eggs, and plant pathogenic bacteria. The timing of sowing or planting may be helpful in avoiding periods particularly favorable for a pathogen or insect. Regulation of soil moisture is another way to minimize infestation, because some diseases are particularly enhanced by excessive soil moisture or by irrigation. Proper and balanced soil fertility levels and correction of deficiencies in trace elements are two other important measures to discourage pathogens and insects.

Table 7. Major diseases of medicinal plants.

Scientific name	Common name	Major diseases	
		Name	Pathogen
Abies balsamea (L.) Mill.	Balsam fir	Needle blight	*Acanthostigma parasiticum* (Hartig) Sacc., *Rhabdogloeum abictinum* Dearn., *Hypodermella abietis-concoloris* (Mayr.) Dearn
		Canker	*Cyptospora pinastri* Fr., *C. Friesii* Sacc., *Dasyscypha resinaria* (Cke. & Phill.) Rehm, *Melampsora abieti-capraearum* Tub., *Melampsorelia cerastii* (Pers.) Schroet
		Rust	*Micropora polypodophila* (Bell) Faull.
Achillea millefolium L.	Yarrow	Crown gall	*Agrobacterium tumefaciens* (E. F. Sm. & Town.) Conn.
		Powdery mildew	*Erysiphe cicharacearum* DC.
		Root rot	*Phymatotrichum omnivorum* (Shear) Dug., *Rhizoctonia solani* Kuehn.
Aconitum napellus L.	Monkshood	Powdery mildew	*Erysiphe polygoni* DC., *Heterodera marioni* (Cornu) Goodey, *Plasmopara pygmaea* (Ung.) Schroet.
		Downy mildew	*Phymatotrichum omnivorum* (Shear) Dug.
		Root rot	*Rhizoctonia solani* Kuehn
		Bacterial leaf spot	*Pseudomonas delphinii* (E.F.Sm) Stapp.
Acorus calamus L. var. Americanus Wolff. *A. tatarinowii* L.	Calamus, sweet flag Shi Chang Pu	Leaf spot	*Cylindrosporium acori* PK., *Ramularia aromatica* (Sacc.) Hoehn., *Septocylindrium* spp.
Actaea alba L. *A. rubra* (Ait.) Willd.	White baneberry Red baneberry	Leaf spot	*Ascochyta actaeae* (Bres.) J. J. Davis, *Ramularia actaeae* Ell. & Hollw

Scientific name	Common name	Major diseases	
		Name	Pathogen
Aesculus hippocastanum L.	Horse chestnut	Root rot	*Armillaria mellea* Vahl. ex Fr., *Phymatotrichum omnivorum* (Shear) Dug., *Collybia velutipes* Curt.
		Wood rot	*Fomes applanatus* (Wallr.) Gill., *Ceronspora aesculina* Ell. & Kell., *Macrisoiruyn baccatyn* Ell. & Ev.
		Leaf spot	*Pyhllactinia paviae* Desm.
		Powdery mildew	*Phyllactinia corylea* Karst., *Uncinula flexuosa* PK.
Agastache foeniculum L. *A. anethrodora* L.	Aniseed	Powdery mildew	*Sphaerotheca humuli* (DC.) Burr.
		Leaf, stem spot	*Ascochyta lophanthi* J. J. Davis, *Ramularia lophanthi* Ell., *Septoria lophanthi* Wint.
		Verticillium wilt	*Verticillium dahliae* Kleb.
Agrimonia euparoria L.	Agrimony	Downy mildew	*Peronospora patentillae* d By., *P. agrimoniae* Syd. ex Gaum.
		Root rot	*Phymatotrichum omnivorum* (Shear) Dug.
		Powdery mildew	*Sphaerotheca humuli* (DC.) Burr.
Agropyron repens (L.) Beauvois	Couch grass	Root rot	*Fusarium culmorum* (W. G. Sm.) Sacc., *Rhizoctonia solani* Kuehn., *Gloeosporium bolleyi* Sprague, *Helminthosporium sativum* Pam, King & Bakke.
		Leaf spot	*Ovularia pulchella* (Ces) Sacc., *Phytophthora* spp.
		Leaf mold	*Cladosporium graminum* Pers. ex Lk.
		Powdery mildew	*Erysiphe graminis* DC., *Phyllactinia corylea* Pers. ex Karst.
		Rust	*Puccinia coronata* Cda., *P. graminis* Pers., *P. rubigo-vera* (DC.) Wint.

Scientific name	Common name	Major diseases	
		Name	Pathogen
Alchemilla vulgaris L.	Lady's mantle	Black root rot Leaf spot	*Thielaviopsis basicola* (Berk. & Broome) Ferraris *Venturia adysta* (Fuckel) E. Mulfer
Allium cepa L.	Onion	Purple blotch Soft rot Gray mold rot Bacteria soft rot Bulb or root rot Damping off	*Alternaria porri* (Ell.) Cif. *Rhizotonia solani* Kuehn, *Sclerotinia* spp. *Botrytis* spp. *Erwinia carotovora* (L. R. Jones) Holland, *Pseudomonas allicola* Starr & Burkh. *Fusarium* spp., *Phytophthora drechsleri* Tacker *Pythium* spp.
A. sativum L. *A. schoenoprasum* L.	Garlic Chives	Gray mold, rot Soft rot Dry rot	*Botrytis allii* Munn. *Erwinia carotovora* (I. R. Jones) Holland, *Sclerotium cepivorum* Berk *Fusarium* spp., *Helminthosporium allii* Campanile
Alnus crispus (Ait.) Pursh. *A. incana* (L.) Moench. subsp. *tenufolia*	Alder	Powdery mildew Wood rot	*Erysiphe aggregata* (Pk.) Farl, *Phyllactinia corylea* Pers. ex Karst. *Fomes igniarius* (L. ex Fr.) Kickx, *Leuzites trabea* Pers. ex Fr., *Polyporus* spp., *Poria* spp.
Aloe vera (L.) Burm.f.	Aloe	Root rot	*Pythium ultimum* Trow.
Althaea officinalis L.	Marshmallow	Powdery mildew Leaf spot	*Erysiphe cichoracearum* DC., *E. polygoni* DC., *Oidium* spp. *Phoma exigua* Desmaz, *Septoria malvicola* Ellis & G. Martin

Scientific name	Common name	Major diseases	
		Name	Pathogen
Amelanchier alnifolia Nutt.	Saskatoon	Rust Fruit or brown rot Die back canker Powdery mildew	*Gymnosporungium biseptatum* Ell. and spp. *Monilinia amelanchieris* (Reade) Honey, *M. fructicola* (Wint.) Honey *Nectria cinnabarina* Tode ex Fr., *Nummularia discincola* (Schw.ex Fr.) Cke. *Phyllactinia guttata* (Fr.) Lev., *Podosphaera axyacanthae* (DC.) d By.
Ananas comosus (L.) Merr.	Pineapple	Brown rot Stem, root rot Yellow spot	*Fusarium* spp. *Phytophthora* spp., *Pythium* spp., *Rhizoctonia* spp. Ananas virus
Anaphalis margaritacea (L.) Bench & Hook	Everlastings	Leaf spot Rust	*Septoria margaritaceae* PK., *Phoma antennariae* Clements *Uromyces amoenus* Syd.
Anethum graveolens L.	Dill	Leaf and stem spot Root and stem rot Yellows Powdery mildew	*Cercospora anethi* Sacc., *Phoma anethi* (Pers ex. Fr.) Sacc. *Phymatotrichum omnivorum* (Shear) Dug., *Sclerotinia sclerotorum* (Lib.) d By. Aster yellow phytoplasma *Erysiphe anethii*
Angelica archangelica L. *A. sinensis* (Oliv.) Diels	Angelica Dong quai	Leaf spot Rust	*Fusicladium angelicae* (Fr.) Lind, *Phyllosticta angelicae* Sacc. *Puccinia angelicae* (Schum.) Fickl., *Puccinia* spp.
Antennaria magellanica Schultz	Everlastings	See *Anaphalis margaritacea*	

Scientific name	Common name	Major diseases	
		Name	Pathogen
Anthemis nobilis L.	Chamomile	See *Chamaemelum nobile*	
Apium graveolens L.	Celery	Damping off Gray mold rot Bacterial soft rot Fusarium yellows Root rot, stem rot Leaf spot Yellow disease Seed rot Celery blight	*Aphanomyces euteiches* Drechs., *Pythium* spp., *Sclerotinia* spp. *Botrytis cinerea* Pers. ex Fr. *Erwinia carotavora* (L. R. Jones) Holland *Fusarium oxysporum* Schlecht. *Rhizoctonia solani* Kuehn. *Septoria apiicola* *Fusarium oxysporum* F. subsp. *apii* *Sclerotinia selerotiarum, Botrytis cineria, Septoria apiicola* *Cercospora apii*
Aralia racemosa L. *A. nudicaulis* L.	Spikenard	Root rot Leaf spot	*Armillaria mellea* Vahl ex Fr. *Colletotrichum* spp., *Phyllosticta calocasiae* Hoehn.
Arctium lappa L.	Burdock	Leaf spot	*Ascochyta lappae* Kab. & Rub., *Cercaspora arctii* F. L. Stevens, *Gloeosporium lappae* Dearn. & House, *Phlyctaena arcuata* Berk., *Septoria lapparum* Sacc.
Arctostaphylos uva-ursi (L.) Spreng	Bearberry, Uva-ursi	Rust Leaf spot Root rot	*Chrypsomyxa arctostaphyli* Diet. *Cercospora gaultheriae* Ell. & Eiv., *Cryptostictis arbuti* (Bonar) Zeller, *Phyllosticta amicta* Ell. & Ev. *Phymatotrichum amnivorum* (Shear) Dug., *Poria ferruginosa* (Schrad ex Fr.) Cke.

Scientific name	Common name	Major diseases	
		Name	Pathogen
Armoracia rusticana Gaertn, Mey & Scherb.	Horseradish	Crown gall Gray leaf spot Powdery mildew Root rot	*Agrobacterium tumefaciens* (E. F. Sm. & Town) Conn. *Alternaria brassicae* (Berk.) Sacc. *Erysiphe polygoni* DC. *Phymatotrichum omnivorum* (Shear) Dug., *Thielaviopsis basicola* (Berk. & Br.) Ferr.
Arnica latifolia Bong. *A. montana* L. *A. chamissonis* L. subsp. *Foliosa* *A. condifolia* Hook *A. fulgens* Pursh *A. sororia* Greene	Arnica	Powdery mildew Rust	*Erysiphe cichoracearum* DC., *Sphaerotheca humuli* (DC.) Burr. *Puccinia arnicalis* PK., *Uromyces junci* (Desm.) L. Tul.
Artemisia absinthium L. *A. annua* L. *A. vulgaris L.*	Wormwood Quing Hao Mugwort	Rust Leaf spot	*Albugo tragopogonis* Pers., *Puccinia absinthii* (Hedw.f.) DC., *Puccinia* spp. *Cercospora ferruginea* Fekl., *Gloeosporium heterophyllum* Ell. & Ev.
A. dracunculus L. *A. tridentata* Nutt.	Tarragon Sagebrush	Gray-mold blight Leaf spot Rust	*Botrytis cinerea* Pers. ex Fr. *Cylindrosporium artemisiae* Dearn. & Barth., *Heterosporium* spp., *Ramularia artemisiae* J. J. Davis, *Septoria artemisiae* Pass. *Puccinia absinthii* (Hedw.f.) DC., *Uromycea oblongisporus* Ell. & Ev.

Scientific name	Common name	Major diseases	
		Name	Pathogen
Asarum canadensis L.	Wild ginger	Leaf spot	*Ascochyta versicolor* Bub., *Laestadia asarifolia* (Cke.) Sacc.
		Root rot	*Sclerotinia sclerotiorum* (Lib.) d By.
Astragalus americana Bunge. *A. membranaceus* Bunge.	Astragalus	Leaf spot	*Cercospora astragali* Wor., *Didymaria astragali* (Ell. & Holw.) Sacc., *Napicladium astragali* Ell. & Ev., *Polystigma astragali* (Lasch) Hoehn., *Septoria astragalicola* Pk., *Stemphylium* spp.
A. sinensis L.	Huang Qi	Powdery mildew	Erysiphe polygoni DC., *Microsphaera euphorbiae* (Pk.) Berk. & Curt.
Avena sativa L.	Oat	Leaf or head mold	*Cladosporium graminum* Pers. ex Lk., *Fusicladium destruens* Pk., *Gibberella zeae* (Schw.) Petch.
		Blight	*Fusarium* spp., *Helminthosporium sativum* Pam, King & Bakke, *H. victoriae* Meehan & Murphy, *Selerotium rolfuii* Sacc.
		Rust	*Puccinia coronata* Cds., *Puccinia* spp.
		Damping off	*Pythium* spp.
		Root rot	*Rhizoctonia solani* Kuehn., *Wojnowicia graminis* (Mc Alp.) Sacc.
		Smut	*Ustilago avenae* (Pers.) Rostr.
Baptisia tinctora (L.) R.Br. ex Ait.f.	Wild indigo	Leaf spot	*Cerocospora velutina* Ell. & Kell., *Marssonina baptisiae* (Ell. & Ev.) Magn., *Septoria baptisae* Cke., *Stagonospora baptisiae* (Ell. & Ev.) J. J. Davis
		Powdery mildew	*Erysiphe polygoni* DC., *Microsphaera alni* DC. ex Wint.

Scientific name	Common name	Major diseases	
		Name	Pathogen
Bellis perennis L.	Daisy	Root and crown rot	*Phymatotrichum omnivorum* (Shear) Dug., *Pythium mastophorum* Drechsl., *Sclerotinia sclerotiorum* (Lib.) d By.
Berberis aquifolium L.	Oregon grape	See *Mahonia aquifolium*	
Berberis vulgaris L.	Barberry	Rust	*Cumminsiella sanguinea* (Pk.) Arth., *Puccinia graminis* Pers.
		Root rot	*Phymatotrichum omnivorum* (Shear)Dug., *Poria punctata* (Fr.) Cke., *Rhizoctonia solani* Kuehn.
		Damping off	*Pythium debaryanum* Hesse.
Beta vulgaris L.	Beet root	Crown and root gall	*Agrobacterium tumefaciens* (E. F. Sm. & Town) Conn., *Heterodera schachtii* A. Schm.
		Root rot	*Fusarium* spp., *Rhizoctonia crocorum* (Pers.) DC ex Fr., *Phoma betae* Frank., *Phymatotrichum omnivorum* (Shear) Dug., *Physalospora rhodina* (Berk. & Curt.) Cke., *Pythium* spp., *Rhizopus* spp., *Sclerotinia sclerotiorum* (Lib.) d By.
		Damping off	*Aphanomyces cochlioides* Drechs., *Pythium* spp.
		Mosaic, savoy, yellow vein, black heart, black root	Virus
Betula lenta L.	Birch	Wood rot	*Daedalea* spp., *Daldinia concentrica* (Bolt. ex Fr.) Ces & de N., *Fomes* spp., *Hydnoporia fuscescens* (Schw.) Murr., *Polyporus* spp., *Poria* spp., *Schizophyllum commune* Fr., *Ganoderma lucidum* (Leyss. ex Fr.) Karst., *Stereum* spp.
		Root rot	*Armillaria mellea* Vahl.ex Fr., *Phymatatrichum omnivorum* (Shear) Dug.

Scientific name	Common name	Major diseases	
		Name	Pathogen
Borago officinalis L.	Borage	Leaf spot	*Ramularia* spp.
Boswellia sacra Roxb. ex Colebr. *B. carteri* Birdwood	Frankincense	Leaf spot Wood rot	*Fomes fastuosus* (Sw. Fr.) Cocke *Hexagonia hydnoides* (Sw. Fr.) M. Fidalgo, *Trametes hirsuta* (Wullen. Fr.) Quel
Brassica nigra (L.) Kock.	Black mustard	Leaf spot Powdery mildew Mosaic, yellows	*Alternaria brassicae* (Berk.) Sacc., *A. Oleracea* Milbrath., *Cercosporella brassicae* (Fautr. & Room) Hoehn, *Septoria brussicae* Ell. & Ex. *Erysiphe polygoni* DC. Virus
Buxus sempervirens L.	Boxwood	Root rot Blight	*Armillaria mellea* Vahl ex Fr., *Fusarium* spp., *Phytophthora parasitica* Dast., *Rhizoctonia solani* Kuehn. *Hyponectria buxi* (DC. ex Fr.) Sacc., *Phoma conid'ogena* Schnegg.
Calamintha nepeta (L.) Savi *C. ascendens* L. *C. officinalis* L.	Basil thyme Calamint Mountain mint	Rust Downy mildew	*Puccinia menthae* Pers *Plasmopara calaminthae* J. Schrot
Calendula officinalis L.	Marigold, calendula	Leaf spot Root rot Mosaic, spotted wild, yellows	*Cercospora calendulae* Sacc., *Colletotrichum gloeosporioides* Penz. *Phymatatrichum omnivorum* (Shear) Dug., *Pythium ultimum* Trow., *Rhizoctonia solani* Kuehn. Virus

Scientific name	Common name	Major diseases	
		Name	Pathogen
Camellia sinensis (L.) Kunize	Green tea	Blight	*Botrytis cinerea* Pers. ex Fr., *Pestalotia guepini* Desm., *Sclerotinia camelliae* Hara.
		Blister blight	*Exobasidium vesans* Massee
		Gray blight	*Pestalotia theae* Sawada, *P. longiseta* Spegazzini
		Brown blight	*Glomerella cingulata* (Stomen) Spanlding et Schrenk.
		Canker	*Botryosphaeria ribis* Gross. & Dug., *Glomerella cingulata* (Ston.) Spauld. & Schrenk
		Anthracnose	*Gloeosporium theae-sinenesis* Miyake
		Scab	*Elsinoe leuospila* Bitancourt et Jenkins
		Root rot	*Poria hypolateritia* Berk.
		Ring, yellow spot	Virus
Cananga odorata (Lam.) J. D. Hook & T. Thompson	Ylang-Ylang	Root rot	*Phytophthora* spp., *Rhizoctonia* spp.
		Leaf blight	*Sclerotium rolfsii* Sacc.
Cannabis sativa L.	Hemp	Stem rot	*Fusarium* spp., *Gibberella saubinetii* (Mont.) Sacc., *Sclerotinia sclerotiorum* d By.
		Leaf spot	*Cylindrosporium* spp., *Septoria cannabis* (Lasch) Sacc.
Capsella bursa-pastoris (L.) Medik.	Shepherd's purse	Leaf spot	*Cylindrosporium capsellae* Ell. & Ev., *Pseudomonas tabaci* (Wolf & Foster) F. L. Stevens, *Ramularia armoraciae* Fckl.
		Downy mildew	*Peronospora lepidii* (McAlp.) G.W.Wils., *P. parasitica* Pers. ex Fr.
		Curly top, mosaic, yellows	Virus

Scientific name	Common name	Major diseases	
		Name	Pathogen
Capsicum annuum L. var. Annuum *C. annum* L. var. Grossum *C. annum* L. var. Acuminatum	Cayenne pepper Sweet pepper Chilli pepper	Fruit and root rot	*Alternaria solani* (Ell. & G.Martin) Sor., *Colletotrichum nigrum* Ell. & Halst., *Crvularia lunata* (Wakk.) Boed., *Diaporthe phaseolorum* (Cke. & Ell.), *Fusarium* spp. *Gloeosporium piperatum* Ell. & Ev., *Glomerella cingulata* (Ston.) Spauld. & Schrenk., *Macrophomina phaseoli* (Maubl.) Ashby., *Phoma destructiva* Plowr., *Phytophthora capsici* Leonian., *Pythium* spp., *Sclerotinai sclerotiorum* (Lib.) d By.
		Mosaic, spotted wilt, ring spot	Virus
Carbenia benedicata L.	Blessed thistle	See *Cnicus benedictus*	
Carduus benedita L.	Blessed thistle	See *Cnicus benedictus*	
Carica papaya L.	Papaya	Blossom-end rot	*Alternaria* spp.
		Fruit, stem, root rot	*Ascochyta caricae* Pat., *Colletotrichum* spp., *Diplodia* spp., *Fusarium* spp., *Phymatotrichum omnivorum* (Shear) Dug, *Pythium aphanidermatum* (Edson) Fitzp.
		Powdery mildew	*Erysiphe cichoracearum* DC., *Oidium caricae* Noack., *Asterina caricarum* Rehm.
		Damping off	*Pythium debaryanum* Hesse., *Rhizoctonia solani* Kuehn.
Carthamus tinctorius L.	Safflower	Leaf spot	*Alternaria* spp., *Septoria carthami* Murashkinsky
		Stem rot	*Sclerotinia sclerotiorum* (Lib.) d By.
Carum caroi L.	Caraway	Stem rot	*Sclerotinia sclerotiorum* (Lib.) d By.

Scientific name	Common name	Major diseases	
		Name	Pathogen
Cassia senns L. *C. angustifolia* Vahl.	Alexandria senna Tinnevelly senna	Powdery mildew Blight	*Microsphaera alni* DC. ex Wint., *Asterina elaeocarpi* Svd. *Rhizoctonia solani* Kuehn.
Castanea satuva Mill	Chestnut	Root rot	*Phytophthora cambivora* (Petri) Buisman, *P. cinnamomi* Rands, *P. nicotianae* Breda de Haan, *Pythium* spp., *Armillria mellea* (Vahl.Fr.) P. Kumm., *Phanerochaete* spp., *Phymatotrichopsis omnivora* (Duggar) Hennebert
		Canker	*Amphiporhe castanea* (Tul.) Barr., *Bonyosphaeria* spp., *Ceratocystis microspora* (arx.) R. W. Davidson
		Leaf spot	*Cylindrosporium castaneae* (Lev.) Krenner, *Discella ochroleuca* (Berk. & M.A.Curtis) Arix, *Monochaetia concentrica* (Berk. & Broome) Sacc.
Catharanthus roseus (L.) Don.	Periwinkle	See *Vinca minor*	
Caulophyllum thalietroides (L) Michx. *C. giganteum* (F.) Loconte & W. H. Blackwell	Blue cohosh	Leaf, stem spot	*Botrytis* spp., *Calloria caulophylli* (Ell.& Ev.) Rehm, *Cercospora caulophylli* Pk., *Vermicularia hysteriiformis* Pk.
Cedrus libani subsp. *atlantica*	Cedarwood	Root rot	*Armillaria meliea* Vahl ex Fr., *Clitocybe tabescens* (Scop.ex Fr.) Bres., *Fomes pini* (Brot.ex Fr.) Karst., *Phymatotrichum omnivorum* (Shear) Dug.

Scientific name	Common name	Major diseases	
		Name	Pathogen
Centaurea calcitrapa L.	Star thistle	Stem, root rot	*Phymatotrichum omnivorum* (Shear) Dug., *Phytophthora halstedii* (Leb. & Cohn) Schroet., *Rhizoctonia solani* Kuehn., *Sclerotium rolfsii* Sacc.
		Powdery mildew	*Erysiphe cichoracearum* DC.
Centella asiatica (L.) Urb.	Gotu Kola	Leaf spot	*Septoria asinticae* Speg.
Chamaelirium luteum (L.) A. Gray	False unicorn root, fairywand	Not available	
Chamaemelum nobile (L.) All.	Roman chamomile	Damping-off	*Rhizoctonia solani* Kuehn.
Chamaenerion angustifolium (L) Scop.	Firewood	See *Epilobium angustifolium*	
Chenopodium ambrosioides L.	Wormseed	Leaf spot	*Cercospora anthelmintica* Atk., *Hendersonia bliti* Clements, *Phyllosticta ambrosjoides* Thuem., *Stagonospora atriplicis* (West) Lind.
		Downy mildew	*Peronospora effusa* (Grev. ex Desm.) Ces.
C. album L.	Lamb's quarter	Root rot	*Aphanomyces cochlioides* Drechs., *Phymatotrichum omnivorum* (Shear) Dug., *Rhizoctonia solani* Kuehn., *Sclerotinia sclerotiorum* (Lib.) d By.
		Rust	*Albugo bliti* (Biv. Bern.) Kuntze., *Uromyces peekianus* Farl.

Scientific name	Common name	Major diseases	
		Name	Pathogen
Chrysanthemum cinerarifolium (Trevir) Vis. *C. parthenium* (L) Berhn.	Pyrethrum Feverfew	Gray mold blight Leaf spot	*Botrytis cinerea* Pers.ex Fr. *Cercospora chrysanthemi* Heald & Wolf, *Cylindrosporium chrysanthemi* Ell. & Dearn
C. vulgare L.	Tansy	See *Tanacetum vulgare*	
Cichorium intybus L.	Chicory	Root rot Powdery mildew	*Phoma* spp., *Phymatotricum cichorii* (D.B.Swing) Stapp., *Rhizoctonia solani* Kuehn., *Sclerotinia sclerotiorum* (Lib.) d By. *Erysiphe cichoracearum* d By.
Cimicifuga racemosa (L.) Nutt.	Black cohosh	Leaf spot	*Ascochtya actaeae* (Bres.) J. J. Davis, *Ectostroma afflatum* (Schw.) Fr.
Cinnamomum verum J. Presl. *C. camphora* (L.) Nees & Eberm.	Cinnamon Camphor	Canker Powdery mildew Rust Leaf blight	*Diplodia camphorae* Tassi, *D. natalensis* P. Evans, *D. tubericola* (Ell. & Ev.) Taub., *Gloeosporium camphorae* Sacc., *G. ochraceum* Patterson *Erysiphe cichoracearum* DC. *Puccinia menthae* Pers. *Glomerella cingulata*

Scientific name	Common name	Major diseases	
		Name	Pathogen
Citrus limon (L.) Burm. *C. bergamia* Risso & Poit *C. reticulata* Blanco *C. aurantifolia* Swingle *C. aurantium* L. *C. paradisi* Macfad	Lemon Bergamot orange Mandarin Lime Bitter orange Grapefruit	Leaf and fruit spot Gummosis Diplodia rot Brown rot Collar rot Black fruit rot Anthracnose Green mold Root rot Damping-off Cottony fruit rot	*Diaporthe citri* Wolf *Diplodia natalensis* P. Evans *Phytophthora citrophthora* (R. E. Sm. & E. H. Sm.) Leonian *P. parasitica* Dast. *Alternaria citri* Ell. & Pierce *Diplodia natalensis* P. Evans *Septoria lla cingulata* (Ston.) Spauld. & Schrenk. *Penicillium* spp. *Phymatatrichum omnivorum* (Shear) Dug. *Rhizoctonia solani* Kuehn. *Sclerotinia sclerotiorum* (Lib.) D By.
Cnicus benedictus L.	Blessed thistle	Southern blight	*Sclerotium rolfsii* Sacc.
Cochlearia amorucia L.	Horseradish	See *Armoracia rusticana*	
Codonopsis pilosula (Franch.) Nannfeldt. *C. tangshen* Oliver	Codonopsis, Dang Shen, Bell flower	Powdery mildew Gray mold blight Leaf spot Root rot	*Erysiphe cichoracearum* DC. *Botrytis cinerea* Per. ex Fr. *Cercosporma minuta* (J. J. Davis) Chupp, *Ascochyta bohemica* Kab. & Bubak, *Phyllosticta allarifollae* Allesch., *Septoria campanulae* (Lev.) Sacc. *Rhizoctonia solani* Kuehn
Coffea arabica L.	Coffee	Anthracnose Twig blight Leaf spot Damping-off	*Colletatrichum coffeanum* Noack. *Fusarium lateritium* Nees. *Glomerella cingulata* (Ston.) Spauld. & Schrenk. *Rhizoctonia solani* Kuehn.

Scientific name	Common name	Major diseases	
		Name	Pathogen
Colchicum autumnale L.	Crosus, Meadow saffron	Leaf spot Leaf smut	*Botrytis elliptiva* (Berk.) Cke. *Urocystis colchici* (Schlecht.) Rab.
Collinsonia canadensis L.	Pilewort	Stem spot	*Dendryphiella interseminata* (Berk. & Rav.) Bub. & Banoj., *Leptosphaeria collinsoniae* Dearn & House, *Spondylocladium tenellium* Pk.
Commiphora molmol Engl. ex Tschirch *C. myrrha* (Nees) Engl.	Myrrh	Not available	
Convallaria majalis L.	Lily of the valley	Rhizome rot Leaf spot	*Botrytis paeoniae* Oud. *Gloeosporium convallariae* Allesch., *Kabatiella microsticta* Bub.
Convolvulus arvensis L. *C. sepium* L.	Bindweed	Rust Black mildew	*Coleosporium ipomoeae* (Schw.) Burr., *Puccinia convolvuli* (Pers.) Cast. *Parodiella paraguayensis* Speg.
Coriandrum sativum L.	Coriander, Cilantro	Stem gall Powdery mildew Wilt Stem rot Blight Anthracnose Root knot nematode	*Protomyces macrosporus* *Erysiphe polygoni* *Fusarium oxysporum* f. sp. *corianderii* *Sclerotinia sclerotium, Rhizoctonia solani* *Colletotrichum gloeosporoides* *Gloeosporium* spp. *Meloidogyne* spp.

Scientific name	Common name	Major diseases	
		Name	Pathogen
Cornus canadensis L.	Bunchberry	Leaf spot	*Ceratobasidium anceps* (Bres. & Syd.) Jacks., *Discohainesia oenotherae* (Cke. & Ell.) Naunf., *Septoria canadensis* Pk., *Ramularia* spp.
C. florida L.	Dogwood	Root rot	*Armillaria mellea* Vahl ex Fr., *Corticium galactinum* (Fr.) Burt., *Phymatotrichum omnivorum* (Shear) Dug.
		Wood rot	*Daedalea confragosa* Bolt. ex Fr., *Daldinia vernicosa* (Schw.) Ces. & De N., *Hypoxylon rubiginosum* Pers. ex Fr., *Polyporus* spp.
		Powdery mildew	*Phyllactivia corylea* Pers. ex Karst.
Corylus cornuta Marsh. *C. rostrata* Marsh. *C. avellana* L.	Hazelnut	Canker	*Apioporthe anomala* (Pk.) Hoehn.
		Leaf spot	*Cylindrosporium vermiformis* J. J. Davis., *Gloeosporium coryli* (Desm.) Sace., *Septogloeum profusum* (Ell. & Ev.) Sacc.
		Root rot	*Armillaria mellea* Vahl ex Fr., *Phymatotrichum omnivorum* (Shear) Dug.
Crataegus monogyna Jacq. *C. pinnatifida* Bge.	English hawthorn Chinese hawthorn	Fruit rot	*Botrytis cinerea* Pers. ex Fr.
		Leaf spot	*Cercospora apiifoliae* Tharp., *Cylindrosporium brevispina* Dearn., *Septoria crataegi* Kickx.
		Leaf blight	*Fabraea maculata* Atk., *Sclerotium rolfsii* Sacc.
		Rust	*Gymnosporangium* spp.
Crocus sativus L.	Saffron	Corm rot	*Fusarium oxysporum* Schlecht.
		Scab	*Pseudomonas marginata* (McCull.) Stapp.

Scientific name	Common name	Major diseases	
		Name	Pathogen
Cryptotaenia japonica Hassk.	Japanese parsley, Mitsuba	See *Petroselinum crispum*	
Cucurbita pepo L.	Pumpkin	Leaf blight	*Alternaria cucumerina* (Ell. & Ev.) J. A. Elliott., *Sclerotium rolfsii* Sacc.
		Gray mold rot	*Botrytis cinerea* Pers. ex Fr.
		Soft rot	*Erwinia aroideae* (Town.) Holland
		Powdery mildew	*Erysiphe cichoracearum* DC.
		Damping-off	*Pythium ultimum* Trow.
Cuminum cyminum L.	Cumin	Leaf spot	*Alternaria* spp.
		Alternaria blight	*Alternaria bulnsii*
		Powdery mildew	*Erysiphe polygoni* DC.
		Root rot	*Fusarium* spp., *Rhizoctonia* spp.
		Bacteria wilt	*Bacterium cumini* E. F. Sm
		Wilt	*Fusarium oxysporum* f. sp. *cuminii*
Curcuma domestica L. *C. xanthorrhiza L.* *C. longa* L.	Curcuma, Jiang Huang	Fruit rot	*Colletotrichum capsici* (Duf.) R. J. Bulter & Blsby
		Wood rot	*Phyllosticta zingiheri* F. Stevens & Ryan
		Root rot	*Pythium batleri* Drechs, *P. graminicola* Subramanian
Cuscuta epithymu Murr.	Dodder	Fruit rot	*Colletotrichum destructivum* O'Gara
		Stem canker	*Phoma* spp.
		Stem gall	*Phomopsis cuscutae* H. C. Greene

Scientific name	Common name	Major diseases	
		Name	Pathogen
Cymbopogon cĩtratus (DC. ex Nees) Stapf. *C. nardus* L. *C. winterianus* Jowitt *C. martinii* (Roxb) Wats	Lemon grass Citronella Palmarosa, gingergrass	Eye and leaf spot Web blight Tangle-top Grassy shoot Leaf blight Smut	*Helminthosporium sacchari* (B. de Haan) Butler, *Carvalaria verruciformis, Cochliobolus* spp. *Himantia stellifera* Johnston *Myriogenospora paspali* Atk. *Balansia sclerotica* *Colletotrichum graminicola*, *Rhizoctonia* spp. *Tolysporium christensenii*
Cynara scolymus L. *C. cardunculus* L.	Artichoke Wild artichoke	Gray-mold blight Powdery mildew Root rot	*Botrytis cinerea* Pers. ex Fr. *Erysiphe cichoracearum* DC. *Phymatotrichum omnivorum* (Shear) Dug., *Phytophthora megasperma* Drechs
Cypripedium calceolus L.	Ladyslipper	Leaf spot Rust	*Aecidium graebnerianum* P. Henn., *Cercospora cypripedii* Ell. & Dearn., *Fusicladium aplectri* Rll. & Ev., *Nycosphaerella cypripedii* (Pk.) Lindau *Puccinia cypripedii* Arth. & Holw., *Pucciniastrum goodyerae* (Tranz.) Arth.
Cytisus scoparius (L.) Link.	Broom	Spot on branches	*Gloeosparium garganicum* Sacc.& D. Sacc., *Nectria coccinea* Pers. ex. Fr.

Scientific name	Common name	Major diseases	
		Name	Pathogen
Daucus carota L.	Carrot	Leaf blight	*Alternaria dauci* (Kuehn.) Groves & Skolko.
		Gray-mold rot	*Botrytis cinerea* Pers. ex. Fr.
		Root rot	*Phymatotrrichum omnivorum* (Shear) Dug., *Phytophthora megasperma* Drechs., *Rhizoctonia crocorum* (Pers.) DC. ex. Fr.
		Damping-off	*Phythium* spp.
		Soft rot	*Sclerotinia sclerotiorum* (Lib.) d By.
		Mosaic	Western celery mosaic viurs
Digitalis purpurea L.	Foxglove	Anthracbise	*Colletotrichum juscum* Laub.
		Root rot	*Fusarium* spp., *Rhizoctonia solani* Kuehn.
		Wilt	*Fusarium* spp., *Verticillium albo-atrum* Reinke & Berth.
Dioscorea oppositae Thunb. *D. villosa* L.	Chinese yam (Shen yao) Wild or Mexican yam	Leaf spot	*Cereospora carbonacea* Miles., *Phyllachora ulei* Wint., *Phyllosticta dioscoreae* Cke.
		Root rot	*Diplodia theobromae* (Pat.) Nowell, *Phymatotrichum omnivorum* (Shear) Dug
		Rust	*Sphenospora pallida* (Wint.) Diet., *Uredo dioscoreicola* Kern, Cif. & Thurston

Scientific name	Common name	Major diseases	
		Name	Pathogen
Echinacea angustifolia DC. *E. pallida* (Nutt.) Nutt. *E. pururea* (L.) Moench.	Narrowleaf echinacea Pale-flower echinacea Purple coneflower	Leaf spot Root rot Yellows diseases Stem rot Botrytis blight Wilt Virus diseases	*Cercospora rudbeckii* Pk., *Septoria lepachydis* Ell. & Ev., *Fusarium* spp., *Alternaria* spp. *Phymatotrichum omnivorum* (Shear) Dug., *Penicillium* spp., *Pythium* spp., Bacteria Phytoplasmas *Sclerotinia sclerotiorum* (Lib.) de Bary *Botrytis cinerea* Pers. ex. Fr. *Fusarium oxysporum* Schlecht. Cucumber mosaic, broad bean wilt, mosaic
Echinopanax horridus (Sm) Decne. & Planch. ex. H. A. T. Harms	Devil's club	See *Oplopanax horridus*	
Elaeagnus angustifolia L.	Russian olive	Crown gall Powdery mildew Canker Root rot	*Agrobacterium tumefaciens* (E. F. Sm. & Town.) Conn. *Phyllactinia corylea* Pers. ex. Karst. *Phytophthora cactorum* (Leb. & Cohn.) Schroet. *Phymatotrichum omnivorum* (Shear) Dug.
Eleutherococcus senticosus (Rupr. ex Maxim) Maxim.	Siberian ginseng	Leaf blight, spot Root rot	*Alternria panax* Whet., *Alternaria* spp. *Phymatotrichopsis omnivorum* (Shear) Dug.
Elymus repens L.	Couch grass	See *Agropyron repens*	

Scientific name	Common name	Major diseases	
		Name	Pathogen
Ephedra distachya L. *E. nevadensis* Wats. *E. sinica* Stapf.	Ephedra Mormon tea Ma Huang	Rust gall Root rot	*Peridermium ephedrae* Cke. *Phymatotrichum omnivorum* (Shear) Dug.
Epilobium angustifolium L. *Epilobium parviflorum* Schreb.	Fireweed Small-flowered willow herb	Gray-mold blight Leaf spot Powdery mildew	*Botrytis cinerea* Pers. ex. Fr. *Cercospora montana* (Speg.) Sacc., *Discosia bubakii* Kab., *Ramularia cercosporoides* Ell. & Ev. *Sphaeratheca humuli* (DC.) Burr.
Equisetum arvense L.	Horsetail	Root and stem rot Blight	*Rhizoctonia solani* Kuehn. *Gloeosparium equiseti* Ell. & Ev.
Erythroxylum coca Lam.	Coca	Rust	*Bubakia erythroxylonis* (Graz.) Cumm.
Eschscholtzia california Cham.	California poppy	Collar rot Gray-mold blight Powdery mildew	*Alternaria* spp. *Botrytis cinerea* Pers. ex. Fr. *Erysiphe polygoni* DC.
Eucalyptus citriodora Hool. *E. globulus* Labill.	Gum tree Eucalyptus	Crown gall Root rot Wood rot Blight Leaf spot Damping off	*Agrobacterium tumefaciens* (E. F. Sm. & Towns.) Conn. *Phymatotrichum omnivorum* (Shear) Dug. *Polyparus gilvus* (Schw.) Fr. *Cylindrocladium scoparium* *Physalospora* spp., *Mycospherella* spp. *Fusarium solani, Rhizoctonia solani*
Eugenia caryophyllata L.	Clove	See *Syzygium aromaticum*	

Scientific name	Common name	Major diseases	
		Name	Pathogen
Eupatorium perfoliatum L.	Boneset	Gray-mold blight Rust Powdery mildew Root rot	*Botrytis cinerea* Pers. ex Fr. *Cionohria praelonga* (Wint.) Arth., *Coleosporium eupatorii* Arth. *Erysiphe cicharacearum* DC. *Phymatotrichum omnivorum* (Shear) Dug., *Rhizoctonia solani* Kuehn.
Euphrasia officinalis L.	Eyebright	Rust	*Coleosporium euphrasiae* Arth, *C. pinicola* Arth.
Fagopyrum tataricum L.	Buckwheat	Seed mold Damping-off	*Alternaria tenuis* Auct. *Rhizoctonia solani* Kuehn.
Fagus grandifolia Ehrb. *F. sylvatica* L.	Beech European beech	Root rot Heart rot Powdery mildew Trunk canker	*Armillaria mellea* Vahl.ex Fr. *Fomes pinicola* (Sw. ex Fr.) Cke., *F. igniarius* (L. ex Fr.) Kickx. *Microsphaera alni* DC.ex Wint., *Phyllactinia corylea* Pers. ex Karst. *Strumella coryneoidea* Sacc. & Wint., *Ustulina deusta* (Hoffm. ex Fr.) Petr.
Filipendula ulmaria (L.) Maxim.	Meadow sweet	Powdery mildew Rust	*Sphaerotheca humuli* (DC.) Burr. *Triphragmium ulmariae* (Hedw.f.) Lk.

Scientific name	Common name	Major diseases	
		Name	Pathogen
Foeniculum vulgare Mill.	Fennel	Gray-mold rot	*Botrytis cinerea* Pers. ex Fr.
		Bacterial soft rot	*Erwinia carotovora* (L. R. Jones) Holland
		Root rot	*Phymatotrichum omnivorum* (Shear) Dog.
		Withering stem	*Phomopsis foenienli*
		Necrotic leaf spot	*Cosporidium punctum*
		Bulb rot	*Sclerotina sclerotiorum*
		Powdery mildew	*Leveillula taurica*
		Leaf spot, blight	*Cercospora foeniculi, Alternaria* spp., *Ramularia* spp.
Forsythia suspensa (Thunb.)Vahl.	Forsythia, Lian qiao	Crown gall	*Agrobacterium tumefaciens* (E. F. Sm. & Towns.) Conn.
		Anthracnose	*Gleosporium* spp.
		Stem gall	*Phomopsis* spp.
		Root rot	*Phymatotrichum omnivorum* (Shear) Dug.
Fragaria vesca L.	Strawberry	Blight	*Botrytis cinerea* Pers. ex Fr., *Dendrophama obscuruns* (Ell. & Ev.) H. W. Anderson
		Root rot	*Cylindrocladium scoparium* Mort., *Phytophthora fragariae* Hickman, *Ramularia* spp., *Rhizoctonia solani* Kuehn.
		Leather rot (fruit)	*Phytophthora cactorum* (Leb. & Cohn) Schroet.
		Soft rot	*Rhizopus* spp.
		Powdery mildew	*Sphaerotheca humuli* (DC.) Hurr.

Scientific name	Common name	Major diseases	
		Name	Pathogen
Fraxinus americana L. *F. excelsior* L.	White ash	Leaf spot	*Cylindrosporium fraxini* (Ell. & Kell.) Ell. & Ex., *Gloeosporium aridum* Ell. & Ev., *Mycosphaerella* spp., *Piggotia fraxini* Berk. & Curt.
		Twig canker	*Cytospora annularis* Ell. & Ev., *Dathiarella fraxinicola* Ell. & Ev.
		Canker	*Diplodia infuscans* Ell. & Ev., *Sphaeropsis* spp.
		Powdery mildew	*Phyllactinia corylea* Pers. ex Karst.
		Heart rot	*Polyporus hispidus* Bull.ex Fr.
Fucus vesiculosus L.	Kelp	Gall	*Halenchus dumnonicus*
Galium aparine L.	Cleavers	Leaf spot	*Cercospora galii* Ell. & Holw., *Melasmia galii* Ell. & Ev., *Pseudopeziza repanda* (Fr.) Karst., *Septoria eruciatae* Rob. & Desm.
		Powdery mildew	*Erysiphe cicharacearum* DC.
		Downy mildew	*Peronospora calotheca* d By.
		Root rot	*Phymatotrichum omnivorum* (Shear) Dug.
		Rust	*Pucciniastrum galii* (Lk.) E. Fisch., *Uromyces galii-californici* Linder
Gaultheria procumbens L.	Wintergreen, teaberry	Black mildew	*Meliola* spp.
		Powdery mildew	*Microsphaera alni* DC. ex Wint.
		Leaf spot	*Discohainesia oenotherae* (Cke. & Ell.) Nannf., *Lachnum gaultheriae* (Ell. & Ev.) Zeller, *Postalopezia brunnea-pruinosa* (Zeller) Seaver.

Scientific name	Common name	Major diseases	
		Name	Pathogen
Geranium macrorrhizum L.	Geranium, American cranesbill	Leaf spot Stem rot Downy mildew Powdery mildew	*Botrytis cinerea* Pers. ex Fr., *Cylindrosporium geranii* Ell. & Ev., *Pestalaziella subsessilis* Sacc. & Ell., *Ramularia geranii* (West.) Fckl. *Botrytis cinerea* Pers. ex Fr. *Plasmopara geranii* Cke. & Harkn. *Sphaerotheca humuli* (DC.) Burr.
Ginkgo biloba L.	Ginkgo	Sapwood or wound rot Leaf spot, anthraenose	*Fomes connatus* (Weinum. ex Fr.) Gill, *Oxyparus papulinus* (Schum. ex Fr.) Donk. *Glomerella cingulata* (Ston.) Spauld. & Schrenk.
Glechoma hederacea L.	Ground ivy	Leaf spot	*Glomerella cingulata* (Ston.) Spauld. & Schrenk., *Macrophoma* spp.
Glycine max (L.) Merrill.	Soybean	Leaf spot Stem rot Pod and stem blight Powdery mildew Seed rot	*Alternaria* spp., *Cercospora canescens* Ell. & G. Martin., *Helminthosporium vignae* L. S. Olive, *Mycosphaerella cruenta* (Sacc.) Latham., *Phyllosticta glycines* Tehon & Daniels. *Cephalasporium gregatum* Allington & Chamberlain, *Phythium* spp., *Selerotinia sclerotiorum* (Lib.) d By. *Diaporthe sajae* (Lehman) Wehm. *Erysiphe polygoni* DC. *Penicillium* spp.
Glycyrrhiza glabra L. *G. uralensis* Fisch ex DC	Licorice Chinese licorice	Powdery mildew Leaf spot	*Erysiphe polygoni* DC., *Microsphaera diffusa* Cke. & Pk. *Cylindrosporium glycyrrhizae* Harkn., *Septoria glycyrrhizae* Ell. & Kell.

Scientific name	Common name	Major diseases	
		Name	Pathogen
Gossypium hirsutum L.	Cotton	Leaf blight	*Alternaria* spp., *Pellicularia filamentosa* (Pat.) Rogers
		Boll rot	*Aspergillus* spp., *Botryosphaeria ribis* Gross. & Dug., *Fusarium moniliforme* Sheldon, *Glomerelia gossypii* Edg., *Nematospora coryli* Pegl., *Physalospora rhodina* (Berk. & Curt.) Cke., *Rhizopus stolonifer* (Ehr. ex Fr.) Lind.
		Root and stem rot	*Schizophyllum commune* Fr., *Thielaviopsis basicola* (Berk. & Br.) Ferr.
		Stem and leaf blight	*Macrophomina phaseoli* (Maubl.) Ashby., *Pellicularia filamentosa* (Pat.) Rugers.
		Wilt	*Verticillium albo-atrum* Reinke & Berth.
Grindelia robusta Nutt. *G. squarrosa* (Pursh) Donal.	Gumweed	Rust	*Coleosporium solidaginis* (Schw.) Thuem., *Puccinia extrnsicola* Plowr., *Uromyces junci* (Desm.) L. Tul.
		Powdery mildew	*Erysiphe cichoracearum* DC.
Hamamelis virginiana L.	Witch hazel	Crown gall	*Agrobacterium tumejaciens* (E.F.Sm. & Towns.) Conn.
		Leaf spot	*Discosia artocreas* Tode ex Fr., *Gonatobotryum maculicola* (Wint.) Sacc., *Phyllosticta hamamelidis* (Cke) G. Martin., *Ramularia hamamelidis* Pk.
		Powdery mildew	*Phyllactinia corylea* Pers. ex Karst., *Podosphaera biuncinata* Cke.
Harpagophytum procumbens DC. ex Meisn	Devil's claw	Stem and root rot	*Sclerotinia sclerortiorum* (Lib.) de Bary, *Phymatotrichopsis omnivora* (Duggar) Hennebert
		Leaf spot	*Cercospora beticola* Sacc.

Scientific name	Common name	Major diseases	
		Name	Pathogen
Hedeoma pulegiodes (L) Pers.	Pennyroyal	Downy mildew Rust	*Peronospora hedeomatis* Kell. & Swing. *Puccinia menthae* Pers.
Hedera helix L.	English ivy	Anthracnose Leaf spot	*Amerosporium trichellum* (Fr.) Lind., *Colletotrichum trichellum* (Fr.) Duke *Glomerella cingulata* (Ston.) Spauld. & Schrenk., *Macrophoma* spp., *Ramularia hedericola* Heald & Wolf., *Sphaeropsis hedericola* (Speg.) Sacc.
Helianthus annuus L.	Sunflower	Rust Powdery mildew Root and stem rot	*Coleosporium helianthi* (Schw.) Arth., *Uromyces junci* (Desm.) L. Tul. *Erysiphe cichoracearum* DC. *Phymatotrichum omnivorum* (Shear) Dug., *Selerotinia sclerotiorum* d By.
Helichrysum angustifolium (Lam) DC	Curry plant	Wilt Stem rot	*Verticillium albo-atrum* Reinke & Berth. *Fusarium* spp.
Helonias dioica L.	False unicorn root	See *Chamaelirium luteum*	
Heracleum maximum Bartr. *H. lanatum* Michx. *H. sphondylium* L.	Cow parsley Cow parsnip Cow parsnip	Leaf spot Root rot	*Cylindrosporium heraclei* (Fr.) Ell. & Ev., *Fusicladium angelicae* (Fr.) Ell. & Ev., *Phyllosticta heraclei* Ell. & Dearn., *Ramularia heraclei* (Oud.) Sacc. *Phymatotrichum omnivorum* (Shear) Dug

Scientific name	Common name	Major diseases	
		Name	Pathogen
Hibiscus sabdariffa L. *H. rosa-sinensis* L.	Hibiscus	Leaf spot	*Cercospora hibisci* Tracy & Earle, *C. Malayensis* F. L. Stevens & Solheim, *Phyllosticta hibiscina* Ell. & Ev.
		Root and stem rot	*Fusarium solani* (Mart.) Appel. & Wr., *Phytophthora parasitica* Dast., *Rhizoctonia solani* Kuehn, *Thielaviapsis basicola* (Berk. & Br.) Ferr.
Hierochloe odorata (L.) Beauv.	Sweet grass	Crown and stem rust	*Puccinia coronata* Cds., *P. graminis* Pers.
		Root rot	*Rhizoctonia solani* Kuehn.
Hippophae rhamnoides L	Sea buckthorn	Verticillium wilt	*Verticillium albo-atrum* Reinke & Berth.
		Fusarium wilt	*Fusarium* spp.
		Damping off	*Fusarium*, *Botritis*, *Rizoctonia*, *Phoma* spp.
		Brown rot	*Monillia altaica* (Sacc.) Reade.
		Bacteria leaf spot, gum	*Pseudomonas* spp.
		Leaf spot	*Septoria elaeagni*
Humulus lupulus L.	Hops	Leaf spot	*Cercospora* spp., *Cylindrosporium humuli* Ell. & Ev., *Phyllosticta decidua* Ell. & Kell., *P. humuli* Sacc. & Speg.
		Powdery mildew	*Sphaerotheca humuli-americani* Fairm., *Erysiphe cichoracearum* DC.
		Downy mildew	*Pseudoperonospora humuli* (Miy. & Tak.) G. W. Wils.
		Wilt	*Verticillium albo-atrum* Reinke & Berth.

Scientific name	Common name	Major diseases	
		Name	Pathogen
Hydrangea arborescens L.	Hydrangea	Gray-mold blight	*Botrytis cineriea* Pers. ex Fr.
		Powdery mildew	*Erysiphe polygoui* DC.
		Root and stem rot	*Phymatotrichum omnivorum* (Shear) Dug., *Rhizoctonia solani* Kuehn.
		Rust	*Pucciniastrum hydrznpene* (Perk. & Curt.) Arth.
Hydrastis canadensis L.	Goldenseal	Leaf blight	*Alternaria* spp., *Botrytis hydrastis* Whet.
		Wilt	*Fusarium* spp., *Heterodera marioni* (Cornu) Goodey
		Root rot	*Phymatotrichum omnivorum* (Shear) Dug., *Rhizoctonia solani* Kuehn.
Hypericum perforatum L.	St. John's wort	Leaf spot	*Cercospora hyperici* Tehon & Daniels., *Cladosporium gloeosporioides* Atk.
		Powdery mildew	*Erysiphe cichoracearum* DC.
		Rust	*Mesopsora hypericorum* (Wint.) Diet., *Uromyces hyperici* (Spreng.) Curt.
Hyssopus officinalis L.	Hyssop	Root knot nematodes	*Meloidogyne* spp.
Ilex aquifolium L.	English holly	Canker	*Boydia insculpta* (Oud.) Grove
		Leaf spot	*Phyllosticta* spp., *Physalospora ilicis* (Schleicher ex Fr.) Sacc.
Inula helenium L.	Elecampane	Powdery mildew	*Erysiphe cichoracearum* DC.
Iris versicolor L.	Blue flag	Leaf spot	*Alternaria iridicola* (Ell. & Ev.) J. A. Elliott, *Didymellina macrospora* Kleb., *Phyllasticia iridis* Ell. & G. Martin.
		Rust	*Puccinia iridis* (DC.) Wallr., *P. sessilis* Schneid.

Scientific name	Common name	Major diseases	
		Name	Pathogen
Jasminum grandiflorum L. *J. auriculatum* Vahl.	Royal jasmine	Crown gall Blossom blight Root rot	*Agrobacterium tumefaciens* (E. F. Sm. & Towns.) Conn. *Choanephora infundibulifera* (Curr.) Sacc. *Clitocybe tabescens* (Scop. ex Fr.) Bres., *Corticium galactinum* (Fr.) Burt.
Juglans regia L. *J. nigra* L.	Walnut Black walnut	Root rot Leaf blight Wood rot Powdery mildew	*Armillaria mellea* Vahl. ex Fr. *Cylidrosporium juglandis* Wolf. *Daedalea confragosa* Bolt. ex Fr., *Fomes conchatus* (Pers. ex Fr.) Gill., *Schizophyllum commune* Fr. *Erysiphe polygoni* DC., *Phyllactinia corylea* Pers. ex Karst., *Microsphaera alni* DC. ex Wint.
Juniperus communis L.	Juniper	Crown gall Needle blight Wood and root rot Rust gall Damping-off	*Agrobacterium tumefacieus* (E. F. Sm. & Town.) Conn. *Chlorocyha juniperina* (Ell.) Seaver, *Pestalotia funerea* Desm. *Coniophora corrugis* Burt., *Daedalea juniperina* Murr., *Lenzitea saepiaria* Wulf ex. Fr., *Phymatotrichum omnivorum* (Shear) Dug. *Gymnosporangium* spp. *Rhizoctonia solani* Kuehn.
Laminaria digitata (Hudss.) Lamk. *L. saccharina* (L.) Lamk.	Kelp	See *Fucus vesiculosus*	

Scientific name	Common name	Major diseases	
		Name	Pathogen
Lamium album L.	Nettle	Powdery mildew Downy mildew Root rot	*Erysiphe cicharacearum* DC. *Peronospora lamii* A. Braun *Rhizoctonia solani* Kuehn
Larrea tridentata (Sesse. & Moc. ex DC.) Coville.	Chapatral	Leaf blight	*Omphalia* spp.
Laurus nobilis L.	Bay laurel	Thread blight	*Corticinu koleroga* Cke.
Lavandula angustifolia Mill. *L. spica* L.	Lavender	Root rot Leaf spot Scab Die back	*Armillaria mellea* Vahl. ex Fr. *Septoria lavandulae* Desm. *Phoma lavandulae* *Coniothyrium lavandulae*
Ledum latifolium Jacq.	Labrador tea	Rust Leaf spot Leaf gall	*Chrysomyxa ledi* (Alb.& Schw.) d By. *Ascochyta ledi* Rostr., *Cryptostietis arbuti* (Bonar) Zeller *Exohasidium vaccinii* Wor., *E. ledi* Karst., *Synchytrium vaccinii* Thomas
Lemna minor L.	Duckweed	Parasitic on epidermal	*Reessia amoeboides* C. Fisch
Leonurus cardiaca L.	Motherwort	Leaf spot	*Ascochyta leonuri* Ell. & Dearn., *Dinemasporium hispidulum* (Schrad.ex Fr.) Curt., *Phyllosticta decidua* Ell. & Kell., *Septoria lamii* Pass.

Scientific name	Common name	Major diseases	
		Name	Pathogen
Levisticum officinale W. Koch.	Lovage	Chlorotic mottle Leaf spot	Celery mosaic virus *Ramularia schroeteri* Ell. & Ev.
Ligustrum lucidium W. T. Aiton. *L. vulgare* L.	Ligustrum fruit Privet	Crown gall Root rot Cander, dieback	*Agrobacterium tumefaciens* (E. F. Sm. & Towns.) Conn. *Clitocybe tabescens* (Scop. ex Fr.) Bres., *Phymatotrichum omnivorum* (Shear) Dug. *Glamerella cingulata* (Ston.) Spauld. & Schrenk.
Linum usitatissimum L.	Flax	Leaf spot, blight Root rot Rust Stem rot	*Alternaria* spp., *Helminthosporium sativum* Pam, King, & Bakke. *Curvularia geniculata* (Tracy & Earle) Boed., *Phythium* spp., *Rhizoctonia solani* Kuehn, *Thielaviopsis basicola* (Berk. & Br.) Ferr. *Melampsora lini* (Pers.) Lev. *Sclerotium delphinii* Welch, *Sclerotinia sclerotiorum* (Lib.) d By
Liquidambar styraciflua L. *L. orientalis* L.	Gum tree, storax Oriental sweet gum	Leaf spot Pink wood stain Root rot	*Cerocospora liquidambaris* Cke. & Ell., *Discosia artoereas* Tode ex Fr. *Fusarium moniliforme* Sheldon, *Graphium rigidum* (Pers.) Sacc. *Phymatatrichum omnivorum* (Shear) Dug., *Polyporus sector* Ehr. ex Fr.
Lobelia inflata L. *L. siphilitica* L. *L. pulmonaria* L.	Lobelia Blue lobelia Lungwort	Leaf spot Rust	*Cereospora effusa* (Berk. & Curt.) Ell. & Ev., *C. lobeliae* Kell. & Swing, *Gloeosporium hawaiencs* Thuem. *Puccinia labeliae* Gerard

Scientific name	Common name	Major diseases	
		Name	Pathogen
Lomatium dissectum (Nutt.) Math. & Const.	Lomatium	Downy mildew Rust	*Plasmopara nivea* (Unger) Schroet *Puccinia asperior* Ell. & Ev., *P. jonessi* Pk., *P. ligustiei* Ell. & Ev.
Lonicera caerulea L. *L. caprifolium* L.	Mountain fly honeysuckle Dutch honeysuckle	Gray-mold blight Collar rot Leaf blight Powdery mildew	*Botrytis cinerea* Pers. ex Fr. *Fomes ribis* (Schum. ex Fr.) Cke. *Herpobasidium deformans* Gould *Microsphaera alni* DC. ex Wint, *Erysiphe polygoni* DC.
Lycium barbarum L. *L. chinense* Mill. *L. pallidum* L.	Lycium Chinese wolfberry Wolfberry	Leaf spot Powdery mildew Rust	*Alternaria* spp., *Cercospoe* Ell. & Halst, *Phyllosticta lycii* Ell. & Kell. *Erysiphe polygoni* DC., *Microsphaera diffusa* Cke. & Pk., *Sphaerotheca pannosa* (Wallr.) Lev. *Puccinia globosipes* Pk.
Lycopodium clavatum L.	Clubmoss	Root rot Black rot	*Phytophthora cinamomi* Rands, *Phaeosphaeria* spp. *Ceuthospora lycopodii* Lind
Lysimachia vulgaris L.	Loosestrife	Stem and leaf necrosis Leaf spot Rust	*Ceratobasidium anceps* (Bres.& Syd.) Jacks. *Cercospora lysimachiae* Ell. & Halst., *Cladosporium lysimachiae* Guba., *Rumularia lysimachiae* Thuem., *Septoria conspicua* Ell. & G. Martin *Puccinia limosae* Magn.
Lythrum salicaria L.	Purple loosestrife	Leaf spot Root rot	*Cercospora lythri* (West.) Niessl., *Pezizella oenotherae* (Cke. & Ell.) Sacc., *Septoria lythrina* Pk. *Rhizoctonia solani* Kuehn.

Scientific name	Common name	Major diseases	
		Name	Pathogen
Macropiper excelsum G. Forst	Kava-Kava	See *Piper methysticum*	
Mahonia aquifolium (Pursh) Nutt.	Oregon grape	Rust Leaf spot	*Cammiusiella sanguinea* (Pk.) Arth., *Puccinia graminis* Pers. *Heterodera marieni* (Cornu) Goodey, *Phomopsis* spp., *Phyllosticta* spp.
Marrubium vulgare L.	Horsehound	Leaf spot Leaf gall	*Cercospora marrubii* Therp. *Synchytrium marrubii* Tobler
Matricaria chamomilla L. *M. recutita* L.	Chamomile German chamomile	Powdery mildew Yellow	*Erysiphe cichoracearum* DC., *Sphaerotheca humuli* (DC.) Burr. Virus
Matteucia struthiopteris (L.) Tod.	Ostrich fern	Leaf blister Root rot	*Taphrina hiratsukae* *Sclerotium deciduum* Sacc.
Medicago sativa L.	Alfalfa	Root rot Black stem Stem canker Leaf spot	*Aphanomyces euteiches* Drechs., *Corticium praticola* Kotila, *Corynebacterium insidiosum* (McCull) H. L. Jens, *Fusarium* spp., *Phymatotrichum omnivorum* (Shear) Dug., *Rhizoctonia* spp., *Sclerotium rolfsii* Sacc. *Ascochyta imperfecta* Pk. *Phoma* spp. *Pleosphaerulina hyalospora* (Ell. & Ev.) Berl., *Pleospora herbarum* (Pers. ex Fr.) Rah.

Scientific name	Common name	Major diseases	
		Name	Pathogen
Melaleuca alternifolia (Maid. & Bet.) Cheel	Melaleuca, medicinal tea tree	Sooty mold	*Limacinia fuliginodes* Bonar., *Phycopsis* spp., *Teichospora anstrale* Fuckel.
Melilotus officinalis (L.) Lamk. *M. arvensis* L.	Sweet clover Melilot	Stem canker Anthracnose Stem rot Root rot	*Ascochyta caulicola* Laub. *Colletotrichum destructivum* O'Gara, *C. trifolii* Bain & Essary, *Kabatiella caulivora* (Kirchn.) Karak. *Sclerotinia sclerotiorum* (Lib.) d By., *S. trifoliorum* Eriks. *Phytophthora cactarum* (Leb. & Cohn) Schroet
Melissa officinalis L.	Lemon balm, balm	Gray-mold blight Leaf spot	*Botrytis cinerea* Pers. ex Fr. *Phyllosticta decidua* Ell. & Kell.
Mentha piperita L. *M. spicata* L. *M. pulegium* L.	Peppermint Spearment Pennyroyal	Stem canker Powdery mildew Leaf spot Verticillium wilt Rust Anthracnose Stolon rot	*Alternaria* spp. *Erysiphe cichoracearium* DC., *Sphaerotheca humuli* (DC.) Burr. *Cercospora menthivola* Tehon & Daniels, *Phyllosticta decidua* Ell. & Kell., *Ramularia menthicola* Sacc., *Septoria menthicola* Sacc. & Letendre. *Verticillium albo-atrum* Reinke & Berth. *Puccinia menthae* *Sphaceloma menthae* *Fusarium sloani, Rhizoctonia* spp.
Monarda odoratissima Benth. *M. punctata* T.J. Howell	Horsebalm Horsemint	Leaf spot Rust	*Phyllosticta decidua* Ell. & Ke., *Ramularia brevipes* Ell. & Ev. *Puccinia angustata* Pk., *P. menthae* Pers.

Scientific name	Common name	Major diseases	
		Name	Pathogen
Myosotis scorpiodes L.	Lily of the valley	See *Convallaria majalis*	
Myrica penxylvanica Lois.	Bayberry	Leaf spot Root rot Seedling blight	*Cercospora disporsa* Ell. & Ev., *Ramularia monilioides* Ell. & G. Martin, *Septoria myricae* Ell. & Ev. *Clitoeybe tabescens* (Scop. ex Fr.) Bull., *Phymatotrichum omniivorum* (Shear) Dug. *Rhizoctonia solani* Kuehn.
Myristica fragrans Houtt.	Nutmeg	Leaf spot	*Phyllosticta* spp.
Myrtus communis L.	Myrtle	Leat spot Stem rot	*Pestalotia decolarata* Speg. *Solerotium rolfsii* Sacc.
Narcissus pseudonarcissus L.	Daffodil	Scale necrosis Root and bulb rot Neck rot	*Aphelenchaides folasistus* (Ritz.-Bos) Steiner & Buhrer *Armillarjz meltea* Vahl ex Fr. *Botrytis* spp., *Fusarium azysporum* Schlecht., *Sclerotinia narcissicola* Gregory
Nasturtium officinale L.	Watercress	Leaf spot Rust Root rot Damping-off	*Cercospora nasturtii* Pass. *Puocinia aristidae* Tracy *Pythium debaryanum* Hesse *Rhizoctonia solani* Kuehn.
Nepeta cataria L.	Catnip, catmint	Leaf spot Root rot	*Ascochyta nepetae* J. J. Davis, *Cercospora nepetae* Tehon, *Phyllosticta decidua* Ell. & Rell, *Septoria alabamensis* Atk. *Rhizoctonia solani* Kuehn.

Scientific name	Common name	Major diseases	
		Name	Pathogen
Nymphaea alba L.	White water lily	Leaf spot	*Alternaria* spp., *Cerecospora exotica* Ell. & Ev., *Helicoceras nymphaearum* (Rand) Linder, *Ovularia nyumphaearum* Allesch., *Phyllosticta fatiscens* Pk.
		White smut	*Entyloma nymphaeae* (D. D. Cunn.) Setch.
		Leaf and stem rot	*Pythium* spp.
Ocimum basilicum L.	Basil	Powdery mildew	*Sphaerotheca humuli* (DC.) Burr.
		Leaf spot	*Corynespora cassicola*
		Leaf blight	*Alternaria* spp.
		Die back	*Collectotrichum capsici*
		Scab	*Elsinoe arxii*
		Wilt	*Fusarium* spp.
Ocotea bullata (Birch) E. May	Stinkwood	Black mildew	*Ivenina (Meliola) glabroides* F. L. Stevens, *Lembosia microspora* Chardon
		Green scurf	*Cephaleuros virescens* Kunze
Oenothera biennis L.	Evening primrose	Rust	*Aecidium anograe* Arth., *Uromyces plumbarius* Linder
		Leaf spot	*Alternaria tenuis* Auct., *Cercospora cenotherae* Ell. & Ev., *Pezizella oenotherae* (Cke. & Ell.) Sacc., *Septoria ocnotherae* West.
		Powdery mildew	*Erysiphe polygoni* DC.
		Downy mildew	*Peronospora arthuri* Farl.
		Root rot	*Phymatotrichum omnivocum* (Shear) Dug.

Scientific name	Common name	Major diseases	
		Name	Pathogen
Olea europaea L.	Olive	Root rot Leaf spot	*Armillaria mellea* Vahl., *Phymatotrichum omnivorum* (Shear) Dug. *Asterina oleina* Cke., *Cycloclonium oleaginum* Cast.
Oplopanax horridus (Sm.) Miq.	Devil's club	Gray-mold blight Leaf spot	*Botrytis cinerea* Pers. *Cercospora dacmonicola* Sprague
Opuntia fragilis (Nutt.) Haw.	Prickly pear	Cladodes rot	*Aspergilius alliaceus* Thom. & Church, *Collectotrichum dematium* (Fr.) Grove, *Gloeosporium opuntiae* Ell. & Ev., *Macrophoma opuntiicola* (Speg.) Sacc. & Syd., *Phoma* spp.
Origanum majorana L. *Origanum vulgare* L. subsp. *hirtum* (Link) Ietswaart.	Sweet marjoram Marjoram, Oregano	Root rot Leaf spot	*Pythium* spp. *Alternaria* spp., *Botrytis* spp., *Helminthosporium* spp., *Stemphylium* spp.
Paeonia lactiflora Pall. *P. officinalis* L. *P. suffruiticosa* Andr.	White peony, Bia shao European peony Tree peony, Mu Dan Pi	Root and stem rot Leaf blotch Stem canker Wilt	*Armillaria mellea* Vahl ex. Fr., *Botrytis* spp., *Fusarium* spp., *Phymatotrichum omnivorum* (Shear) Dug, *Rhizoctonia solani* Kuehn, *Thielaviopsis basicola* (Berk. & Br.) Ferr. *Cladosporium paeaniae* Pass, *C. herbarum* Lk. ex Fr. *Coniothyrium* spp. *Verticillium albo-atrum* Reinke & Berth.

Scientific name	Common name	Major diseases	
		Name	Pathogen
Panax ginseng C. A. Meyer. *P. notoginseng* (Burk.) F. H. Chen *P. quinquefolium* L.	Asian ginseng Tian Qi American ginseng	Alternaria blight Root rot Blight, stem and seed rot Damping off Wilt	*Alternaria panax* Whet. *Armillaria mellea* Vahl ex Fr., *Fusarium scirpi* Lambotte & Fautr., *Phytophthora cactorum* (Leb. & Cohn) Schroet., *Ramularia* spp., *Rhizoctonia solani* Kuehn., *Botrytis cinerea* Pers. ex Fr. *Pythium debarryanum* Hesse., *Rhizoctonia solani* Kuehn *Verticillium albo-atrum* Reinke & Berth
Papaver braceatum L. *P. rhoens* L.	Iranian poppy Poppy, corn poppy	Gray-mold blight Leaf and seed pod spot Powdery mildew Root and stem rot	*Botrytis cinerea* Pers. ex Fr. *Corcospora papaveri* Muller & Chupp., *Septoria* spp. *Erypsiphe polygoni* DC. *Rhizoctonia solani* Kuehn
Passiflora incarnata L.	Passion flower	Leaf spot Root rot	*Cercospora biformis* Pk., *Glocosporium fructigenum* Berk. *Phymatotrichum omnivorum* (Shear) Dug.
Pelargonium capitatum (L.) L'Her *P. crispum* L'Her. ex. Aiton. *P. graveolens* (L.) L'Her ex.Aiton	Wild rose geranium Lemon geranium Rose geranium	Crown gall Leaf spot Bacteria leaf spot Root and stem rot	*Agrobacterium tumefaciens* (E. F. Sm. & Town) Conn. *Ascochyta* spp., *Botrytis cinerea* Pers. ex Fr. *Pseudomonas erodii* Lewis, *Xanthomonas pelargonii* (N. A. Brown) Starr & Burkh. *Pythium* spp., *Rhizoctonia solani* Kuehn, *Thielaviopsis basicola* (Berk. & Br.) Ferr.
Peltigera canina L.	Ground liverwort, Dog lichen	Not available	

Scientific name	Common name	Major diseases	
		Name	Pathogen
Petroselinum crispum (Mill.) Nym.	Parsley	Leaf blight	*Alternaria dauci* (Kuehn) Groves & Skolko., *Septoria petruselini* Desm.
		Bacteria soft rot	*Erwinia aroideae* (Town.) Holland
		Wilt	*Fusarium* spp., *Heterodera marioni* (Cornu) Goodey
		Root rot	*Phymatotrichum omnivorum* (Shear) Dug., *Rhizoctonia solani* Kuehn, *Selerotinia scierotiorum* (Lib.) d By.
Phataris canariensis L.	Canary creeper	See *Tropaeolum majus*	
Phellodendron chinensis Schneid.	Chinese corktree (Huang Bai)	Leaf spot	*Cercospora* spp.
Phoradendron leucarpum (Raf.) Rev. & M. C. Johnst.	Mistletoe	Leaf spot	*Asterinella phoradendri* Rynn., *Exosporium phoradondri* Tharp
		Leaf and twig blight	*Phyllosstiota phoradendri* Bonar, *Nectria cinnaharina* Tode ex Fr., *Sphaeropsis visci* (West.) Archer.
P. flavescens (Push.) Nutt.	American mistletoe	Rust	*Uredo pharudendri* Jacks
Phyllanthus emblica L.	Myrobalan, emblic	Rust	*Aecidium favaccum* Arth., *Phakopsora fenestrala* Arth.
Physalis alkekengi L.	Chinese lantern	Rust	*Aecidium physalidis* Burr., *Puccinia physalidis* Pk.
		Leaf spot	*Alternaria solani* (Ell. & G. Martin) sor., *C. physalicola* Ell. & Barth., *Leptosphaeria physalidis* Ell. & Ev., *Pseudomonas angulata* (Fromme & Murrary) Stapp.
		White smut	*Entyloma australe* Speg.
		Root rot	*Phymatotrichum omnivorum* (Sher) Dug.

Scientific name	Common name	Major diseases	
		Name	Pathogen
Phytolacca americana L.	Pokeweed	Leaf spot	*Alternaria* spp., *Cercospora flagellaris* Ell. & G. Martin, *Dendryphium nodulosum* Sacc., *Phyllosticta phytolaccae* Cke., *Septoria phlyctaenoides* Berk. & Curt.
		Root rot	*Phymatotrichum omnivorum* (Shear) Dug., *Rhizoctonia crocorum* (Pers.) DC. ex Fr.
Picea mariana (Mill) Black.	Black spruce	Witches' broom	*Arccuthobium pusillum* Pk., *Melampsorella cerastii* (Pers.) Schroet.
P. glauca (Moench) Voss.	White spruce	Needle rust	*Chrysomyxa* spp.
		Canker	*Cytospora chrysosperma* Pers. ex Fr., *C. leucostoma* Pers. ex Sacc.
P. ables (L.) Karst	Norway spruce	Seedling blight	*Ascochyta piniperda* Lindau., *Phytophthora cinnsmomi* Rands.
Pimpinella anisum L.	Anise	Leaf spot	*Cercospora malkoffii* Bub.
		Root and stem rot	*Phymatotrichum omnivorum* (Shear) Dug., *Sclerotinia sclerotiorum* (Lib.) d By.
Pinus mugo Torra var. Pumilio	Dwarf mountain pine	Root and stem rot	*Armillaria mellea* Vahl ex Fr., *Cylindrocladium scoparium* Morg., *Fomes annosus* (Fr.) Cke., *Fusarium* spp.,
P. palustris Mill.	Southern pitch pine		*Phytophthora cinnamomi* Rands.
P. strobus L.		Canker	*Caliciopsis pinea* Pk., *Phoma bacteriophila* Pk.
P. albicaulis Engeim.	White pine	Damping-off	
P. contorta Dougl. ex. Loud	Lodgepole pine	Needle and seedling blight	*Pythium ultimum* Trow., *Rhizoctonia solani* Kuehn, *Septoria spadicea* Patterson & Charies, *Thelephora terrestris* Ehr. ex Fr., *Rhizina undulata* Fr.

Scientific name	Common name	Major diseases	
		Name	Pathogen
Piper methysticum G. Forst	Kava-Kava	Dieback Wilt Nematodes	Cucumber mosaic cucumovirus *Sphaerulina* spp., *Colletotrichum* spp. *Amphisbaenema* spp.
P. nigrum L.	English pepper	Leaf spot Root rot	*Cercoseptoria piperis* (F. L. Stevens & Dalbey) Petr., *Cercospora piperis* Pat., *Cyclodothis pulchella* Syd., *Guignardia pipericola* F. L. Stevens, *Omphalia flavida* (Cke.) Maubl. & Rangel. *Rosellinia bunodes* Berk. & Br.
Plantago asiatica L. *P. psyllium* L. *P. lanceolata* L. *P. major* L.	Psyllium Plantain	Leaf spot Powdery mildew Downy mildew Rust Root and stem rot	*Ascochyta plantaginella* Tehon., *Cercospora plantaginella* Tehon., *Mycosphaerella columbi* Rehm, *Ramularia* spp., *Septoria plantaginea* Pass. *Erysiphe cichoracearum* DC., *Sphaceloma plantaginis* Jenkins & Bitane. *Peronospora alta* Fekl. *Puccinia aristidae* Tracy, *P. pacifica* Blasd. *Phymatotrichum omnivorum* (Shear) Dug., *Sclerotium rolfsii* Sacc.
Podophyllum peltatum L.	Mayapple	Leaf spot	*Cercospora podophylli* Tehon & Daniels, *Discohainesia oenotherae* (Cke. & Ell.), *Phyllosticta podophylli* (Curt.) Wint., *Septoria podophylima* Pk., *Vermicularia podophylli* Ell. & Dearn.
Polygala senega L.	Seneca snakeroot	Rust Leaf spot	*Aecidium renatum* Arth. *Cercps[pra grosea* Cke. & Ell., *Septoria consocia* Pk.

Scientific name	Common name	Major diseases	
		Name	Pathogen
Polygonum bistorta L. *P. multiflorum* Thunb.	Bistort root Flowery knotweed	Leaf spot Rust Leaf smut	*Pseudopeziza bistortae* (DC. ex Fr.) Fckl. *Puccinia bistortae* (Strauss) DC., *P. septentrionalis* Juel *Ustilago bistortarum* (DC.) Koern.
Populus balsamifera L. *P. candicans* L. *P. tremuloides* Michx.	Poplar, Balm of gilead American aspen	Leaf spot Root rot Canker	*Cercospora poputina* Ell. & Ev., *Marssonina castagnoi* (Desm. & Mont.) Magn., *Mycosphaerella populorum* G. E. Thompson, *Phyllosticta maculana* Ell. & Ev. *Fomos applanatas* (Pers. ex Fr.) Gill, *Phymatotrichum omnivorum* (Shear) Dug *Valisa nivea* Hoffm. ex Fr.
Primula vulgaris Huds. *P. veris* L.	Primrose Cowslip	Leaf spot Root rot	*Ascochyta primulae* Trail, *Asteroma garrettianum* Syd., *Cercosporella primulae* Allesch., *Colletotrichum primulae* Halst. *Mycosphaerelia* spp. *Pythium irregulare* Buism, *Rhizoctonia solani* Kuehn.
Prunella vulgaris L.	Heal all, self heal	Powdery mildew Leaf spot Powdery mildew	*Erysiphe cichoracearum* DC. *Gibberidea abundans* (Dobroz) Shear, *Linospora brunellae* Ell. & Ev., *Ramularia brunellae* Ell. & Ev., *Septoria brunellae* Ell. & Holw. *Sphaerotheca humuli* (DC.) Burr.
Prunus africana L.	African prune	Crown gall Canker Root rot	*Agrobacterium tumelaciens* (E. F. Sm. & Towns.) Conn. *Cladosporium carpophilum* Thuem *Phymatotrichum omnivorum* (Shear) Dug.

Scientific name	Common name	Major diseases	
		Name	Pathogen
P. dulcis (Mill.) D.A. Webb.	Almond	Shot hole	*Cercospora circumscissa* Sacc., *Coryneum carpophilum* (Lev.) Jauch
		Root rot	*Phymatotrichum omnivorum* (Shear) Dug.
		Green fruit rot	*Sclerotinia sclerotiorum* (Lib.) d By.
		Bacteria shoot blight	*Pseudomonas syringae* Van Hall.
		Crown gall	*Agrobacterium tumefacicus* (E. F. Sm. & Town) Conn.
		Bacterial leaf spot	*Xabtginibas orybu* (E. F. Sm.) Dows.
P. mume Siebold & Zucc.	Japanese apricot	Crown gall	*Agrobacterium tumelaciens* (E. F. Sm. & Towns.) Conn.
		Blossom end rot	*Alternaria* spp.
		Canker	*Cladosporium carpophilum* Thuem., *Phytophthora cactorum* (Leb. & Cohn) Schroet., *P. cithrophthora* (R. E. & E. H. Sm.) Leonian
		Brown rot	*Montilinia fructicola* (Wint.) Honey
		Root rot	*Phymatotrichum omnivorum* (Shear) Dug.
P. serotina J. F. Ehrb.	Black cherry, wild cherry	Shot hole	*Coecomyces lutescens* Higgins, *Cylindrosporium batescens* Higgins.
		Blight	*Monilia angustior* (Sacc.) Reade., *Monilinia demissa* (Dana) Honey
		Powdery mildew	*Podosphaera oxyacanthae* (DC.) d By.

Scientific name	Common name	Major diseases	
		Name	Pathogen
P. virginiana L.	Chokecherry	Root rot, collar rot	*Armillaria metlea* Vahl ex Fr., *Fomes* spp., *Phymatotrichum omnivorum* (Shear) Dug., *Phytophthora cactorum* (Leb. & Cohn.) Schroet
		Blight	*Botrytis cinerea* Pers. ex Fr.
		Brown rot	*Monilia* spp.
		Canker	*Pseudomonas syringae* Van Hall, *Phytophthora citrophthora* Leonian.
Pueraria lobata (Wild) Ohwi.	Kudzu	Stem and root rot	*Fusarium* spp., *Macrophomina phasoli* (Maubl.) Ashby., *Phymatotrichum omnivorum* (Shear) Dog.
		Damping-off	*Rhizoctonia solani* Kuehn
Pygeum africanum L.	Pygeum	Not available	
Pyrethrum cinerarifolium L.	Pyrethrum	See *Chrysanthemum cinerarifolium*	
Quercus alba L.	White oak	Leaf spot	*Actinopelte dryina* (Sacr.) Hoehn., *Cylindrosporium microspilum* Sacc. & Wint. *Dothiarella phomiformis* (Sacr.) Petr. & Syd., *Gloeasporium* spp., *Marssonina martini* (Sacc. & Ell.) Magn., *Septogloeum querceum* J. J. Davis
		Root rot	*Armillaria mellea* Vahl ex Fr., *Clitoeybe tabescens* (Scop. ex Fr.) Bres., *Phymatotrichum omnivorum* (Shear) Dug., *Phytophthora cinnamomi* Rands
		White heart rot	*Fomes* spp., *Ganoderma curtisii* (Berk.) Murr., *Polyporus* spp.
		Powdery mildew	*Microsphaera alni* DC. ex Wint., *Phyllactinia corylea* Pers. ex Karst

Scientific name	Common name	Major diseases	
		Name	Pathogen
Ranunculus occidentalis Nutt. *R. ficaria* L.	Buttercup Pilewort	Gray mold	*Botrytis cinerea* Pers. ex Fr.
		Leaf spot	*Ceratobasidium anceps* (Bres. & Syd.) Jacks, *Cerospora ranunculi* Ell. & Holw., *Cylindrosporium ficariae* Berk., *Ovularia decipiens* Sacc., *Fabraea ranunculi* (Fr.) Karst., *Septocylindrium ranunculi* Pk., *Ramularia aequivoca* (Ces.) Sacc.
		Powdery mildew	*Erysiphe polygoni* DC., *Sphaerothera humuli* (DC.) Burr.
		Downy mildew	*Peronospora ficariae* Tul.
		Root and stem rot	*Phymatotrichum omnivorum* (shear) Dug., *Sclerotium rolfsii* Sacc., *Sclerotinia solerotiorum* (Lib.) d By.
		Rust	*Puccinia* spp., *Uromyces* spp.
Raphanus sativus L.	Radish	Gray and black leaf spot	*Alternaria brassicae* (Berk.) Sacc., *A. oleracea* Milbrath
		Soft rot	*Erwinia carotovora* (L. R. Jones) Holland
		Powdery mildew	*Erysiphe polygoni* DC.
		Downy mildew	*Perouospora parasitica* Pers. ex Fr.
		Root rot	*Phymatotrichum omnivorum* (Shear) Dug., *Rhizoctonia solani* Kuelin.
		Damping-off	*Pythium debaryanum* Hesse., *P. ultimum* Trow.
Rhamnus catharticus L. *R. frangula* L. *R. purshiaanus* L.	Buckthorn Alder buckthorn Cascara sagrade	Leaf spot	*Cercospora aeruginosa* Cke., *Cylindrosporium rhamni* Ell. & Ev., *Marssonina rhamni* (Ell. & Ev.), *Phyllosticta rhamnigena* Sacc., *Septoria blasdalei* Sacc.
		Heart rot	*Daedalea unicolor* Bull. ex Fr., *Fomes igniarius* (L. ex Fr.) Kickx.
		Crown rust	*Puccinia coronata* Cda.

Scientific name	Common name	Major diseases	
		Name	Pathogen
Rheum palmatum L. *R. officinal* Baill. *R. tanguticum* L.	Chinese rhubarb Rhubarb Rhubarb	Crown gall Root rot Gray-mold rot Soft rot	*Agrobacterium tumefaciens* (E. F. Sm. & Town.) Conn. *Armillaria mellea* Vah. ex Fr., *Rhizoctonia solani* Kuehn., *Phytophthora spp* *Botrytis cinerea* Pers. ex Fr. *Erwinia carotovora* (L. R. Jones) Holland
Rhodiola rosea L.	Roseroot	See *Sedum rosea* subsp. *integrifolium*	
Rhus radicans L.	Poison ivy	Leaf spot Powdery mildew Root rot	*Actinopelte dryina* (Sacc.) Hoehn., *Cercospora bartholomaei* Ell. & Kell, *Cercosporella californica* Bonar, *Cylindrosporium irregulare* (Pd.) Dearn., *Discohainesia oenotherae* (Cke. & Ell.) Natuuf., *Ophiocarpella tarda* (Hark.) Th. & Syd. *Phyllactinia corylea* Pers. ex Karst. *Phymatotrichum omnivorum* (Shear) Dug., *Ehizoctonia crocorum* (Pers.) DC. ex Fr.
R. toxicodendron L.	Roseroot	See *Sedum rosea* subsp. *integrifolium*	
Ribes nigrum L. *R. lacustre* (Pers.) Poir.	Black currant Black gooseberry	Root rot Cane blight Leaf spot Collar rot	*Armilaria metlea* Vahl ex Fr., *Hypholoma perplexum* Pk., *Botryosphaeria ribis* Gross & Dug. *Cercospora angulata* Wint., *Botrytis cinerea* Pers. ex Fr., *Mycosphaerella ribis* (Fckl.) Feltg., *Phyllosticta grossulariae* Sacc. *Fomes ribis* (Schum ex Fr.) Gill.
Robertium macrorrhizum Pic.	Geranium	See *Geranium mactorrhizum*	

Scientific name	Common name	Major diseases	
		Name	Pathogen
Rosa canina L. *R. damascena* L.	Dog rose Damask rose	Blossom rot Crown canker Downy mildew Rust Powdery mildew	*Botrytis cinerea* Pers. ex Fr. *Cylindrocladium scoparium* Morg., *Griphosphaeria corticola* (Fckl.) Hoeh. *Peronospora sparisa* Berk. *Phragmidium* spp. *Sphaerotheca humuli* (DC.) Burr., *S. pannosa* (Wallr.) Lev.
Rosmarinus officinalis L.	Rosemary	Leaf spot Leaf blight	*Ascochyta rosmarini* Lib. *Sclerotium delphinii* tode. Fr.
Rubus chamaemorus L. *R. fruiticosus* L.	Cloudberry Blackberry	Crown gall Fruit rot Rust Leaf spot Powdery mildew	*Agrobacterium rhizogenes* (Riker et al.) Conn. *Botrytis cinerea* Pers. ex Fr. *Kuehneola uredinia* (Lk.) Arth. *Pezizella oeuotherae* (Cke. & Ell.) Sacc., *Mycosphaerella rubi* Roark. *Sphaerotheca humuli* (DC.) Burr.
R. idaeus L.	Raspberry	Crown gall Root rot Gray mold Orange, yellow rust Leaf rust Wilt	*Agrobacterium tumefaciens* (E. F. Sm. & Rown.) Conn. *Armillaria mellea* Vahl. ex Fr., *Rhizoctonia solani* Kuehn. *Botrytis cinerea* Pers. ex Fr. *Gymnoconia peckiana* (Howe) Trott., *Kuehneola uredinis* (Lk.) Arth *Pezizella oenotherae* (Cke. & Ell.) Sacc., *Phragmidium rubi-idaei* (DC.) Karst. *Verticillium albo-atrum* Reinke & Berth.

Scientific name	Common name	Major diseases	
		Name	Pathogen
Rumex acetosella L. *R. obtusifolia* L.	Sorrel Dock	Leaf spot Rust Root rot	*Gloeosporium rumicis* Ell. & Ev., *Phyllosticta acetosellae* A. L. Sm., *Septoria pleosporoides* Sacc. *Puccinia acetosae* (Schum.) Koern. *Rhizoctonia solani* Kuchn.
Ruscus aculeatus L.	Butcher's broom	Wood rot Leaf spot	*Leptosphaeria rusci* Ces. & De Not. *Phyllosticta ruscigena* Pers.
Ruta graveolens L.	Rue, herb of grace	Fruit rot	*Penicillium* spp.
Salix alba L. *S. discolour* Muhlenb. *S. caprea* L.	White willow Pussy willow	Branch and twig canker Leaf spot Rust Powdery mildew	*Botryosphaeria ribis* Gross & Dug., *Discella carbonacea* (Fr.) Berk. & Br., *Physalospora miyabeana* Fukushi, *Valsa sordida* Nits. *Cercospora salicina* Ell. & Ev., *Gloeosporium salicis* West, *Phyllosticta apicalis* Davis, *Marssonina kriegeriana* (Bres.) Magn *Melampsora abieti-capraearum* Tub., *M. bigelowii* Thuem. *Uncinula salicis* DC. ex Wint.
Salvia officinalis L. *S. sclarea* L.	Sage Clary sage	Rust Downy mildew Root rot	*Aecidium subsimulans* Arth. & Mains, *Puccinia ballotaeflorae* Long *Peronospora lamii* A. Braun *Phymatotrichum omnivorum* (Shear) Dug., *Rhizoctonia solani* Kuehn

Scientific name	Common name	Major diseases	
		Name	Pathogen
Sambucus nigra L. *S. canadensis* L. *S. racemosa* L.	Elderberry Elder	Leaf spot Powdery mildew Root rot	*Gloeosporium tinenm* Sacc., *Ramnlaria smbacina* Sacc. *Microsphaera alni* DC. ex Wint., *Phyllactinia corylea* Pers. ex Karst *Phymatotrichum omnivorum* (Shear) Dug., *Rhizoctonia erocorum* (Pers.) DC. ex Fr., *Xylaria* spp.
Sanguinaria canadensis L.	Bloodroot	Gray-mold blight Leaf spot	*Botrytis* spp. *Cercospora sanguinariae* Pk., *Gloeosporium sanguinariae* Ell. & Ev., *Phyllosticta sanguinariae* Wint.
Sanicula marĩlandica L.	Black snakeroot	White smut Rust Leaf gall	*Entyloma saniculae* Pk. *Puccinia marylandica* Lindr. *Urophlyctis pluriannulata* (Berk. & Vurt.) Farl.
Santalum album L.	Sandalwood	Root rot Spike disease	*Armillaria mellea* (Vahl. Fr.) P. Karst, *Rhizoctonia* spp. Mycoplasma
Saponaria officinalis L.	Soapwort	Leaf spot Root rot	*Alternaria saponariae* (Pk.) Neerg., *Cylindrosporium officinale* Ell. & Ev., *Phyllostricta tenerrima* Ell. & Ev. *Phymatotrichum omnivorum* (Shear) Dug.
Sarothamnus scoparius L.	Broom	See *Cytisus scoparius*	
Sassafras albidum (Nutt.) Nees	Sassafras	Leaf spot Root and trunk rot	*Actinothyrium gloeosporioides* Tehon., *Phyllosticta illinoensis* Tehon & Daniels. *Armillaria mellea* Vahl. ex Fr., *Daedalea confragosa* Bolt. ex Fr.

Scientific name	Common name	Major diseases	
		Name	Pathogen
Satureia hortensis L. *S. montana* L.	Savory	Leaf spot Rust	*Phyllosticta docidua* Ell. & Kell. *Puccinia menthae* Pers.
Schisandra chinensis (Turcz.) Baill.	Schizandra, wu wei zi	Root rot	*Fusarium sporotrichioides* Sherb., *F. schizandra* Link.
Scutellaria baicalensis Georgi. *S. lateriflora* L. *S. galericulata* L.	Baical skullcap Virginia skullcap Skullcap	Stem rot Leaf spot Powdery mildew Root rot	*Botrytis cinerea* Pers. ex Fr. *Ceroospora scutellariae* Ell. & Ev., *Phyllosticta decidua* Ell. & Kell., *Rhizoctonia solani* Kuehn. *Erysiphe galeopsidis* DC., *E. cichoracearum* DC. *Phymatotrichum omnivorum* (Shear) Dug.
Sedum acre L. *S. rosea* (L.) Scop. subsp. *integrifolium*	Stonecrop, Small houseleek Roseroot	Stem rot Leaf spot Rust	*Phytophthora* spp., *Rhizoctonia solani* Kuehn. *Pleospora* spp., *Septoria sedi* West *Puccinia rydbergii* Garrett
Sempervivum tectorum L.	Houseleek, hens and chicks	Rust Stem and leaf rot Root rot	*Endophyllum sempervivi* (Alb. & Schw.) d By. *Phytophthora parasitica* Dast. *Pythium* spp.
Senecio vulgris L.	Groundsel	White rust Rust White smut Root and stem rot	*Albugo trogopogonis* Pers. ex S. F. Gray *Baeodromus californicus* Arth., *Coleosporium occidentale* Arth., *Puccinia angostata* Pk. *Entyloma compositarum* Farl. *Phymatotrichum omnivorum* (Shear) Dug., *Phytophthora* spp., *Sclerotinia sclerotiorum* (Lib.) d By

Scientific name	Common name	Major diseases	
		Name	Pathogen
Senna alexandrina L.	Alexandrina	See *Cassia angustifolia*	
Serenoa repens (Bartr.) Small.	Saw palmetto	Leaf spot	*Achorella attaleae* F. L. Stevens., *Diplodia theobromas* (Pat.) Nowell., *Exosporium palmivorum* Sacc., *Phomopsis* spp.
		Leaf blight	*Colletotrichum gloeosparioides* Penz.
		Leaf canker	*Glomerella cingulata* (Ston.) Spauld. & Schrenk
		Root rot	*Thielaviopsis paradoxa* (De Seyn.) Hoehn.
Sesamum indicum L. *S. orientale* L.	Sesame	Leaf spot	*Cercospora sesami* Zimm.
		Bacteria leaf spot	*Pseudomonas sesami* Malkoff.
		Charcoal rot	*Macrophomina phaseoli* (Maubl.) Ashby.
Shepherdia canadensis (L.) Nutt.	Buffalo berry	White heart rot	*Fomes fraxinophilus* (Pk.) Sacc., *F. ellisianus* (F. W. Anderson) Baxter
		Powdery mildew	*Phyllactinia corylea* Pers. ex Karst.
		Root rot	*Phymatotrichum omnivorum* (Shear) Dug.
		Rust	*Puccinia caricis-shepherdiae* J. J. Davis
		Damping-off	*Pythium ultimum* Trow, *Rhizoctonia solani* Kuehn, *Sphaerotheca humilis* (DC.) Burr.
Silybum marianum (L.) Gaertn.	Milk thistle	Powdery mildew	*Erysiphe cichoracearum* DC.
		Virus	Mosaic, tobacco yellow leaf dwarf
Simmondsia chinensis (Link) C. Schneid	Jojoba	Wood rot	*Dendrophora albobadia* (Schwein. Fr.) Chamuris., *Hypochnicium punctulatum* (Cooke) J. Eriksson, *Meruliopsis corium* Ginns
		Leaf spot	*Strumella simmondsiac*, *Coniothyrium* spp.

Scientific name	Common name	Major diseases	
		Name	Pathogen
Solidago virgaurea L.	Goldenrod	Leaf spot	*Ascochyta compositarum* J. J. Davis, *Cercospora parvimaculans* J. J. Davis, *Colletotrichum solitarium* Ell. & Barth., *Macrophoma sphaeropsispora* (Ell. & Ev.) Tassi, *Phyllosticta solidaginicola* Tehon & Daniels, *Ramularia serotina* Ell. & Ev.
		Rust	*Coleosporium solidaginis* (Schw.) Thuem, *Puccinia extensicola* Plowr.
		Powdery mildew	*Erysiphe cichoraccearum* DC., *Phyllactinia corylea* pers. ex Karst.
Sorbus aucuparia L.	Mountain ash	Canker	*Cytospora* spp., *Valsa* spp.
		Root rot	*Armillaria mellea* Vahl ex Fr.
		Fire blight	*Erwinia amylovora* (Burr.) Winslow et al.
		Rust	*Gymnosporangium* spp.
Spirea ulmaria L.	Meadowsweet	See *Filipendula ulmaria*	
Stachys officinalis (L.) Trev.	Betony, woundwort	Leaf spot	*Cercospora stachydis* Ell. & Ev., *Cylindrosporidum stachydis* Ell., *Ovularia bullala* Ell. & Ev., *Phyllasticta decidua* Ell. & Kell, *Ramularia stachydis* (Pass.) Massal., *Septoria stachydis* Rob
		Powdery mildew	*Erysiphe galeopsidis* DC., *Sphaerotheca humuli* (DC.) Burr.

Scientific name	Common name	Major diseases	
		Name	Pathogen
Stellaria media (L.) Vill.	Chickweed	Leaf, flower, seed smut Rust Root and stem rot	*Entyloma alsines* Halst., *Sorosporium saponariae* Rudolphi, *Ustilago alsineae* Clint. & Zundel *Melampsorella cerastii* (Pers.) Schroet, *Puccinia arenariae* (Schum.) Wint *Phymatotrichum omnivorum* (Shear) Dug., *Sclerotinia sclerotiorum* (Lib.) d By
Stevia rebaudiana (Bertoni) Bertoni	Stevia Sweet herb of Paraguay	Powdery mildew Damping-off Stem rot	*Erysiple cichoracearum* DC. *Rhizoctonia solani* Kuehn. *Sclerotium delphinii* Welch.
Symhytum officinale L.	Comfrey	Leaf spot Root rot	*Stemphylium* spp. *Rhizoctonia* spp.
Syringa vulgaris L.	Lilac	Crown gall Gray-mold blight Leaf blotch Powdery mildew Blossom blight Bacterial blight Root rot	*Agrobacterium tumejaciens* (E. F. Sm. & Towns.) Conn. *Botrytis cinerea* Pers. ex Fr. *Cladosporium* spp., *Heterosporium syringae* Oud. *Microsphaera alni* DC. ex Wint. *Phytophthora cactorum* (Leb. & Cohn) Schroet. *Pseudomonas syringae* Van Hall *Thielaviopsis basicola* (Berk.) Ferr.

Scientific name	Common name	Major diseases	
		Name	Pathogen
Syzygium aromaticum (L.) Merr. & Perry	Clove	Leaf spot	*Asteridium eugeniae* Mont., *Asterina colliculosa* Speg., *Asterinella cylindrotheca* (Speg.) Th., *Dictyochorina portoricensis* Chardon., *Phyllachora eugeniae* Chardon., *Phyllosticta eugeniae* Young.
		Root rot	*Clitocybe tabescens* (Scop. ex Fr.) Bres., *Rosellinia bunodes* Berk. & Br.
		Die back	*Cryptosporella eugenear*
		Leaf rot	*Cylindrocladium quinquesepatum*
		Sudden death	Mycoplasma
		Samatra disease	*Pseudomonas solanacearum*
Tagetes lucida Cav.	Mexican mint marigold	Head blight	*Botrytis cinerea* Pers. ex Fr.
		Rust	*Coleosporium madiae* Cke., *Puccinia tageticola* Diet & Holw.
		Stem rot	*Fusarium* spp., *Macrophomina phaseoli* (Maubl.) Ashby., *Phytophthora cryptogea* Pethyb. & Laff., *Rhizoctonia solani* Kuehn, *Sclerotinia sclerotiorum* (Lib.) d By.
Tanacetum parthenium (L.) Schultz.	Feverfew	See *Chrysanthemum parthenium*	
T. cinerarifolium L.	Pyrethum	See *Chrysanthemum cinerarifolium*	
T. vulgare L.	Tansy	Stem, root rot	*Fusarium roscum* Lk., *Phymatotrichum omnivorum* (Shear) Dug.,.*Scleratinia scierotiorum* (Lib.) d By.

Scientific name	Common name	Major diseases	
		Name	Pathogen
Taraxacum officinale G. H. Weber ex Wigg.	Dandelion	Leaf spot	*Ascochyta taraxaci* Grove., *Ramularia lineola* Pk.
		Bacteria leaf spot	*Pseudomonas tabaci* (Wolf & Foster) Stapp., *Xanthomonas taraxaci* Niederhauser
		Powdery mildew	*Erysiphe cicharacearum* DC., *Phyllactinia corylea* Pers. ex Karst.
Taxus x media Rehd. *T. brevifolia* Nutt.	Yew	Root rot	*Armillaria mellea* Vahl ex Fr.
		Needle, twig blight	*Herpotrichia nigra* Hartig, *Pestalotia funerea* Desm, *Phyllostictina hysterella* (Sacc.) Petr., *Phytophthora cinnamomi* Rands, *Sphaeropsis* spp., *Sphaerulina taxi* (Cke.) Mass., *S. taxicola* (Pk.) Berl.
Thymus citriodorus (Pers.) Schreb.	Lemon thyme	Gray mold	*Botrytis cinerea* Pers. ex Fr.
Thymus vulgaris L	Thyme	Root rot	*Rhizoctonia solani* Kuehn.
Tillia cordata Mill. *T. europaea* L.	Linden, small-leaved lime	Trunk and branch canker	*Botryosphaeris* spp., *Strumella* spp., *Cerospora microsora* Sacc.
		Leaf blight	*Daedulea confiragosa* Bolt., *Fomes counatus* (Weinm. ex Fr.) Gill, *Hyduum septentrionale* Fr.
		White spongy rot	*Phlyctacna tiliae* Dearn., *Phyllosticta praetervisa* Bub.
		Leaf spot	*Sphaeropsis* spp.
		Root rot	*Phymatotrichum omnivorum* (Shear) Dug., *Ustulina vulgaris* Tul.
Tribulus terrestris L.	Puncture vine	Root rot	*Rhizoctonia solani* Kuehn

Scientific name	Common name	Major diseases	
		Name	Pathogen
Trifolium incarnatum L.	Clover (crimson)	Gray-mold leaf blight	*Botrytis* spp.
T. pratense L.	Clover (red)	Sooty blotch	*Cymadothea trifolii* (Pers. ex Fr.) Wolf
		Powdery mildew	*Erysiphe polygoni* DC.
		Bacterial leaf spot	*Pseudomonas syringae* Van Hall.
		Rust	*Uromyces minor* Schroet., *U. trifolii* (Hedw. f.) Lev.
		Root rot	*Phymatotrichum omnivorum* (Shear) Dug.
Trigonella foenum-gracecum L.	Fenugreek	Leaf smut	*Entyloma trigonellae* Stevenson
		Root knot nematodes	*Meloidogyne* spp.
Tropaeolum majus L.	Canary creeper, Nasturtium	Gray mold	*Botrytis cinerea* Pers. ex Fr.
		Leaf spot	*Cercospora tropaeoli* Atk., *Heterosporium tropaeoli* T. Bond., *Pleospora* spp., *Pseudomonas aptata* F. I. Stevens
Tussilago farfara L.	Colisfoot	Leaf spot	*Mycosphaerella tussilaginis* (Rehm.) Lindau, *Ramularia bruunea* Pk., *Septoria farfaricola* Dearn.
Ulmus rubra Muhl.	Slippery elm, sweet elm	Leaf spot	*Ceratophorum ulmicola* Ell. & Kell., *Phyllosticta confertissima* Ell. & Ev.
U. procera L.	English elm	Powdery mildew	*Uncinula macrospora* Pk.
		Wilt	*Ceratostomella ulmi* Buis., *Dothiorella ulmi* Verrall & May
		White spongy rot	*Fomes connatus* (Weinm. ex Fr.) Gill

Scientific name	Common name	Major diseases	
		Name	Pathogen
Umbelluslaria californica Nutt.	California laurel Bay laurel	Black mildew White-mottled butt rot Canker, dieback Wood rot	*Asterina anomala* Cke. & Harkn. *Fomes applanatus* (Pers. ex Fr.) Gill, *F. fomentarius* Kickx., *Nectria cinnabarina* Tode ex Fr., *N. coccinea* Pers. ex Fr. *Lenzites betulina* L. ex Fr., *Polyporus versicolor* L. ex Fr., *Stereum albobadium* Schw. ex Fr.
Uncaria tomentosa (Willd.) DC. *U. guianensis* (Aubl.) Gmel.	Cat's claw	White rot	*Fomes lignosus* (Fr) Kickx.
Urtica dioica L. *U. urens* L.	Stinging nettle	Leaf spot Rust Powdery mildew Downy mildew	*Cylindrosporium urticae* Dearn., *Ramularia urticae* Ces., *Septoria urticae* Rob. ex Desm *Aecidium libertum* Arth., *Puccinia caricis* (Schum) Schroest *Erysiphe cichoracearum* DC. *Pseudoperonospora urticae* (Lib.) Salmon & Ware, *Peronospora* spp.
Vaccinium macrocarpon L. *V. vitis-idaea* L.	Cranberry Mountain cranberry, cowberry, foxberry	Berry blotch rot Blossom blight Berry black rot Berry bitter rot Powdery mildew Witches' broom Berry rot	*Acanthorhyncus vaccinii* Shear *Botrytis cinerea* Pers. ex Fr. *Ceuthospora lunata* Shear *Glomerella cingulata* (Ston.) Spauld & Schrenk. *Microsphaera alni* DC. ex Wint. *Naevia axycocci* Dearn. *Sphaeronema pomorum* Shear, *Phyllosticta putrefaciens* Shear

Scientific name	Common name	Major diseases	
		Name	Pathogen
V. mrytilloides Michx. *V. myrtillus* L. *V. oreophilum* Rydb.	Blueberry Bilberry	Leaf spot Leaf and shoot gall Witches' broom	*Ceuthospora latitans* (Fr.) Hoehn., *Pestalotia maculiformis* Cuba & Zeller, *Phyllosticta sparsa* Bonar. *Exobasidium vaccinii* Wor. *Exobasidium vaccinii-aliginosi* Boud., *Peccinastrum goeppertianum* (Kuchn) Kleb.
Valeriana officinalis L.	Valerian	Powdery mildew Rust Stem and root rot	*Erysiphe cichoracearum* DC. *Puccinia commutata* Syd., *P. extensicola* Plowr., *P. valerianae* Car. *Rhizoctonia solani* Kuehn., *Sclerotium delphinii* Welch.
Verbascum thapsus L.	Mullein	Leaf spot Powdery mildew Downy mildew Root rot	*Cercospora verbascicola* Ell. & Ev., *Phyllosticta verbascicola* Ell. & Kell, *Ramularia variabilis* Fckl., *Septoria verbascicola* Berk. *Erysiphe cichoracearum* DC. *Peronospora sordida* Berk. & Br. *Phymatotrichum omnivorum* (Shear) Dug.
Veronica officinalis L.	Speedwell	Leaf smut Root and stem rot Downy mildew Rust	*Entyloma verpnicae* (Wint.) Lagh. *Fusarium* spp., *Phymatotrichum omnivorum* (Shear) Dug. *Peronospora prisea* Ung. *Puccinia albulensis* Magn., *P. probabilis* Arth. & Cumm., *P. rhaetica* E. Fish
Vetiveria zizanioides L. (Nash.)	Vetiver	Leaf spot	*Didymella andropogonis* Ell. & Ev., *Carvularia trifolii*

Scientific name	Common name	Major diseases	
		Name	Pathogen
Viburnum opulus L.	Highbush cranberry	Crown gall Leaf spot Leaf mold Root rot	*Agrobacterium tumefaciens* (E. F. Sm. & Town.) Conn. *Cercospora opuli* (Fckl.) Hoehn., *Helminthosporium beaumontii* Sacc., *Phyllosticta punctata* Ell. & Dearn. *Claudosporium herbarum* Lk. ex Fr. *Clitocyhe tabescens* (Scop. ex Fr.) Bres.
Vinca minor L.	Periwinkle	Leaf spot Rust Root and stem rot	*Alternaria* spp., *Septoria viucae* Desm., *Volutella vincae* Fairm. *Coleosporium opocynaceaum* Cke., *Puccinia vieca* (DC.) Berk. *Rhizoctonia solani* Kuehn.
Viola tricolor L.	Pansy	Leaf spot Rust Powdery mildew Damping-off	*Alternaria violae* Gall & Dorsett, *Cercospora violae* Sacc., *Phyllosticta rafinesquii* H. W. Anderson *Puccinia ellisiana* Thuem, *P. violae* (Schum) DC., *Uromyces audropagonis* Tracy *Sphaerotheca humuli* (DC.) Burr. *Rhizoctonia solani* Kuehn
Viscum album L.	Mistletoe	Black mildew	*Meliola visci* F. Stevens
Vitex agnus-castus L. *V. negundo* L.	Chaste tree Chinese chaste tree	Leaf spot Root rot	*Cercospora viticis* Ell. & Ev. *Phymatotrichum omnivorum* (Shear) Dug.

Scientific name	Common name	Major diseases	
		Name	Pathogen
Vitis labrusca L. *V. vinifera* L.	Grape	Crown gall	*Agrobacterium tumefaciens* (E. F. Sm. & Town.) Conn.
		Root rot	*Armillaria mellea* Vahl ex Fr., *Clitocybe tabescens* (Scop. ex Fr.) Bres.
		Black mold rot	*Aspergillus niger* v. Tiegh, *Guignardia bidwelli* (Ell.) Viala & Ravaz.
		Gray mold rot	*Botrytis cinerea* Pers. ex Fr.
		Bitter rot	*Melanconium fuligineum* (Seribner & Viala) Cav.
		Rust	*Physopella vitis* (Thuem) Arth.
		Downy mildew	*Phasmopara viticola* (Berk. & Curt.) Berl. & de T.
		Powdery mildew	*Uncinula necator* (Schw.) Burr.
Withania somnifera Dunal.	Ashwagandha, Indian ginseng	Witches-broom	Mycoplasma
		Leaf blight	*Alternaria* spp.
Yucca aloifolia L. *Y. glauca* L.	Yucca Soapweed	Leaf spot	*Cercospora cuncentrica* Cke., *Coniothyrium concentricum* (Desm.) Sacc.
		Leaf mold	*Phyllosticta* spp., *Stagonospora gigantea* Heald & Wolf., *Torula maculans* Cke.
Zanthoxylum americanum Mill	Prickly ash, toothache tree	Leaf spot	*Cercospora xanthoxyli* F. L. Stevens, *Septoria pachyspora* Ell. & Holw.
		White heart rot	*Fomes igniarius* (L. ex Fr.) Kickx.
		Black mildew	*Meliola pilocarpi* F. L. Stevens
		Powdery mildew	*Ovulariopsis farinosa* Syd., *Phyllactinia guttata* (Fr.) Lev.

Scientific name	Common name	Major diseases	
		Name	Pathogen
Zea mays L.	Corn	Root and stalk rot	*Alternaria* spp., *Corticium saskii* (Shirai) T. Matsu., *Fusarium* spp., *Phytophthora parasitica* Dast.
		Rust	*Angiopsora zeae* Mains., *Pucciyia polysora* Underw.
		Bacterial wilt	*Bacterium stewartii* E. F. Sm.
		Dry rot of ear and stalks	*Diplodia macrospora* Earle, *Nigrospora* spp.
		Bacteria root rot	*Erwinia dissolvens* (Rosen) burkh.
		Pink ear rot	*Gibberella fujikuroi* (Saw.) Wr.
		Charcoal rot	*Macrophomina phaseoli* (Maubl.) Ashbv.
		Seedling root rot	*Rhizoctonia solani* Kuehn.
		Seed rot	*Rhizopus* spp., *Trichoderma* spp.
		Head smut	*Sphacelotheca relliana* (Kuehn) Clint.
Zingiber officinale Roscoe	Ginger	Leaf spot	*Phyllostieta zingiberi, Coniothyrium zingiber* F. L. Stevens & Atlenza
		Root rot	*Fusarium* spp., *Pythium butleri* Subr.and spp.
		Bacterial wilt	*Pseudomonas solanacearum*
		Root knot nematode	*Meloidogyne incognita, M. arenaria*

*Note: For detailed list of ginseng diseases, see Thomas S. C. Li and Raj S. Utkhede. 1993. Pathological and non-pathological diseases of ginseng and their control. Current Topics in Bot. Res. 1: 101-115.

Table 8. Major insects found in medicinal plants.

Scientific name	Common name	Major insects
Abies balsamea (L.) Mill.	Balsam fir	Wooly adelgid, twig aphid, cone worms, conifer aphids, eripophyid needle mites, spruce spider mite, western spruce budwoom
Achillea millefolium L.	Yarrow	Aphids, root-knot nematodes, *Phytoecia rufiventris*, *Ueolwuxon fovonia*, *Tebenna ishikii*, *Aphis fukii*, *Lixus subilis*, *Dichrampha montaniana*
Aconitum napellus L.	Monkshood	*Delphiniobium junackianum* (aphid), *Brachycaudus napelli, Lepidopterous caterpillar*
Acorus calamus L. var. Americanus Wolff. *A. tatarinowii* L.	Calamus, sweet flag Shi Chang Pu	*Aphis craccivoa*, *Lasioderma serricorne*
Actaea alba L. *A. rubra* (Ait.) Willd.	White baneberry Red baneberry	No information based on the databases searched is available.*
Aesculus hippocastanum L.	Horse chestnut	Scale insects
Agastache foeniculum L.	Aniseed	*Anabrus simplex* (cricket)
Agrimonia euparoria L.	Agrimony	*Acyrthosiphon pelargonii agrimoniae*
Agropyron repens (L.) Beauvois	Couch grass	Salt-marsh caterpillar, leafhoppers, McDaniel mite
Alchemilla vulgaris L.	Lady's mantle	*Acyrthosiphon pelargonii, bactericera femoralis*
Allium cepa L.	Onion	Cutworm, army worms, brown wheat mite, bulb mite, onion thrips, onion maggot, pea leaf miner, wireworm, eelworm

Scientific name	Common name	Major insects
A. sativum L.	Garlic	Cutworm, armyworm, thrips, wireworm, garden symphylan, eelworm
A. schoenoprasum L.	Chives	Eelworm, cutworm
Alnus crispus (Ait.) Pursh *A. incana* (L.) Moench. subsp. *tenufolia* *A. glutinosa* (L.) Gaertn.	Alder	Alder flea beetle, aphids, leafminers
Aloe vera (L.) Burm. *A. barbadensis* L.	Aloe	Mealybugs, scales
Althaea officinalis L.	Marshmallow	Aphids, borers, caterpillars, hollyhock weevils, Japanese beetles, mealybugs, mites, nematodes, red spider mites, slugs, thrips, *Pexicopia malvella*, *Podagrica fuscicomis*
Amelanchier alnifolia Nutt.	Saskatoon	Aphids, borers, caterpillars, curculios, leaf miners, mites, swaflies, scales
Ananas comosus (L.) Merr.	Pineapple	Fruit fly, thrips
Anaphalis margaritacea (L.) Bench & Hook	Everlastings	*Philaenus spumarius*, *Tyria jacobaeae* (cinnibar moth)
Anethum graveolens L.	Dill	Aphids, lygus bug, carrot rust fly, cutworm
Angelica archangelica L. *A. sinensis* (Oliv.) Diels	Angelica Dong quai	*Systole coriandri, S. albipennis,* aphids
Antennaria magellanica Schultz	Everlastings	See *Anaphalis margaritacea*

Scientific name	Common name	Major insects
Anthemis nobilis L.	Chamomile	See *Chamaemelum nobile*
Apium graveolens L.	Celery	Carrot rust fly, caterpillars, loopers, green peach aphid, slugs, wireworm, willow aphids, leaf miner, celery fly, carrot weevil
Aralia racemosa L. *A. nudicaulis* L.	Spikenard	Silkworm, aphids, caterpillars, scales, thrips
Arctium lappa L.	Burdock	*Pantamorus, Phytoecia rufiventris, Uroleucon gobonis, Tebenna ishikii, Aphis fukii, Lixus subtilis*
Arctostaphylos uva-ursi (L.) Spreng	Bearberry, Uva-ursi	Aphids, scales, whiteflies
Armoracia rusticana Gaertn, Mey & Scherb.	Horseradish	Aphids, cabbage aphid, turnip aphid, cutworms, diamond back moth, cabbage flea beetle, wireworm
Arnica latifolia Bong. *A. montana* L. *A. chamissonis* L. subsp. *foliosa* *A. condifolia* Hook *A. fulgens* Pursh *A. sororia* Greene	Arnica	*Amphorophora arnicae* (aphid)

Scientific name	Common name	Major insects
Artemisia absinthium L. *A. annua* L. *A. dracunculus* L. *A. tridentata* Nutt. *A. vulgaris* L.	Wormwood Quing Hao Tarragon Sagebrush Mugwort	Aphids, mealybugs, gall midges, scales, thrips, grasshopper, locust, *Craspedolepa campestrella, Heliococcus acirculus, Nyslus ericae, Plusia arichalcea, Franklinella* spp., *dolycoris indicus, Phytoecia rufiventris, Uroleucon gobonis, Tebenna ishikii, Lixus subtilis, Coleophora solenella, Ostrinia nubilalis, Ostrinia scapulalis, Phobetus comatus, chrysolina aurichalcea, Rhizaspidiotus taiyuensis, Lygocoris spinolae, Szelegiewicziella chamaerhodi*
Asarum canadensis L.	Wild ginger	*Luehodorifia japonica* (butterfly), *Aleurodes asari* (coccid)
Astragalus americana Bunge. *A. membranaceus* Bunge. *A. sinensis* L.	Astragalus Huang Qi	*Acanthoscelides aureolus, Hypera postica* (weevil)
Avena sativa L.	Oat	Oat birdeherry aphid, chinch bug, vetch weevil, greenbug
Baptisia tinctora (L.) R.Br. Ex Ait.f.	Wild indigo	*Apion rostrum*
Bellis perennis L.	Daisy	Root-knot nematodes, *Chasmatopterus* spp.
Berberis aquifolium L.	Oregon grape	See *Mahonia aquifolium*
Berberis vulgaris L.	Barberry	Lecanium scale
Beta vulgaris L.	Beet root	Armworm, cucumber beetles, flea beetle, garden symphylan, slug, spider mites, wireworm, vegetable leaf miners
Betula lenta L.	Birch	Aphids, birch leaf miner, bronze birch borer, oystershell scale, leopard moths, apple and thorn skeletonizer, cicadas

Scientific name	Common name	Major insects
Borago officinalis L.	Borage	*Vanessa cardui*
Boswellia sacra Roxb. ex Colebr. *B. carteri* Birdwood	Frankincense	Termite, *Rhesala imparata*
Brassica nigra (L.) Kock.	Black mustard	Aphids, cabbage aphid, turnip aphid, cabbage maggot, diamond back moth, flea beetles, wireworm
Buxus sempervirens L.	Boxwood	Boxwood leaf miner, boxwood mite, psyllid, Italy scale, holly scale
Calamintha nepeta (L.) Savi *C. ascendens* L. *C. officinalis L.*	Basil thyme Calamint Mountain mint	*Argyrotaenia pulchellana*
Calendula officinalis L.	Marigold, calendola	Leafhoppers
Camellia sinensis (L.) Kunize	Green tea	Black citrus aphid, cottony camellia scale, root weevils, brown soft scale
Cananga odorata (Lam.) J. D. Hook & T. Thompson	Ylang-Ylang	*Samia cynthis ricini* (silkworm), *Mallodon downesi* (longicorn beetle), *Aleurodicus giganteus*
Cannabis sativa L.	Hemp	Wireworm, spidermites, thrips, mustard aphid, beet worm, hemp aphids, stem borers, beetle grub, leafminers, caterpillar, Bertha army worms, *Lygus rugulipennis, Helicoverpa armigera, Aphis gossypii. Nezara* spp., *Mordellistena* spp., *Acherontia atropos, Euteryx atropunctata, Coccinella septempunctata, Drosicha mangiferae*
Capsella bursa-pastoris (L.) Medik.	Shepherd's purse	Flea beetle, moth, *Lygus.* spp.

Scientific name	Common name	Major insects
Capsicum annuum L. var. Annuum *C. annum* L. var. Grossum *C. frutescens* L. *C. annum* L. var. Acuminatum	Cayenne pepper Sweet pepper Chilli pepper	Colorado potato beetles, aphids, beetles, bugs, caterpillars, cutworms, maggots, mealybugs, mites, red spider mites, root-knot nematodes, weevils, whiteflies
Carbenia benedicta L.	Blessed thistle	See *Cnicus benedictus*
Carduus benedita L.	Blessed thistle	See *Cnicus benedictus*
Carica papaya L.	Papaya	Leafhoppers
Carthamus tinctorius L.	Safflower	Safflower aphid, leafminer, aphid, *Acanthiophilus helianthi, Coccinella spetempunctata, Helicoverpa punctigera, Dactynotus compositae, Trichoplusis orichalcea, Perigea capenesis, Heliothis armigera, Helicoverpa armigera, Helicoverpa assulta, Helicoverpa peltigera*
Carum caroi L.	Caraway	*Systole coriandri, S. albipennis*
Cassia senns L. *C. angustifolia* Vahl.	Alexandria senna Tinnevelly senna	Lacebugs, root-knot nematodes, scales, thrips, mealybug, *Caryedon serratus, Lepidopteran, Zabrotes interstitlalis, Thrips tabaci, Dendrothrips bisponosus, Zabrotes chavesi, Catopsilla crocale*
Castanea satuva Mill	Chestnut	San Jose scale, aphids, beetles, borers, caterpillars, scales
Catharanthus roseus (L.) G. Don	Periwinkle	See *Vinca minor*

Scientific name	Common name	Major insects
Caulophyllum thalietroides (L.) Michaux. *C. giganteum* (F.) Loconte & W. H. Blackwell	Blue cohosh	*Epilachna yasutimii*
Cedrus libani subsp. *atlantica*	Cedarwood	Scales, weevils, moth, caterpillar, aphids, sawfly, *Acleris undulana, Dendrolimus pini, Pandemis cerasana, Argyrotaenia ljungiana, Phloeosinus pfefferi, Trichoferus fasciculatus*
Centaurea calcitrapa L.	Star thistle	Oblique-banded leaf roller
Centella asiatica (L.) Urb.	Gotu Kola	*Lasiodema serricome* (beetle)
Chamaelirium luteum (L.) A. Gray	False unicorn root	No information based on the databases searched is available.*
Chamaemelum nobile (L.) All.	Roman chamomile	No information based on the databases searched is available.*
Chamaenerion angustifolium (L.) Scop.	Firewood	See *Epilobium angustifolium*
Chenopodium album L. *C. ambrosioides* L.	Lamb's quarters Wormseed	Cotton leafworm, huauzontle borer, aphid, painted bug, *Cassida nebulosa, Polymerus cuneatus, Colaspis brunnea, Prodenia, Desmis tajes,Pholisora catullus*
Chrysanthemum cinerarifolium (Trevir.) Vis. *C. parthenium* (L.) Berhn.	Pyrethrum Feverfew	Aphids, beetles, borers, caterpillars, cutworms, galls, leaf hoppers, leaf miners, leaf nematodes, mealybugs, mites, orthezias, scales, slugs, tarnished plant bugs, thrips, weevils, whiteflies
C. vulgare L.	Tansy	See *Tanacetum vulgare*

Scientific name	Common name	Major insects
Cichorium intybus L.	Chicory	Aphids, leaf-mining fly, whitefly, *Pemphigus bursarius*
Cimicifuga racemosa (L.) Nutt.	Black cohosh	Root-knot nematodes
Cinnamomum verum J. Presl. *C. camphora* (L.) Nees & Eberm.	Cinnamon Camphor	Mites, scales, termite, whitefly, moth, thrips, leaf miner, sawfly, *Thiymiatris* spp. *Reticulitermes chinensis, Oncideres captiosa, Mymarothrips flavidonotus, Triozidus hangzhouicus, Schizotetranychus lushanensis, Aulacaspis tubercularis, Sorolopha archimedias, Mehteria hemidoxa, Hiratettix niger, Rhachisphora koshunensis, Moriella rufonata*
Citrus limon (L.) Burm. *C. bergamia* Risso & Poit *C. reticulata* Blanco *C. aurantifolia* Swingle *C. aurantium* L. *C. paradisi* Macfad	Lemon Bergamot orange Mandarin Lime Bitter orange Grapefruit	Aphids, beetles, spotted bug (*Clovia bipunctata*), caterpillars, mealybugs, mites, nematodes, thrips, weevils, whiteflies, leafminer, brown aphid, orange fruit borer, foraging ants, Mediterranean fruit fly, citrus gray weevil, cotton aphid, arrowhead scale (*Duplachionapsis divergens*), *Tetraleurodes semi-leunanaria*, *Pantomorus cervinus, Aleurotuberculatus takahashi, Aonidiella aurantii, Dialeurodes citri, Lopholeucaspis japonica, Disprepes abbreviatus*
Cnicus benedictus L.	Blessed thistle	*Ostrinia kasmirica, Altica carduorum, Ceutorhynchus litura, Urophora cardui, Lygaeus equestris*
Cochlearia amorucia L.	Horseradish	See *Armoracia rusticana*
Codonopsis pilosula (Franch.) Nannfeldt. *C. tangshen* Oliver	Codonopsis, Dang Shen, bellflower	No information based on the databases searched is available.*

Scientific name	Common name	Major insects
Coffea arabica L.	Coffee	Mealybugs, red spider mites, scales, whiteflies, coffee berry borer, green scale, white coffee stem borer, leafminer, coffee bug, antestia bugs, *Odontotermes obesus, O. redemanni, O. homi, Archips occidentalis, Tortrix dinota, Icerya pattersoni, Leucoptera meyrick, Macrotermes estherae, Nasutitermes indicola, Aleurothrixus floccosus*
Colchicum autumnale L.	Crocus, meadow saffron	Tulip-bulb aphids
Collinsonia canadensis L.	Pilewort	No information based on the databases searched is available.*
Commiphora molmol Engl. ex Tschirch *C. myrrha* (Nees) Engl.	Myrrh	*Paoliella hystrix* (aphid)
Convallaria majalis L.	Lily of the valley	Slugs, snails, root-knot nematodes, weevils
Convolvulus arvensis L. *C. sepium* L.	Bindweed	Pained bug (*Bagrada cruciferarum*, *B. hirta*), aphids, beetles, caterpillars, leaf miners, scales, weevils, whiteflies
Coriandrum sativum L.	Coriander, Cilantro	*Systole coriandri, S. albipennis*, aphids
Cornus canadensis L.	Bunchberry	Aphids, beetles, borers, caterpillars, cicadas, galls, leaf hoppers, scales, whiteflies, saddleback caterpillar, stinging caterpillar, hagmoth caterpillar, coccid
C. florida L.	Dogwood	Oyster shell scale, brown soft scale, lecanium scale, tree borers, San Jose scale, apple grain aphid, walnut scale

Scientific name	Common name	Major insects
Corylus cornuta Marsh. *C. rostrata* Marsh. *C. avellana* L.	Hazelnut	Lecanium scale, filbert aphid, apple mealybug, eye spotted bud moth, filbert leafroller, tent caterpillars, filbertworm, ominivorous leaftier
Crataegus monogyna Jacq. *C. pinnatifida* Bge.	English hawthorn Chinese hawthorn	Apple-and-thorn skeletonizer, hawthorn aphid, leaf rollers, pear slug, scale insects, spider mites leopard moths, oyster shell scale, wolly elm aphid, apple grain aphid
Crocus sativus L.	Saffron	Aphids, bulb mites
Cryptotaenia japonica Hassk.	Japanese parsley, Mitsuba	See *Petroselinum crispum*
Cucurbita pepo L.	Pumpkin	Bean aphid, melon aphid, potato aphid, nitidulid beetles, seed corn maggot, spider mites, cucumber beetles, squash bug, wireworm, vine borers
Cuminum cyminum L.	Cumin	Aphids, cigarette beetle
Curcuma domestica L. *C. xanthorrhiza* L. *C. longa* L.	Curcuma, Jiang Huang Turmeric	*Lamprosema charesalis, Conogethes punctiferalis, Udaspes folus, Aspidiella hartii, Lema fulvimana, Mimegralla coeruleifrons, Smicronyx roridus*
Cuscuta epithymum Murr. *C. chinensis* Lam.	Dodder	*Smicronyx jungermanniae, Melanagmyza* spp., *Eupoecilla ambiguella, Celastrina argiolus*

Scientific name	Common name	Major insects
Cymbopogon citratus (DC. ex Nees) Stapf. *C. nardus* L. *C. winterianus* Jowitt *C. martinii* (Roxb) Wats	Lemon grass Citronella Palmarosa, gingergrass	*Chilo infuscatellus, Fossispa lutena, Mythimna separata, Microtemes obesi, Stemmatophora fuscibasalis, Diatraea saccharalis*
Cynara scolymus L. *C. cardunculus* L.	Artichoke Wild artichoke	Aphids, thistle aphid, artichoke plume moth, cutworm, variegated cutworm, painted lady butterfly, slugs
Cypripedium calceolus L.	Ladyslipper	*Melanagromyza tokunagai, Pseudoparlatoria parlatorioides, Tenthecoris bicolor* (orchid bug), *Neoneella zikani* (capsid bug), *Neofurius carvalhoi*
Cytisus scoparius (L.) Link.	Broom	Aphids, mealybugs, red spider mites, scales, *Lampides boeticus*
Daucus carota L.	Carrot	Aphids, carrot rust fly, cut worms, garden symphylan, wireworms, leafhopper, slugs
Digitalis purpurea L.	Foxglove	Aphids, beetles, mealybugs, nematodes, thrips, *Dicyphus palidicornis*
Dioscorea oppositae Thunb. *D. villosa* L.	Chinese yam, shen yao Wild or Mexican yam	*Lasioptera* spp., *Heteroligus meles, Coccobius comperei, Adelencyrtus moderatus, Crioceris livida, Cosmopolites sordidus*
Echinacea angustifolia DC. *E. pallida* (Nutt.) Nutt. *E. pururea* (L.) Moench.	Narrowleaf echinacea Pale-flower echinacea Purple coneflower	*Chlosyne gorgone* (nymphalid) *Ligyrocoris barberi* (seedbug) *Pratylenchus penetrans* (nematode)
Echinopanax horridus (Sm.) Decne. & Planch. ex. H. A. T. Harms	Devil's club	See *Oplopanax horridus*

Scientific name	Common name	Major insects
Elaeagnus angustifolia L.	Russian olive	Aphids, scale insect, beetle, *Curculio flavoscutellatus*
Eleutherococcus senticosus (Rupr. ex Maxim) Maxim.	Siberian ginseng	Silkworm
Elymus repens L.	Couch grass	See *Agropyron repens*
Ephedra distachya L. *E. nevadensis* Wats. *E. sinica* Stapf.	Ephedra, horse tail Mormon tea Ma Huang	Parasitic *Hymenoptera, Lasioptera ephedricola*
Epilobium angustifolium L.	Fireweed	*Aphis oenotherae, Macrosiphum euphorbiae*
Epilobium parviflorum Schreb.	Small-flowered willow herb	*Altica tombacina* (bronze flea beetle), *Mompha lacteella, Jalysus spinosus, Aphis epilobii*
Equisetum arvense L. *E. telmatecia* Ehrb.	Horsetail	*Ceresa bubalus*
Erythroxylum coca Lam.	Coca	Aphids, caterpillar, *Hylesia lineata*
Eschscholtzia california Cham.	California poppy	Root-knot nematodes
Eucalyptus citriodora Hool. *E. globulus* Labill.	Gum tree Eucalyptus	Aphids, borers, caterpillars, mealybugs, mites, scale, snout beetle, leaf beetle, brown inchworm, longhorn borer, termite, *Psaltoda moerens, Acromyrmex subterraneus molestans, Blera* spp., *Psorocampa denticulata, Eupseudosoma aberrans, E. Involuta, Glena* spp., *sarsina violascens, Gonipterus scutellatus, Cardiaspina crawford*
Eugenia caryophyllata L.	Clove	See *Syygium aromaticum*

Scientific name	Common name	Major insects
Eupatorium perfoliatum L.	Boneset	Aphids, leaf miners, root-knot nematodes, scales, thrips, *Argyrotaenia amatana* (moth)
Euphrasia officinalis L.	Eyebright	*Phytomyza affinis*
Fagopyrum tataricum L.	Buckwheat	*Thysanoptera* (thrips), *Aphis gossypii* (aphid), *Acyrthosiphon rubi* (aphid)
Fagus grandifolia Ehrb. *F. sylvatica* L.	Beech European beech	Aphids, borers, caterpillars, leaf hopers, scale
Filipendula ulmaria (L.) Maxim.	Meadow sweet	*Cladius pectinicornis, Aulacorthum solani* (aphid), *Pegomya rubivora*
Foeniculum vulgare Mill.	Fennel	*Systole coriandri, S. albipennis*, aphid
Forsythia suspensa (Thunb.) Vahl.	Forsythia, Lian qiao	Leaf bugs
Fragaria vesca L.	Strawberry	Aphids, garden symphylan, slug, spittlebug, strawberry leafroller, root weevils, two spotted mites, red spider mites
Fraxinus americana L. *F. excelsior* L.	White ash	Aphids, boreres, bugs, caterpillars, flower galls, nematodes, scales, hagmoth caterpillar, beetle, moth, ash lopper, ash weevil, ash bark beetle, *Macriogta hispana, Zeuzera pyrina, Cossus cossusas, Strictocepthala bisonia, Tettigella viridis, Prays fraxinella*
Fucus vesiculosus L.	Kelp, bladderwrack	No information based on the databases searched is available.*
Galium aparine L.	Cleavers	*Myzus cerasi*, *Dysaphis reaumuri* (aphid) *Dysaphis pyri* (aphid), *Aroga argutiola, Macrosiphum miscanthi* (aphid)

Scientific name	Common name	Major insects
Gaultheria procumbens L.	Wintergreen, teaberry	*Poliaspis gaultheriae*
Geranium macrorrhizum L.	Geranium, American cranesbill	Leaf hopper
Ginkgo biloba L.	Ginkgo	*Mahasena aurea, Etiella zinckenella, Brachytrupes portentosus, Argolis ipsilon, Metabolus flavescens, Gulcula panterinaria*
Glechoma hederacea L.	Ground ivy	No information based on the databases searched is available.*
Glycine max (L.) Merrill.	Soybean	Bean stem miner, bean leaf roller, soybean aphid, sweet potato whitefly, cutworm, pod borer, bean stink bug, green stink bug, bean bug
Glycyrrhiza glabra L. *G. uralensis* Fisch ex DC	Licorice Chinese licorice	*Porphyrophora ningxiana, Acanthoscelides aureolus*
Gossypium hirsutum L.	Cotton	Cotton aphids, boll weevil, bollworm, cotton fleahopper, tarnished plant bug, rapid plant bug, conchuela, southern green stink bug, spider mites, grasshoppers, thrips, white fly
Grindelia robusta Nutt. *G. squarrosa* (Pursh) Donal.	Gumweed	*Helipodus ventralis* (weevil), *Dactynotus richardsi* (aphid), *Trionymus grindeliae* (mealybug)
Hamamelis virginiana L.	Witch hazel	Aphids, mealybugs, red spider mites
Harpagophytum procumbens DC. ex Meisn	Devil's claw	No information based on the databases searched is available.*
Hedeoma pulegiodes (L.) Pers.	Pennyroyal	No information based on the databases searched is available.*

Scientific name	Common name	Major insects
Hedera helix L.	English ivy	Aphids, Japanese holly mites, broad mites, cyclamen mites, soft scale
Helianthus annuus L.	Sunflower	Oblique-banded leaf roller, aphids, beetles, bugs, caterpillars, maggots, mealybugs, nematodes, scales
Helichrysum angustifolium (Lam.) DC.	Curry plant	Leaf hoppers, root-knot nematodes
Helonias dioica L.	False unicorn root	See *Chamaelirium luteum*
Heracleum maximum Bartr. *H. lanatum* Michx. *H. sphondylium* L.	Cow parsley Cow parsnip Cow parsnip	Aphids, caterpillars
Hibiscus sabdariffa L. *H. rosa-sinensis* L.	Hibiscus	Aphids, beetles, caterpillars, root-knot nematodes, scales, whiteflies, mealybug, plant bug, chysomelid beetle, *Aulacophora temoralis, Spermophagus pygopubens, Earias vittella, Dysdercus cingulatus, Pseudagrilus subbius, Diaeretiella rapae, Trioxys angelicae, Bemisia tabaci, Acontia transversa, Pectinophora endema, Pyroderces falcatella, Oxycarenus laetus*
Hierochloe odorata (L.) Beauv.	Sweetgrass	No information based on the databases searched is available.*
Hippophae rhamnoides L	Sea buckthorn	Aphids, thrips, two-spoted mites, metallic wood boring beetles
Humulus lupulus L.	Hops	Armyworms, cutworm, loopers, caterpillars, hop looper, corn earworm, European earwig, hop aphid, leafrollers, omnivorous leaftier, spider mites, western spotted cucumber beetle, root weevils, prionus beetles

Scientific name	Common name	Major insects
Hydrangea arborescens L.	Hydrangea	Aphids, leaf tiers, red spider mites, stem and root-knot nematodes, rose chafer, tarnished plant bugs, thrips
Hydrastis canadensis L.	Goldenseal	No information based on the databases searched is available.*
Hypericum perforatum L.	St. John's wort	Root-knot nematodes, aphids, rose leaf roller, gypsy moth, gall tick, comma-shaped scale, sea buckthorn fly, caterpillars
Hyssopus officinalis L.	Hyssop	Root-knot nematodes, scales
Ilex aquifolium L.	English holly	Aphids, potato aphid, cottony camellia scale, holly bud moth, holly leafminer, heooly scale, lecanium scale, oblique-banded leafroller, filbert leafroller, brown soft scale, orange tortrix
Inula helenium L.	Elecampane	*Cassida ferruginea, Myopites blotii*
Iris versicolor L.	Blue flag	Aphids, beetles, iris borers, caterpillars, flies, mites, nematodes, slugs, thrips, weevils, mealy bug, *Aphtona semicyanea, Ceratilis capitata*
Jasminum grandiflorum L. *J. auriculatum* Vahl.	Royal jasmine	Root-knot nematodes, scales, whiteflies, *Aleurotuberculatus takahashi*
Juglans regia L. *J. nigra* L.	Walnut Black walnut	Walnut aphids, lecanium scale, frosted scale, walnut husk fly, walnut caterpillars
Juniperus communis L.	Juniper	Aphids, cypress tip moth, juniper scale, juniper tip midge, juniper webworm, leaf miners, spider mites, conifer mites
Laminaria digitata (Hudss.) Lamk. *L. saccharina* (L.) Lamk.	Kelp	See *Fucus vesiculosus*

Scientific name	Common name	Major insects
Lamium album L.	Nettle	Aphids, *Meligethes haemorrhoidalis*
Larrea tridentata (Sesse. & Moc. ex DC.) Coville.	Chapatral	*Asphondylia* (gall midges), *Ligurotettix coqulletti* (grasshopper), *Paranthus linsley* (beetle), *Psiloptera cuperopunctata*, *Bootettix argentatus* (grasshopper), *Cibolacris parviceps* (grasshopper), *Psiloptore drummondi, Psiloptera webbii*
Laurus nobilis L.	Bay laurel, laurel	Lacebugs, holly scale, soft scale, caterpillars, psyllids
Lavandula angustifolia Mill.	Lavender	*Thomsimana lavandulae, Sophoronia humeralla,* spittlebugs or froghoppers, root-knot nematodes
Ledum latifolium Jacq.	Labrador tea	*Masonaphis pinawae, Lepidosaphes ulmi* (coccid)
Lemna minor L.	Duckweed	*Hydrozeles lemnae*
Leonurus cardiaca L.	Motherwort	No information based on the databases searched is available.*
Levisticum officinale W. Koch.	Lovage	Aphids
Ligustrum lucidium W. T. Aiton. *L. vulgare* L.	Ligustrum fruit Privet	Lilac leafminer, thrips, holly scale, privet aphids
Linum usitatissimum L.	Flax	Sawfly, root-knot nematodes
Liquidambar styraciflua L. *L. orientalis* L.	Gum tree, storax Oriental sweet gum	Caterpillars, scales, termite, silk moth, *Passalida* spp., *Longistigma liquidambarus*

Scientific name	Common name	Major insects
Lobelia inflata L. *L. siphilitica* L. *L. pulmonaria* L.	Lobelia Blue lobelia Lungwort	Aphides, caterpillars, leaf hoppers, root-knot nematodes, wireworms
Lomatium dissectum (Nutt.) Math. & Const.	Lomatium	*Papilio zelicaon* (butterfly)
Lonicera caerulea L. *L. caprifolium* L.	Mountain fly honeysuckle Dutch honeysuckle	Aphids, honeysuckle aphid, bugs, caterpillars, flea beetles, four-lined plant bugs, mealybugs, root-knot nematodes, sawflies, scales, whiteflies, *Heliothis virescens, Helicoverpa zea, Metcalfa pruinosa, gigantomiris jupiter*
Lycium barbarum L. *L. chinense* Mill. *L. pallidum* L.	Lycium Chinese wolfberry Wolfberry	Aphids, borers, caterpillars, galls, leaf beetle, Chinese wolfberry aphid
Lycopodium clavatum L.	Clubmoss	No information based on the databases searched is available.*
Lysimachia vulgaris L.	Loosestrife	*Galerucella nymphaeae, Mordwilkoja vagabunda, Aceria laticincta, Aphis frangulae*
Lythrum salicaria L.	Purple loosestrife	Aphids
Macropiper excelsum G. Forst	Kava-Kava	See *Piper methysticum*
Mahonia aquifolium (Pursh) Nutt.	Oregon grape	Holly scale, aphids, root-knot nematodes, whiteflies
Marrubium vulgare L.	Horsehound	Root-knot nematodes

Scientific name	Common name	Major insects
Matricaria chamomilla L. *M. recutita* L.	Chamomile German chamomile	Thrips
Matteuccia strothiopteris (L.) Tod	Ostrich fern	*Chirosia betuleti* (gallformer)
Medicago sativa L.	Alfalfa	Alfalfa weevil, caterpillar, looper, aphids, armyworms, blister beetles, clover leaf weevil, clover root cureullo, cutworm, grasshopper, meadow spittlebug, pea leaf weevil, western spotted cucumber beetle, slugs
Melaleuca alternifolia (Maid. & Bet.) Cheel	Melaleuca, medicinal tea tree	*Melaleuca quinquenervia, Paropsisterma tigrina,* weevil
Melilotus officinalis (L.) Lamk. *M. arvensis* L.	Sweet clover Melilot	*Polyommatus icarus* (blue butterfly), *Thysanoptera* spp., *Tychius meliloti, Apion meliloti, Therioaphis riehmi*
Melissa officinalis L.	Lemon balm, balm	*Choreutis marzoccai*
Mentha piperita L. *M. spicata* L. *M. pulegium* L.	Peppermint Spearmint Pennyroyal	Mint aphid, cabbage looper, alfalfa looper, variegated cutworm, spotted cutworm, painted lady (thistle butter fly), mint cutworm, red backed cutworm, mint root borer, mint flea beetle, root weevils, slugs, wireworm, spider mites, garden symphylan, mint stem borer
Monarda odoratissima Benth. *M. punctata* T. J. Howell	Horsebalm Horsemint	Borers, scales
Myosotis scorpiodes L.	Lily of the valley	See *Convallaria majalis*

Scientific name	Common name	Major insects
Myrica penxylvanica Lois.	Bayberry	Caterpillars, *Myrciamyia maricaensis, Neomitranthella robusta, Lepidosaphes cupressi, Bactrocera cucurbitae, Bactrocera dorsalis, Phyllonorycter* spp.
Myristica fragrans Houtt.	Nutmeg	*Leptocorisa acuta, Dacryphalus sumatranus, Xyleborus myristicae*
Myrtus communis L.	Myrtle	Aphids, mealybugs, trunk borer, scales, *Heliothrips haemorrhoidalis, Tetraleurodes bicolor.*
Narcissus pseudonarcissus L.	Daffodil	Aphids, bulb flies, bulb mites, caterpillars, mealybugs, millipedes, nematodes, thrips
Nasturtium officinale L.	Watercress	Bean aphid, cabbage flea beetle, beet armyworm, cabbage looper, diamondback moth, imported cabbage worm
Nepeta cataria L.	Catnip, catmint	Caterpillars, leaf hoppers
Nymphaea alba L.	White water lily	Water lily beetles, aphids, beetles, leaf cutters
Ocimum basilicum L.	Basil	*Spodoptera littoralis* (moth), *Helicoverpa armigera, Cochliochila bullita*
Ocotea bullata (Birch) E. May	Stinkwood	*Cerambycidae, Scolytidae, Platypodidae, Coptotermes* spp.
Oenothera biennis L.	Evening primrose	*Altica oleracea* (chrysomelid), *Pseudatomoscelis seriatus* (leafhopper), *Jalysus spinosus*
Olea europaea L.	Olive	Beetles, shot hole borers, caterpillars, root-knot nematodes, scales, thrips, fruit fly, *Prays oleae, Lepidosaphes ulmi, Euphyllura olivina, Saissetia oleae, Prolasioptera beriesiana, Dacus oleae*

Scientific name	Common name	Major insects
Oplopanax horridus (Sm.) Miq.	Devil's club	No information based on the databases searched is available.*
Opuntia fragilis (Nutt.) Haw.	Prickly pear	Root mealybugs, corky scab
Origanum majorana L.	Sweet marjoram	*Eupteryx atropunctata, Chamaesphecia acrifrons*
Origanum vulgare L. subsp. *hirtum* (Link) Ietswaart.	Marjoram, Oregano	*Eupteryx atropunctata, Thrips origani*
Paeonia lactiflora Pall. *P. officinalis* L. *P. suffruiticosa Andr*.	White peony, Bia shao European peony Tree peony, Mu Dan Pi	Ants, beetles, bugs, curculios, nematodes, scales, thrips
Panax ginseng C. A. Meyer. *P. notoginseng* (Burk.) F. H. Chen *P. quinquefolium* L.	Asian ginseng Tian Qi American ginseng	Cutworm, slug, *Gryllotalpa unispina, G. africana, Holotrichia diomphalia, Pleonomus canaliculatus, Agrotis ypsilon, Agriotes fusicollis, Hylemyia platura, Gryllus testaceus, Loxostege sticticalis*
Papaver braceatum L. *P. rhoens* L.	Iranian poppy Poppy, corn poppy	Aphids, four-lined plant bugs, leaf hoppers, leaf nematodes, mealybugs, root-knot nematodes, rose chafers, tarnished plant bugs, *Phytomyza atricomis, Ceuthorhyrichus denticulatus* (weevil), green peach aphid, *Ormyrus capsalis*
Passiflora incarnata L.	Passion flower	Caterpillars, mealybugs, root-knot nematodes, scales, mite, *Phitonis crucfier, Trigona spinipes, Lasiodiplodia theobromae, Iridia viridiphaga, Azamora penicillana, Nymphalidae* spp., *Pseudaulacaspis pentagona.*

Scientific name	Common name	Major insects
Pelargonium capitatum (L.) L'Her *P. crispum* L'Her. ex. Aiton. *P. graveolens* (L.) L'Her ex. Aiton	Wild rose geranium Lemon geranium Rose geranium	Aphids, beetles, caterpillars, mealybugs, mites, scales, slugs, weevils, whiteflies, *Cacyareus marshalli*
Peltigera canina L.	Ground liverwort, Dog lichen	No information based on the databases searched is available.*
Petroselinum crispum (Mill.) Nym.	Parsley	Aphids, armyworms, loopers, diamondback moth, carrot rut flies, willow aphids
Phataris canariensis L.	Canary creeper	See *Tropaeolum majus*
Phellodendron chinensis Schneid.	Chinese corktree, Huang Bai	*Calopha nigra, Acariformes, Tetrapodili* (mites)
Phoradendron leucarpum (Raf.) Rev. & M. C. Johnst. *P. flavescens* (Push.) Nutt.	Mistletoe American mistletoe	*Pereute* spp., *Pseudothysanoes phoradendri*, *Atlides halesus*, *Spodoptera eridania* (armyworm), *Pseudodiaspis multipora*
Phyllanthus emblica L.	Myrobalan, emblic	Whitefly, *Betousa stylophora, Resseliella seitner*
Physalis alkekengi L.	Chinese lantern	Aphids, beetles, caterpillars, nematodes, weevils, tobacco aphid, *Heliothis subflexa, Lema trineata daturaphila, Trichobaris mucorea, Systena blanda, Henosepilachna vigintioctopunctata, Epitrix* spp.
Phytolacca americana L.	Pokeweed	*Cephisus siccifolius, Pyrausta ainsliei, Tiracola plagiata*

Scientific name	Common name	Major insects
Picea mariana (Mill) Black. *P. glauca* (Moench) Voss. *P. ables* (L.) Karst	Black spruce White spruce Norway spruce	Aphid, budworm, colley spruce gall adelgid, eastern spruce gall adelgid, coneworms, douglas-fir tussock moth, pine weevil, pine needle scale, spider mites, bud scale, needleminer, gall aphids.
Pimpinella anisum L.	Anise	*Lastoderma serricome* (cigarette beetle)
Pinus mugo Torra var. Pumilio *P. palustris* Mill. *P. strobus* L. *P. albicaulis* Engeim. *P. contorta* Dougl. ex. Loud	Dwarf mountain pine Southern pitch pine White pine Lodgepole pine	Black pineleaf scale, coneworms, eriophyid mites, European pine shoot moth, pandora moth, pine shoot moths, mountain pine beetle, pine butterfly, pine needle scale, spider mites, pine aphids, powdery pine needle aphid, pine bark adelgid, sequoia pitch moth, pine sheath miner. Douglas fir webworm, silver spotted tiger moth
Piper methysticum G. Forst	Kava-Kava	Stem borer, mealybugs, red spider mites
P. nigrum L.	English pepper	Armyworms, cutworms, loopers, green peach aphid, garden symphylan, flea beetles, spider mites, wireworm, stem borer, Colorado potato beetles
Plantago asiatica L. *P. psyllium* L. *P. lanceolata* L.	Psyllium Plantain	*Nymphalidae lavae, Arctiidae larvae, Mniotype solieri, Hyadaphis coriandri* (aphid), *Eurythecta zelaea, Spilosoma congrue*, *Junonia coenia, epiphyas postvitana, Aphis gossypii, Chrysolina banksi, Diacrisis virginica*, Rosy apple aphid
Podophyllum peltatum L.	Mayapple	No information based on the databases searched is available.*
Polygala senega L.	Seneca snakeroot	*Tetranchus telarius* (mite)
Polygonum bistorta L. *P. multiflorum* Thunb.	Bistort root Flowery knotweed	Aphids, painted bug (*Bagrada cruciferarum*, *B. hirta*)

Scientific name	Common name	Major insects
Populus balsamifera L. *P. candicans* L. *P. tremuloides* Michx.	Poplar, Balm of gilead American aspen	Oystershell scale, poplar and willow borer, lettuce root aphid, carpenter worm, satin moth, cicadas, soft scale
Primula vulgaris Huds. *P. veris* L.	Primrose Cowslip	Vine and flower weevils, root mealybugs, leaf hopper, aphids, beetles, black vine weevils, nematodes, red spider mites, slugs, weevils, whiteflies
Prunella vulgaris L.	Heal all, selfheal	*Otiorhynchus porcatus* (weevil)
Prunus africana L.	African prune	Scale, aphid, peach twig borer, leafroller, thrips, spider mites, rust mites, apple and thorn skeletonizer, earwigs, pear slug, featherd borer, plum curealios, lesser appleworm
P. dulcis (Mill.) D. A. Webb.	Almond	Peach leaf curl, aphids, borers, caterpillars, leaf hoppers, mites, nematodes, scales, thrips
P. mume Siebold & Zucc.	Japanese apricot	San Jose scale, mite eggs, peach twig borer, fruit tree leafroller, spider mites, earwigs, aphids, lecanium scale
P. serotina J. F. Ehrb.	Black cherry, wild cherry	San Jose scale, black cherry aphid, leafrollers, climbing cutworms, cherry fruit fly, earwig, pear slug, spider mites, shothole borer
P. virginiana L.	Chokecherry	Leopard moths, plum curculios, San Jose scale, black cherry aphid, leafrollers, climbing cutworms, cherry fruit fly, earwig, pear slug, spider mites, shothole borer
Pueraria lobata (Wild.) Ohwi.	Kudzu	*Maruca testulalis*, *Cerotoma tingomarianus*
Pygeum africanum L.	Pygeum	See *Chrysanthemum cinerarifolium*
Pyrethrum cinerarifolium L.	Pyrethrum	See *Chrysanthemum cinerariifolium*

Scientific name	Common name	Major insects
Quercus alba L.	White oak	Jumping oak gall, leaf galls, stem galls, pit scale, gall wasps, oak skeletonizer, twig gall, western oak looper, flatheaded apple tree borer, yellow-necked caterpillar
Ranunculus occidentalis Nutt. *R. ficaria* L.	Buttercup Pilewort	Aphids, caterpillars, red spider mites, *Rhopalosiphum nymphaeae, Thripidae fallaciosus*
Raphanus sativus L.	Radish	Cabbage aphid, turnip aphid, green peach aphid, cabbage maggot, cutworms, armyworm, diamond back moth, flea beetles, wireworms
Rhamnus catharticus L. *R. frangula* L. *R. purshiaanus* L.	Buckthorn Alder buckthorn Cascara sagrade	Buckthorn aphids, borers, caterpillars, scales, whiteflies
Rheum palmatum L. *R. officinal* Baill. *R. tanguticum* L.	Chinese rhubarb Rhubarb Rhubarb	Armyworm, cutworm, slugs, garden symphylan
Rhodiola rosea L.	Roseroot	See *Sedium rosea*
Rhus radicans L.	Poison ivy	Aphids, boreres, caterpillars, flea beetles, mites, psyllids, scales, *Aceria zhejiangensis, Schlechtendalia chinensis*
R. toxicodendron L.	Roseroot	See *Sedum rosea* subsp. *integrifolium*
Ribes nigrum L. *R. lacustre* (Pers.) Poir.	Black currant Black gooseberry	Cottony maple scale, cambium miner, tent caterpillars, two spotted spider mites, leaf bugs, gooseberry sawflies, currant aphid, currant borer, fruit fly, imported currant worm, San Jose scale, Virginia creeper leafhopper
Robertium macrorrhizum Pic.	Geranium	See *Geranium macrorrhizum*

Scientific name	Common name	Major insects
Rosa canina L. *R. damascena* Mill.	Dog rose Damask rose	Leaf cutting bees, leafrollers, raspberry cane maggot, rose aphid, red humped caterpillar, root weevils, rose curculio, rose leaf hopper, midge, slug, stem miner, small carpenter bees, thrips, spider mites, western spotted cucumber beetles, gall wasp, rose chafers
Rosmarinus officinalis L.	Rosemary	*Frankiniella occidentalis*
Rubus chamaemorus L. *R. fruiticosus* L.	Cloudberry Blackberry	Beetles, borers, caterpillars, galls, leaf hoppers, leaf miners, maggots, mealybugs, mites, sawflies, scales, weevils, whiteflies
R. idaeus L.	Raspberry	Raspberry bettles, aphids, borers, caterpillars, leaf hoppers, maggots, mites, sawflies, scales, spittlebugs, weevils, whiteflies
Rumex acetosella L. *R. obtusifolia* L.	Sorrel Dock	Onion thrips, caterpillar, moth, *Pegomya nigretarsis, Estigmene acraea, Eurythecta zelaea*
Ruscus aculeatus L.	Butcher's broom	*Dynaspidiotus britannicus, Otiorrhynchus sulcatus* (weevil), *Ceroplastes rusci* (coccid)
Ruta graveolens L.	Rue, herb of grace	*Amblyseius chergui* (mite), *Neomargarode erythrocephala, Rhinocola succincta*
Salix alba L. *S. discolour* Muhlenb. *S. caprea* L.	White willow Pussy willow	Giant willow aphid, poplar and willow borer, spiny elm caterpillar, satin moth, carpenter worm, willow bean gall sawflies, gall mites, gall wasp, yellow-necked caterpillar
Salvia officinalis L. *S. sclarea* L.	Sage Clary sage	Aphids, miner, beetles, borers, caterpillars, leaf hoppers, leaf nematodes, mites, orthezias, root-knot nematodes, tarnished plant bugs, whiteflies
Sambucus nigra L. *S. canadensis* L.	Elderberry	San Jose scale, striped alder sawfly, tent caterpillars, gall wasp, aphids, bettles, borers, bugs, mealybugs, thrips

Scientific name	Common name	Major insects
S. racemosa L.	Elder	San Jose scale, striped alder sawfly, tent caterpillars
Sanguinaria canadensis L.	Bloodroot	No information based on the databases searched is available.*
Sanicula marilandica L.	Black snakeroot	*Papilio polyxenes asteruis* (butterfly)
Santalum album L.	Sandalwood	White scales, defoliating caterpillars, green scales, black scales, sandalwood beetle (*Mylabris pustulata*), mealybugs, branch borers, fruit pentatomid bugs, lac insect, *Amata passalis,* jassids (*Pentacephala nigrilinea*), *Sarina migrachypeata, Eubracys tomentosa*
Saponaria officinalis L.	Soapwort	*Subcoccinella vigintiquattuorpuncata* (beetle)
Sarothamnus scoparius L.	Broom	See *Cytisus scoparius*
Sassafras albidum (Nutt.) Nees	Sassafras	Caterpillars, Japanese beetles, scales, *Faraleprodera itzingeri*
Satureja hortensis L. *S. montana* L.	Savory	No information based on the databases searched is available.*
Schisandra chinensis (Turcz.) Baill.	Schizandra (wu wei zi)	*Arboridia erythrina* (leaf hopper), *Arboridia suwai* (leaf hopper)
Scutellaria baicalensis Georgi. *S. lateriflora* L. *S. galericulata* L.	Baical skullcap Virginia skullcap Skullcap	*Phyllobrotica* (leaf beetles)

Scientific name	Common name	Major insects
Sedum acre L. *S. rosea* (L.) Scop. subsp. *integrifolium*	Stonecrop (small houseleek) Roseroot	Aphids, mealybugs, root-knot nematodes, scales
Sempervivum tectorum L.	Houseleek, hens and chicks	Mealybugs
Senecio vulgris L.	Groundsel	Aphids, root-knot nematodes, whiteflies, *Tyria jacobaeae, Homoeosoma deltaeparanensis*
Senna alexandrina L.	Alexandrina	See *Cassia angustifolia*
Serenoa repens (Bartr.) Small.	Saw palmetto	*Litoprosopus futilis* (moth)
Sesamum indicum L. *S. orientale* L.	Sesame	Capsule borer, leaf roller, milkweed bug, aphid, *Asphondylia sesami, Orosius albicinctus*
Shepherdia canadensis (L.) Nutt.	Buffalo berry	*Choristineura rosaceana* (leaf roller)
Silybum marianum (L.) Gaertn.	Milk thistle	*Nezara viridula*
Simmondsia chinensis (link) C. Schneid	Jojoba	*Braconidae*
Solidago virgaurea L.	Goldenrod	Stem-galling fly (*Eurosia solidaginis*), *Epiblema scudderiana, Uroleucon tissoli, Uroleucon nigrituberculatum*

Scientific name	Common name	Major insects
Sorbus aucuparia L.	Mountain ash	Aphids, leaf blister mite, woolly aphids, pear leaf blister, mites, apple and thorn skeletonizer
Spirea ulmaria L.	Meadowsweet	See *Filipendula ulmaria*
Stachys officinalis (L.) Trev.	Betony, woundwort	*Amblyseius victoriensis* (mite), *Eurrhypara hortulata, Tingis stachydis*
Stellaria media (L.) Vill.	Chickweed	*Ytislrutofrd bspotstiotum* (whitefly)
Stevia rebaudiana (Bertoni) Bertoni	Stevia, sweet herb of Paraguay	Aphids, mealybugs, red spider mites, whiteflies
Symphytum officinale L.	Comfrey	*Acleris latifasciana, Aphis rhamni* (aphid)
Syringa vulgaris L.	Lilac	Lilac leafminer, oyster shell scale, aphids, beetles, borers, caterpillars, giant hornet, whiteflies
Syzygium aromaticum (L.) Merr. & Perry	Clove	*Trioza eugeniae, Lymantria xylina, Curculic c-album, Bemisis tabaci, Drosicha mangiferae, Polychrosis cellifera, Acrocercops* spp., *Mylklocerus undecimpustulatus, Poekilocerus pictus, Orgyia posica, Trioze eugenlae,* weevil
Tagetes lucida Cav.	Mexican mint marigold	Beetles, borers, bugs, caterpillars, cutworms, leaf hoppers, mites, root nematodes, root-knot nematodes, slugs, scale, leafminer, *Bemisis tabaci, Hydatothrips samayunkur, Heliothis armigera, Bellicositermis bellicosus*
Tanacetum parthenium (L.) Schultz-Bip.	Feverfew	See *Chrysanthemum parthenium*
T. cinerarifolium L.	Pyrethum	See *Chrysanthemum cinerarifolium*

Scientific name	Common name	Major insects
T. vulgare L.	Tansy	Root-knot nematodes, green peach aphid, *Chasmatopterus* spp.
Taraxacum officinale G. H. Weber ex Wigg.	Dandelion	Weevil
Taxus x media Rehd. *T. brevifolia* Nutt.	Yew	False spider mite, root weevils, lecanium scale, cotton camellia scale, vine weevils
Thymus citriodorus (Pers.) Schreb. *T. vulgaris* L	Lemon thyme Thyme	Mealybugs
Tillia cordata Mill. *T. europaea* L.	Linden, small-leaved lime	Aphids, gall mite, spider mites, Japanese beetles, flatheaded apple tree borer
Tribulus terrestris L.	Puncture vine	*Schistocerca gregaria* (hoppers of locusts), *Amsacta moorei* (red hairy caterpillar)
Trifolium incarnatum L. *T. pratense* L.	Clover (crimson) Clover (red)	Loopers, aphids, webworms, clover leaf weevil, clover root borer, clover root curcullo, cutworm, armyworms, grasshoppers, lesser clover leaf weevil, meadow spittlebug, pea leaf weevil, western spotted cucumber beetle, slugs
Trigonella foenum-gracecum L.	Fenugreek	Aphid, *Empoasca kerri*
Tropaeolum majus L.	Canary creeper Nasturtium	Aphids, bugs, caterpillars, flea beetles, leaf miners, red spider mites, root-knot nematodes, thrips, *Chromatomyia horticola, Pieris rapae, Plutella xylostella (*diamondback moth)
Tussilago farfara L.	Colisfoot	*Acidia cognata*

Scientific name	Common name	Major insects
Ulmus rubra Muhl. *U. procera* L.	Slippery elm, sweet elm English elm	Aphids, leaf beetles, borers, bugs, caterpillars, lacebugs, leaf hoppers, leaf miners, red spider mites, scales, weevils, *Cacopsylla ulmi, Megalopyge opercularis, Lyctus cavicollis, Anoplophora glabripennis, Agenocimbex ulmusvora, Bucclatria ulmella, Zeuzera pyrina, Cossusas cossus, Scolytus multistriatus, Tinocallis saltans, Epinotia abbreviana*
Umbelluslaria california Nutt.	California laurel Bay laurel	Aphids, caterpillars, scales, thrips, whiteflies
Uncaria tomentosa (Willd.) DC. *U. guianensis* (Aubl.) Gmel.	Cat's claw	*Helopeltis sumatranus* (tea mosquito bug), *Areta camea* (drepanid), *Margaronia marginata*, *Tricentrus caliginosus*, *Ebhul varius*, *Centrotypus* spp., *Anaca* spp., *Paranthrene cyanogama*
Urtica dioica L. *U. urens* L.	Stinging nettle	*Flatid metcalfa pruinosa, Frankliniella occidentalis*
Vaccinium macrocarpon L. *V. vitis-idaea* L.	Cranberry Mountain cranberry, cowberry, foxberry	Black headed fire worm, cranberry girdler, cranberry tipworm, cranberry fruit worm, lecanium scale, cutworms, root weevils, strawberry root weevil
V. mrytilloides Michx. *V. myrtillus* L. *V. oreophilum* Rydb.	Blueberry Bilberry	Aphids, lecanium scale, obscure root weevil, leaf rollers, root weevils
Valeriana officinalis L.	Valerian	Japanese beetles
Verbascum thapsus L.	Mullein	Root-knot nematodes, *Campylomma verbasci, Aphis verbasci, Euschistus servus, Euschistus variolarlus*
Veronica officinalis L.	Speedwell	Caterpillars, root-knot nematodes, aphid

Scientific name	Common name	Major insects
Vetiveria zizanioides L. (Nash.)	Vetiver	*Locusta migratoria migratoriodes* (locust), *Lachnosterna serrata* (white grub)
Viburnum opulus L.	Highbush cranberry, crampbark	Aphids, beetles, borers, bugs, caterpillars, galls, leaf hoppers, mites, plant hoppers, root-knot nematodes, scales, thrips, whiteflies
Vinca minor L.	Periwinkle	Aphids, beetles, leaf hoppers, orthezias, root-knot nematodes, scales
Viola tricolor L.	Pansy	Violet leaf gall midges, aphids, bettles, caterpillars, cutworms, mealybugs, mites, root nematodes, root-knot nematodes, sawflies, slugs, sowbugs, wireworms
Viscum album L.	Mistletoe	*Carulaspis* spp.
Vitex agnus-castus L. *V. negundo* L.	Chaste tree Chinese chaste	*Hyblaea puera*
Vitis labrusca L. *V. vinifera* L.	Grape	Erineum mite, European fruit lecanlum scale, thrips, black vine weevil, polycaon borer, stinkbugs, cutworms, grape leaf hopper, grape mealybug, cottony maple scale, twig borer, Japanese beetle, Virginia creeper leaf hopper
Withania somnifera Dunal.	Ashwagandha, Indian ginseng	*Epilachna* spp., painted bug (*Bagrada cruciferarum*, *B. hilaris*)
Yucca aloifolia L. *Y. glauca* L.	Yucca Soapweed	Aphids, borers, bugs, caterpillars, mealybugs, mites, root-knot nematodes, scale, thrips
Zanthoxylum americanum Mill	Prickly ash, toothache tree	*Comopsylla zanthoxylae, Podagricomela cuprea, Alcidodes sauteri, Nortia luteosignata,* flea beetle

Scientific name	Common name	Major insects
Zea mays L.	Corn	Aphids, corn earworm, root worms, cutworms, armyworms, garden symphylan, grasshopper, slugs, seedcorn maggot, spider mites, thrips, wireworm, leaf miner, beetle grubs
Zingiber officinale Roscoe	Ginger	*Dichocrosis punctiferalis* (borer), *Chalidomyia atricornis* (rhizome maggots), *Conogethes punctiferalis, Udaspes folus, Aspidella hartii, Lema fulvimana* (chrysomelid), *Neochetina eichhorniae* (weevil), *Udonga montana* (stink bug), *Holotrichia coracea* (white grub), *Cockerelliela zingiberai*

* Databases searched: Biological Abstracts on compact diskette. 1985-present; Botanical Abstracts. 1918-1922; Bradley, Mary A., Index to publications of the United States Department of Agriculture. 1901-1940; CAB Abstracts on compact diskette. 1972-present; Horticultural Abstracts 1961-present; AGRICOLA. 1984-1993; Review of Applied Mycology. Plant host-pathogen index to volumes 1-40 (1922-1961).

References

1. Abraham, C. C. and K. S. R. Mony. 1978. Occurrence of *Leptocorisa acuta* Fahr. (Coreidae, Hemiptera) as a pest of nutmeg trees. J. Bombay Natural History Soc. 74: 553.

2. Adams, J. 1927. Medicinal Plants and Their Cultivation in Canada. Dominion Expt. Farm Publication, Ottawa, Canada. 29 p.

3. Ahn, S. B., S. H. Lee and W. S. Cho. 1990. The insect pest species on the composite herb plants and their damages. Res. Reports Rural Development Administration Crop Protection 32: 26-31.

4. Annoymous. 1952. Insects, The Yearbook of Agriculture 1952. U.S.D.A., Washington, DC. 780 p.

5. Anonymous. 1970. Index of Plant Diseases in the United States. Agriculture Handbook No. 165. U.S.D.A. Washington, DC. 532 p.

6. Bandra, K. A., N. P. Peries, I. D. R. Kumar and V. Karunaratne. 1990. Insecticidal activity of *Acarus calamus* and *Glycosmis mauritata* Tanaka against *Aphis craccivora*. Tropical Agri. 67: 223-228.

7. Bellardi, M. G., C. Rubies-Autonell and S. Biffi. 1997. Virus diseases of officinal crops in Emilla-Romagna. Informatore-Fitopatologica 47: 28-34.

8. Bello, A. 1967. Data on the geographical distribution of *Halenčhus dumnonicus* (Nematoda). Boletin-de-la-Real-Cociedad-Espanola-de-Historia-Natural-Biologica 65: 1-2, 81-82.

9. Bodkin, G. E. 1915. Report of the economic biologist. Rept. Dept. Sci. Agric. 1914-1915, 17th Sept. 1915, 11 p.

10. Brown, J. F. 1989. Kava and kava diseases in South Pacific. ACIAR-Working-Paper No. 24, 70 p.

11. Brown, J. F. 1998. Reflections of a traditional plant pathologist. Aust. Plant Pathology 27: 1-14.

12. Bucher, G. E. and P. Harris. 1961. Food-plant spectrum and elimination of disease of cinnabar moth larvae, *Hypocrita jacobacea* (L.). Canad. Ent. 93: 931-936.

13. Bunney, S. 1992. The Illustrated Encyclopedia of Herbs, Their Medicinal and Culinary Uses. Chancellor Press, London. 320 p.

14. Campbell, C. D. and W. D. Hutchison. 1995. Sugarbeet resistance to Minnesota populations of sugarbeet root aphid (Homoptera: Aphididae). J. Sugar Beet Res. 32: 37-46.

15. Chang, K. F., R. J. Howard and R. G. Gaudiel. 1997. First report of *Sclerotinia sclerotiorun* on coneflower. Plant Dis. 81: 1093.

16. Chang, K. F., R. J. Howard and S. F. Hwang. 1997. First report of Botrytis blight caused by *Botrytis cinerea*, on coneflower. Plant Dis. 81: 1461.

17. Chang, K. F., R. J. Howard, S. F. Hwang, R. G. Gaudiel and S. F. Blade. 1998. Diseases of *Echinacea* in Alberta in 1997. Can. Plant Dis. Surv. 78: 92-94.

18. Chapman, R. F., E. A. Bernays and T. Wyatt. 1988. Chemical aspects of host-plant specificity in three *Larrea*-feeding grasshoppers. J. Chem. Ecology 14: 561-579.

19. Chemsh, S. I. and V. A. Lukhtanov. 1981. Adaptation to damage in the silkworm *Bombyx mori* L. (Lepidoptera, Bombycidae). I. Influence of ecdysterone and some adaptogens on the resistance of larvae to formaldehyde intoxication. Entomologicheskoe-Obozrenie 60: 1, 21-33.

20. Chevallier, A. 1996. The Encyclopedia of Medicinal Plants. Dorling Kindersley Ltd., London. 336 p.

21. Clarke, A. R., D. Shohet, V. S. Patel and J. L. Madden. 1998. Overwintering sites of *Chrysophtharta bimaculata* (Olivier)(Coleoptera: Cgrysomelidae) in commerically managed *Eucalyptus obliqua* forests. Aust. J. Ento. 37: 149-154.

22. Conte, L., A. Zazzerini and L. Tosi. 1989. Changes in composition of sunflower oil extracted from achenes of *Sclerotium bataticola* infected plant. J. Agric. Food Chem. 37: 36-38.

23. Corbett, G. H. 1934. Division of entomology. Annual report for the year 1933. Gen. Ser. Dep. Agric. S.S. & F. M. S. no. 19: 38-54.

24. Da Costa Lima, A. 1942. Bugs infesting orchids. Orguidea 4: 100-109.

25. Dube, S., P. D. Upadhyay and S. C. Tripathi. 1991. Fungitoxic and insect repellent efficacy of some spices. Indian Phytopathol. 44: 1, 101-105.

26. Dudas, P., B. Galmbosi and G. Bujaki. 1985. Aphids damaging umbellate volatile oil plants. Novenyvedelem 21: 5, 196-198.

27. Davis, R. L. and J. F. Brown. 1996. Epidemiology and management of kava dieback caused by cucumber mosaic cucumovirus. Plant Dis. 80: 917-921.

28. Dusek, J. and M. Rezae. 1967. *Pegomya rubivora,* a new pest on cultivated raspberry in Czechoslovakia. Sb. Vys. Sk. Zemed. Brne (A) 15: 289-295.

29. Everett, T. H. 1960. New Illustrated Encyclopedia of Gardening. Vol. 14. Greystone Press, New York. 192 p.

30. Farr, D. F., G. F. Bills, G. P. Chamuris, and A. Y. Rossman. 1989. Fungi, on Plants and Plant Products in the United States. APS Press, St. Paul, MN. 1252 p.

31. Farrell, B. and C. Mitter. 1990. Phylogenesis of insect/plant interactions: have *Phyllobrotica* leaf beetles (Chrysomelidae and the lamiales) diversified in parallel? Evolution 44: 1389-1403.

32. Fisher, G., J. DeAngelis, and D. M. Burgett. 1993. Pacific NorthWest Insect Control Handbook. Washington State Univ., Pullman, WA. 352 p.

33. Gabriel, B. P. 1982. Prospects for industrial entomology in the Philippines. Philippine Entomologist 4: 525-534.

34. Ghiuta, M. 1971. Contributions to the study and distribution of Eriophyid gall mites in the Bacau district. Contributium la studiul si raspindirea Eriophyidocecidiilor in judetul Bacus Muzeul de Stintele Naturi Bacau Studii si Comunicari (1971). p. 147-161.

35. Glendenning, R. 1926. Some new aphids from British Columbia. Canad. Ent. lviii. no. 4, 95-98.

36. Goldsmith, S. K. 1987. Resource distribution and its effect on the mating system of a longhorned beetle, *Perarthrus linsleyi* (Coleoptera: Cerambycidae). Oecologia 73: 317-320.

37. Golfari, L. 1963. Observations on *Cephisus siccifolius* on *Eucalyptus* trees in Misiones. Idia 189: 9-14.

38. Gour, T. B., T. V. K. Singh, A. Sathe and S. N. Pasha. 1991. *Stemmatophora fuscibasalis* Snellen—a new record as a pest of citronella. Indian J. Plant Protection 19: 220.

39. Graves, H. W. 19943. The Mormon cricket in California. Bull. Dept. Agric. Calif. 32: 201-205.

40. Green, E. E. 1920. Observations on British coccidae. Ento. Mthly-Mag., London, nos. 672-673. Third Ser. nos. 65 and 66, May-June 1920, p. 114-130.

41. Griffiths, G. C. D. 1984. The Alysiinae Hym. Braconidae parasites of the Agromyzidae (Diptera) VII. Supplement. Beitrage zur Ento. 34: 343-362.

42. Grzbowska, T. 1986. Chemical control of brown leaf spot of lovage (*Levisticum officinale* Koch.). Herba-Polonica 32: 3-4, 225-231.

43. Gupta, B. M. and C. P. S. Yadava. 1992. Chemical control of the aphid, *Myzus persicae* (Sulzer) on cumin in semi-arid rajasthan. Indian Cocoa Arecanut and Spices J. 15: 109-113.

44. Hameed, S. F., V. K. Sud and S. P. Giamzo. 1977. New records of aphids from Kulu and Lahaul valleys (Himachal Pradesh). Indian J. Ento. 37: 203-205.

45. Haunold, A. 1993. Agronomic and quality characteristics of native North American hops. Am Soc. Brew. Chem. J. 51: 133-137.

46. Henkel, A. and G. F. Klugh. 1908. The cultivation and handling of goldenseal. U.S.D.A. Bur. Plant Ind. Circ. No. 6. 19 p.

47. Hille Ris Lambers, D. The *Rubus* aphids of the Netherland. Tijdschr. PlZiekt. 56: 253-261.

48. Hoffman, G. D. and P. B. McEvoy 1985. Mechanical limitations on feeding by meadow spittlebugs *Philaenus spumarius* (Homoptera: Cercopidae) on wild and cultivated host plants. Ecological Ento. 10: 415-426.

49. Hwang, S. F., K. F. Chang and R. Howard. 1996. Yellows Diseases of Echinacea, Monarda and Caraway. Agri-fax 630-1, Alberta Agri., Food and Rural Development. 4 p.

50. Hwang, S. F., K. F. Chang, R. J. Howard, A. H. Khadhair, R. G. Gaudiel and C. Hiruki. 1997. First report of a yellow phytoplasma disease in purple cone-flower (*Echinacea* spp.) in Canada. Zeitschrift-fur-pflanzenkrankheiten-und-pflanzenschutz. 104: 182-192.

51. Hyche, L. L. 1997. Some stinging caterpillars on shade and ornamental trees. Highlights Agri. Res. Alabama Agr. Expt. Sta. 44: 8-10.

52. Katsarska, P. and D. Desev. 1976. Study on the essential oil (concrete and absolute) of *Mahonia aquifolia* with a view to its utilization in the perfumery and cosmetic industry. Perfumer Favorist 2: 62.

53. Kennedy, D. M. 1987. Verticillium wilt of sea buckthorn (*Hippophae rhamnoides*). Plant Pathol. 36: 420-422.

54. Ko, C. C., T. C. Hsu and W. J. Wu. 1992. Aleyrodidae of Taiwan. Part I. *Rhachisphora quaintance* et Baker. Japanese J. Ento. 60: 243-260.

55. Lallemand, V. 1918. Un Membracid nouveau pour la Faune francaise. Bull. Soc. Ento. France, Paris, 1920, no.3, p. 53.

56. Langor, D. W. and C. R. Hergert. 1993. Life history, behaviour, and mortality of the western ash bark beetle, *Hylesinus califoricus* (Swaine) (Coleoptera: Scolytidae), in southern Alberta. Canadian Ento. 125: 801-814.

57. Lazorko, W. *Earrhypara hortulata* L. (*articata* L.) on the Pacific coast (Lepidoptera: Pyralidae). J. Ento. Soc. British Columbia. 74: 31.

58. Lepage, H. S. 1942. The orchid bug. Biologica 8: 67-72.

59. Li, T. S. C. 1995. Asian and American ginseng—a review. HortTechnology 5: 27-34.

60. Li, T. S. C. 1998. *Echinacea:* cultivation and medicinal value. HortTechnology 8: 122-129.

61. Li, T. S. C. and W. R. Schroeder. 1996. Sea buckthorn (*Hippophae rhamnoides* L.): a multipurpose plant. HortTechnology 6: 370-380.

62. Li, T. S. C. and R. S. Utkhede. 1993. Pathological and non-pathological diseases of ginseng and their control. Current Topics in Bot. Res. 1: 101-115.

63. MacCreary, D. and H. E. Milliron. 1952. Occurrence of the smartweed borer and the European corn borer in apples. J. Econ. Ent. 45: 348.

64. MacFarlane, H. H. 1968. Review of Applied Mycology. 1-40 Indexes. CAB Press, England. 820 p.

65. McKeown, A. W. and J. Potter. 1994. Native wild grasses and flowers: new possibilities for nematodes. Agri-Food Res. Ontario (Dec., 1994): 20-25.

66. Matsumoto, K. 1989. Effects of aggregation on the survival and development on different host plants in papilonid butterfly, *Leuhdorfia japonica* Leech. Japanese J. Entomology 57: 853-860.

67. Michaud, J. P. 1990. Observations on the biology of the bronze flea beetle *Altica tombacina* (Coleoptera: Chrysomelidae) in British Columbia. J. Entomological Soc. British Columbia 87: 41-49.

68. Mier Durante, M. P. 1983. *Brachycaudus* (*Acaudas napelli*) (Hom. Aphididae): first record for the Iberian Peninsula). Rev. Applied Ento. 1984: 5147.

69. Miller, N. C. E. 1930. Division of entomology. Ann. Report. 1929. Bull. Dept. Agric. S. S. & F. M. S., Gen. Ser. no. 3, 50-61.

70. Muller, H. M., H. Kleinhempel, D. Spaar and H. J. Muller. 1973. Mycoplasma-like

organisms in ornamental plants with flower phyllody symptoms. Archiv. Fur Phytopath. Und Pflanz. 9: 95-104.

71. Nagy, F. and P. Tetenyi. 1986. Modern and well integrated techniques for protecting some medicinal plants grown in Hungary. Parasitica 42: 1, 17-24.

72. Neilson, C. L. 1957. Handbook of the Main Economic Insects of British Columbia. Part 4. Tree Fruit Insects. 68 p.

73. Neilson, C. L. and J. C. Arrand. 1961. Handbook of the Main Economic Insects of British Columbia. Part 7. Shade Tree and Shrub Insects. 60 p.

74. Orton-Williams, K. J. 1982. A new genus and four new species of Criconematidae (Nematoda) from the Pacific. Systematic-Parasitology 4: 3, 239-251.

75. Palaniswami, M. S. 1991. Yam scale insects *Aspidiella hartii* Ckll. and its parasitoids. J. Root Crops. 17: 75-76.

76. Pennacchio, F. and E. Tremblay. 1988. A new species of *Aphidius* Nees from Italy (Hymenoptera, Braconidae, Aphidiinae). Bollettino del Laboratorio di Entomologia Agraria Filippo Silverstri, Italy 45: 167-169.

77. Pereira, J. M. M., J. C. Zanuncio and J. H. Schoereder. 1994. Faunistics indexes of the major lepidopterous pests in the Lassance and Sao Bento Abade Regions, in Minas Gerais, Brazil. Revista Arvore. 18: 79-86.

78. Protsenko, N. G. 1980. Prospects of using the method of biological control against dodder in the conditions of the Chu lowlands of Kirgizia. Entomologicheskie Isledovaniya v Kirgizii 14: 104-109.

79. Rawlinson, C. J. and P. A. Dover. 1986. Pests and diseases of some new and potential alternative arable crops for the United Kingdom. Brighton Crop Prot. Conf. Pests Dis. Surrey: British Crop Protection Council 1986. Vol. 2: 721-732.

80. Redfern, M. and R. R. Askew. 1992. Plant galls. Naturalists' Handbooks. No. 17. Richmond Publishing, Slough, UK. 99 p.

81. Remaldiere, G. 1989. The genus *Therioaphis* Walker in the Middle East. Ann. Soc. Ento. de France 25: 171-198.

82. Robinson, A. G. 1973. Two new Macrosiphini (Homoptera: Aphididae) from Manitoba. Can. Entomologist 105: 813-815.

83. Rogiers, S. Y. and N. R. Knowles. 1997. Physical and chemical changes during growth, maturation, and ripening of saskatoon. Can. J. Bot. 75: 1215-1225.

84. Rubies-Autonell, C. and M. G. Bellardi. 1996. *Levisticum officinale*: natural host of celery mosaic virus. Phytopathologia mediterrancea 35: 58-61.

85. Saeed, M. Q., M. Shahid and H. Gul. 1997. Efficacy of different insecticides against *Epilachna* spp. and their most preferred host plant. Pakistan J. Zoology 29: 94-95.

86. Schedl, K. E. 1939. *Scolyphalus sumatranus* sp. n., *Xyleborus myristicae* sp. n., both from nutmeg (*Myristica fragrans*) in Sumatra. Tijdschr. Ent. 82: 30-53.

87. Scheider, F. 1939. A comparison, based on some observations on the east coast of Sumatra, between primeval forest and monoculture as regards their peril from insect, pests. Schweiz. Z. Forstwesen. 1939. 22 p.

88. Schmelzer, K. R., M. Gippert, M. Weisenfels, and L. Beczner. 1975. Symptoms of host plants of broad bean wilt virus. I. Communication. Zentralblatt Bakteriol. Parasitenkde Infektionskr. Hygiene 130: 696-703.

89. Scriber, J. M. 1986. Local food plant specialization in natural field populations of the southern armyworm, *Spodoptera eridania* (Lepidoptera: Noctuidae). Ento. News. 97: 183-185.

90. Shoji, J. and Y. Tsukitani. 1972. On the structure of senegin—3 of *Senegae radix*. Chem. Pharm. Bull. (Tokyo) 20: 424-426.

91. Singh, H., V. S. Malik and H. Singh. 1993. Biology of painted bug (*Bagrada cruciferarum*). Indian J. Agri. Sci. 63: 672-674.

92. Singh, R. P., S. S. Tomar, C. Devakumar, B. K. Goswami and D. B. Saxena. 1991. Nematicidal efficacy of some essential oils against *Meloidogyne incognita*. Indian Perfumer 35: 35-37.

93. Small, E. (ed.). 1997. Culinary Herbs. NRC Research Press, Ottawa. 710 p.

94. Small, E. and P. M. Catling (eds.). 1999. Canadian Medicinal Crops. NRC Research Press, Ottawa. 240 p.

95. Stanosz, G. R. and M. F. Heimann. 1997. Purple conflower is a host of the aster yellows phytoplasm. Plant Dis. 81: 424.

96. Stockberger, W. W. 1922. Drug plants under cultivation. U.S.D.A. Farmers' Bull. No. 663.

97. Stusak, J. M. 1968. Notes on the bionomics and immature stages of *Tingis stachydis* (Fieber) (Heteroptera, Tingidae). Acta Ento. Bohemoslov. 65: 412-421.

98. Suprapto. 1986. Host range for the stem borer of pepper. Pemberitaan 12: 1-2, 1-11.

99. Sweet, M. H. 1986. *Ligrocoris barberi* (Heteroptera: Lygaeidae), a new seedbug from the southeastern United States with a discussion of its ecology, life cycle, and reproductive isolation. J. NY Ento. Soc. 94: 281-290.

100. Tabatadze, E. S., V. A. Yasnosh and V. Vacante. 1997. Control measures of *Lophọleucaspis japonica* Cockerell (Homotera: coccinea) through integrated citrus pest management. Proc. Integrated Control in Citrus Fruit Crops. Florence, Italy. Aug. 29, 1996. Bull. OILB SROP 20: 45-51.

101. Thakur, R. N., P. Singh and S. Gupta. 1996. Studies on leaf blight disease of *Withania somnifera* (L.) Dunal. Agri. and Equipment International 48: 134-136.

102. Thapa, V. K. 1989. Some higher Himalayan Typhlocybine leafhopper (Homoptera, Cicadellidae) of Nepal. Insecta Marsumurana 42: 93-110.

103. Teobald, F. V. 1928. Aphididae from Italian Somaliland and Eritrea. Bull. Ent. Res. xix. pt. 2: 177-180.

104. True, R. H. 1903. Cultivation of drug plant in the United States. U.S. Dept. Agric. Year Book 1903. p. 337-346.

105. Tryapitsyn, V. A., E. S. Sugonyaev and V. A. Trjapitzin. 1972. *Microterys eleutherococci* sp.n. (Hymenoptera, Encyrtidae) a parasite of *Eupulvinaria pulchra* (Homoptera, coccidae) on *Eleutherococcus* in the Maritime District. Zoologicheskii-Zhurnal 51: 4, 615-617.

106. Van Fleet, W. 1914. Golden seal under cultivation. U.S.D.A. Farmers' Bull. No. 613.

107. Van Fleet, W. 1915. The cultivation of peppermint and spearmint. U.S. Dept. Agric. Farmers' Bull. No. 694.

108. Vasu, H. D. 1971. *Vetiveria zizanioides* (L.) Nash as a new host record for the white grub, *Holotrichia serrata* F. (Coleoptera: Melolonthidae). Indian J. Ento. 32: 272-273.

109. Walker, J. C. 1969. Plant Pathology. McGraw-Hill Book Co., New York. 819 p.

110. Weddell, J. A. 1930. Field notes on the banana fruit-eating caterpillar (*Tiracola plagieta* Walk.) Queensland Agric. J. xxxiii. pt. 3, 186-201.

111. Westcott, C. 1960. Plant Disease Handbook. 2nd ed. D. Van Nostrand Co., Inc., NewYork.

112. Williams, A. H. 1995. New larvel host plant and behaviour of *Chlosyne gorgone* (Lepidoptera: Nymphalidae). Great Lakes Ento. 28: 93-94.

113. Williams, L. E. and A. F. Schmitthenner. 1962. Effect of crop rotation on soil fungus population. Phytopath. 52: 241-247.

114. Wilson, G. F. 1941. Some *Delphinium* pests. *Delphinium* Yearbook 1941. p. 7-9.

115. Winder, R. S. and A. K. A. Watson. 1994. Potential microbial control for fireweed (*Epilobium angustifolium*). Phytoprotection 75: 19-33.

116. Zaim, M. and A. Samad. Association of phytoplasmas with a witches-broom disease of *Withania somnifera* (L.) Dunal in India. Plant Sci. Limerick 109: 225-229.

117. Zheng, M. S., G. Z. Zhao and A. L. Hu. 1987. Studies on the bionomics of *Plebejus argus* L. and its control. Insect Knowledge 24: 232-234.

APPENDIX 1: MAJOR CONSTITUENTS AND THEIR SOURCE

Major Constituents	Source
1,8-Cineole	California laurel
1,8-Dihydroxy-anthracene derivatives	Aloe
11-α,13-Dihydrohelenalin	Arnica
1-Homostarchydrine	Alfalfa
2-3,4-Dihydroxyphenyl-ethanol	Royal jasmine
2-β-glucuronosyl	Licorice
2-Vinyl-4H-1,3-dithin	Garlic
3-n-Pentadecylcatechol	Poison ivy
3-O-β-D-Glucuronide	Fireweed
4'-O-Methylpyridoxine	Ginkgo
α-Acids	Hops
β-Acids	Hops
Abrotamine	Chinese wormwood (Qing Hao)
Absinthol	Wormwood
Acacetin	Gumweed
Acetate	Bay laurel
Acetopenone glucoside	Speedwell
Acetylcholine	Shepherd's purse
Acetylenic compounds	Asian ginseng, Tian Qi, American ginseng
Acetylinic	Anise
Achilleine	Yarrow
Aconitine	Monkshood
Acoric acid	Calamus, Sweet flag, Shi Chang Pu
Acrylic acid	Pineapple
Actein	Black cohosh
Acylphloroglucinal derivatives	Curry plant
Aesculin	Horse chestnut
Agglutinin	Stinging nettle
Agktcibes cgrtsiogabikm	Chinese rhubarb, Rhubarb
Aglycone	English ivy, Everlastings
Agnuside	Chaste tree
Ajoene	Garlic
Alanine	Peppers
Albiflorin	White peony, Mu Dan Pi
Albitocin	Horsebalm, Horsemint, Lily of the valley
Albuminoides	Chinese yam
Aldehydes	Periwinkle

Alkaloides	Blue lobelia, Passion flower, Alfalfa, Chinese lantern, Mistletoe, Wild or Mexican yam, Periwinkle, Ephedra, Horsetail, Ashwagandha, Kava-Kava, Lobelia, Ma Huang, Oregon grape, Boxwood, Daffodil, Cat's claw, Wild indigo, Colisfoot, Valerian
Alkenyl glycoside	Codonopsis, Dang Shen, Bellflower
Alkyl-phthalides	Lovage
Alkylamides	Purpule coneflower, Narrow leaf echinacea
Allantoin	Black snakeroot, Comfrey
Allicin	Onion
Alliin	Chives, Garlic, Onion
Alnulin	Alder
Aloe-emodin	Tinnevelly senna, Alexandria senna
Aloeresin	Aloe
Aloin	Aloe
Alpha-bisabolol	Chamomile
Alpha-peltatin	Mayapple
Alpha-pinene	Coriander, Cilantro
Amidendiol	Arnica
Amines	Shepherd's purse, English hawthorn, Chinese hawthorn
Amygdalin	Almond, English hawthorn, Chinese hawthorn
Amyrin	Peppers
Anagyrine	Blue cohosh
Anahygrine	Ashwagandha, Indian ginseng
Anemonin	Pilewort, Buttercup
Anethole	Fennel, Aniseed, Anise
Angelicide	Angelica
Anisaldehyde	Aniseed
Anthaquinone glycosides	Aloe
Anthemidin	Chamomile, German chamomile
Anthocyanins	Elderberry, Saskatoon, European peony, Chicory, Hibiscus, Dwarf mountain pine, Southern pitch pine, White pine, Lodge pole pine, Raspberry, Grape
Anthocyanosides	Blueberry, Bilberry, Cranberry, Mountain cranberry, Cowberry, Foxberry, Black currant, Black gooseberry
Anthoxanthine	Arnica
Anthraquinone	Buckthorn, Alder buckthorn, Cascara sagrada, Lungwort, Sorrel, Dock, Bistort root, Flowery knotweed (Fu-ti), Garlic, Meadow sweet, Chinese

Anthraquinone *(continued)*	rhubarb, Rhubarb
Anti-protease	Black currant, Black gooseberry
Apigernin	Rosemary, Thyme
Apiin	Parsley, Celery
Apiole	Parsley
Arabinose	Borage
Arachic acid	Sunflower
Araliasides	Devil's club
Arborinine	Rue, Herb of grace
Arbutin	Blueberry, Bilberry, Bearberry, Uva-ursi, Labrador tea
Arctic acid	Burdock
Arctiin	Burdock
Arctiol	Burdock
Aretylcholine	Stinging nettle
Arginine	Jojoba, Radish, Duckweed
Arislolochic acid	Wild ginger
Arnicin	Arnica
Artemisinin	Chinese wormwood (Qing Hao)
Asarone	Calamus, Sweet flag, Shi Chang Pu
Ascaridole	Wormseed
Ash	Eyebright
Asiatic acid	Gotu Kola
Asiaticoside	Gotu Kola
Asparagine	Asparagus, Marshmallow, Astragalus, Horseradish
Aspirin	Bearberry, Uva-ursi
Astragalin	White peony, Mu Dan Pi, Ostrich fern
Astrogen	Soybean
Atlantol	Cedarwood
Aucubigenin	Eyebright
Aucubin	Chaste tree, Heal all, Selfheal, Plantain, Eyebright
Aucuboside	Speedwell
Authraquinone	Tinnevelly senna, Alexandria senna
Azulene	Roman Chamomile, Curry plant, Chamomile, German chamomile
Baldrianic acid	Elder
Barbaloin	Aloe
Benzaldehyde	Black cherry, Wild cherry
Benzoic acid	Tree peony (Mu Dan Pi), Baical skullcap, Virginia skullcap, Skullcap, White peony (Bia shao), Ylang-Ylang, Cloudberry, Vetiver, *Citrus* spp.
Benzoly-aconitine	Monkshood

Benzophenanthridine alkaloids	Bloodroot
Benzoquinone	Loosestrife
Benzoylecgonine	Coca
Berbamine	Barberry
Berberastine	Goldenseal
Berberine	Barberry, Oregon grape, Goldenseal
Berberubine	Barberry
Bergapten	Celery
Bergegin	Dodder
Beta-boswellic acid	Frankincense
Beta-carotene	Saffron
Beta-elemene	English ivy
Beta-peltatin	Mayapple
Beta-phenylethylamine	Mistletoe, American mistletoe
Beta-sitosterol	White peony, Mu Dan Pi, Alder, Lovage, Slippery elm, Sweet elm, English elm, Ashwagandha, Indian ginseng, Ostrich fern
Betaine	Lycium, Chinese wolfberry, Wolfberry, Beet root, Betony, Woundwort
Betanin	Beet root
Betonicine	Horsehound, Betony, Woundwort
Betulinic acid	Heal all, Selfheal
Bilobalide	Ginkgo
Bilobetin	Ginkgo
Bioflavonoids	Lemon, Bergamot orange, Mandarin, Lime, Bitter orange, Grapefruit
Bishomophinolenic acid	Dwarf mountain pine, Southern pitch pine, White pine, Lodgepole pine
Boldine	Sassafras
Borneol	Dwarf mountain pine, Southern pitch pine, White pine, Lodgepole pine, Sage, Clary sage, Rosemary, Wild ginger, Juniper, Lavender
Bornyl acetate	Rosemary
Brefeldin A	Angelica
Bromelain	Pineapple
Bursine	Shepherd's purse
Butane	esame
Butanolide	Sesame
Butyl phthalidine	Lovage
Caffeic acid	Heal all, Selfheal, Artichoke, Wild artichoke, Stevia, Sweet herb of Paraguay, Gum tree, Eucalyptus, Monkshood, Russian olive,

Caffeic acid *(continued)*	Narrowleaf echinacea, Pale-flower echinacea, Purple coneflower, Onion, Peppers, Soybean, Burdock, Foxglove, Digitalis, Sweet marjoram, Linden, Small-leaved lime, Clover (red), European mistletoe, Ostrich fern
Caffeine	Green tea, Coffee, English holly
Calcium	Borage, Watercress, Prickly pear
Calcium oxalate	Chinese rhubarb, Rhubarb
Calcyosm	Astragalus
Campesterol	Slippery elm, Sweet elm, English elm, Ostrich fern
Camphene	Camphor
Camphor	Lavender, Rosemary, Sassafras, Sage, Clary sage, Feverfew, Peppers, Camphor, Yarrow
Canadine	Goldenseal
Capsaicin	Cayenne pepper, Sweet pepper, Chilli pepper
Carbohydrates	Hemp, Soybean
Carbosylic acid	Lobelia
Cariaester	Goldenrod
Carotenes	Plantain, Japanese apricot, Eyebright, Lyçium, Chinese wolfberry, Wolfberry, Sweet marjoram, Dog rose, Damask rose, Oat, Carrot, Sea buckthorn, Arnica, Calendula, Marigold, Ginkgo, Onion, Green tea, Papaya, *Citrus* spp.
Carotenoids	European mistletoe
Carthamin	Celery
Carthamone	Safflower
Carvacrol	Lemon thyme, Thyme, Savory, Catnip, Catmint
Carvone	Dill, Caraway, Peppers
Caryophyllene	Pokeweed
Casticin	Chaste tree
Catalpol	Baical skullcap, Virginia skullcap, Skullcap
Catechins	English hawthorn, Chinese hawthorn, Witch hazel, Motherwort, Chinese rhubarb, Rhubarb
Caulophylline	Blue cohosh
Caulosaponin	Blue cohosh
Centapicrin	Star thistle
Chamaelirin	False unicorn root
Chamazulene	Yarrow
Chelidonine	California poppy
Chlorogenic acid	Black snakeroot, Artichoke, Wild artichoke, Olive, Stevia, Sweet herb of Paraguay, St. John's wort, Burdock, Foxglove, Digitalis, Plantain, Ostrich fern
Cholesterol	Slippery elm, Sweet elm, English elm

Choline	Roman Chamomile, Buckwheat, Mistletoe, Dandelion, Hemp, Foxglove, Digitalis, Motherwort, Betony, Woundwort, Fenugreek, Stinging nettle
Chrysophanic acid	Aloe
Chymopapain	Papaya
Cimicifugin	Black cohosh
Cimin-aldehyde	Cumin
Cincole	Mugwort
Cineole	Chaste tree, Sage, Clary sage, Bay laurel, Rosemary, Melaleuca, Medicinal tea tree. Gum tree
Cinerins	Pyrethrum
Cinnamic acid	Lycium, Chinese wolfberry, Wolfberry, Chinese rhubarb, Rhubarb, Dodder, Curry plant, Gum tree, Storax, *Citrus* spp.
Cinnamolaurine	Sassafras
Cinnamyl cinnamate	Gum tree, Storax
Cinnamyl cocaine	Coca
Cis-aconitate	Saskatoon
Cis-sabinene hydrate	California laurel
Citral	Lemon balm, Balm
Citrate	Saskatoon, Mountain fly honeysuckle, Dutch honeysuckle, Black currant, Black gooseberry, Dog rose, Damask rose, Japanese apricot, Horse chestnut, Hibiscus
Citric acids	Dog rose, Damask rose
Citronellal	Lemon grass, Citronella, Palmarosa, Gingergrass, Lemon balm, Balm
Cocaine	Coca
Codeine	Poppy
Colchicine	Crocus, Meadow saffron
Columbamine	Oregon grape
Convolvuline	Bindweed
Cornic acid	Bunchberry
Cornine	Bunchberry
Coumaric	Gum tree, Eucalyptus, Mexican mint, Marigold, Angelica, Dill, Celery, Agrimony, Dong quai, Sweetgrass, Chamomile, German chamomile, Clover (red)
Coumarins	Cinnamon, Camphor, Highbush cranberry, Chickweed, Sweet clover, Melilot, Alfalfa, White ash, Lemon, Bergamot orange, Mandarin, Lime, Bitter orange, Grapefruit, Black cherry, Rue, Herb of grace, Tarragon, Wild indigo, English hawthorn,

Coumarins *(continued)*	Chinese hawthorn, Siberian ginseng, Lavender, Lovage, Chamomile, German chamomile, Parsley, Anise, Kudzu, Ge gen, Fenugreek
Crataegas acid	English hawthorn, Chinese hawthorn
Creosol	Anise
Creosote	Beech, European beech
Crocine glycosides	Saffron
Cryptopine	California poppy
Cucurbitin	Pumpkin
Curcumin	Sandalwood
Curcuminoids	Curcuma, Jiang Huang
Cutins	White oak
Cyanide	Japanese apricot, Elderberry, Passion flower, Hydrangea, Almond
Cyanogenic glycoside	Kudzu, Ge gen
Cyclolignans	Sesame
Cymene	Savory, Juniper
Cynarin	Artichoke, Wild artichoke
Cynaroside	Artichoke, Wild artichoke, Chamomile, German chamomile
Cypripedin	Lady's slipper
d-Pseudoephedrine	Ephedra
Daidzein	Kudzu
Dammaranedienol	Elecampane
Deanolic acid	Heal all, Selfheal
Dehydrofukinone	Burdock
Dendrolasin	Sandalwood
Dianthrone glucosides	Senna
Dicoumarol	Sweet clover, Melilot
Digitoxin	Foxglove
Digoxin	Foxglove
Dihydro-beta-agarofuran	Sandalwood
Dihydrokawain	Kava-Kava
Dihydrolycopodine	Clubmoss
Dihydromethysticin	Kava-Kava
Dimeric indole alkaloids	Periwinkle
Dinsmin	Rosemary
Diosgenin	Wild yam
Diterpene acids	Spikenard
Ecdysteroids	Lamb's quarters
Ecgonine	Coca

Egin	Arnica
Eleutherosides	Siberian ginseng
Elixen	English ivy
Emblicanins A and B	Myrobalan, emblic
Emodin	Chinese rhubarb, Rhubarb, Buckthorn, Alder buckthorn, Cascara sagrada, Aloe
Emulsin	English hawthorn, Chinese hawthorn
Enicin	Blessed thistle
Ephedrine	Ephedra, Horsetail, Ma Huang
Equisitine	Horsetail
Estragole	Tarragon
Estrogens	Pygeum
Eucalyptol	California laurel
Eudesmic acid	Black cherry
Eugenol	Clove
Eupatorin	Boneset
Fatty acids	Russian olive, Anise, Sea buckthorn, Soybean, Corn, Almond, Pumpkin
Fenchone	Fennel
Ferric oxide	Hazelnut
Ferulic acid	Pygeum, Angelica, Ostrich fern
Fiber	Psyllium
Flavone glycoside	Alder, Broom
Flavonoids	Baical skullcap, Virginia skullcap, Skullcap, Chinese chaste tree, Witch hazel, Boneset, Elderberry , White ash, Sweet clover, Melilot, Clover (crimson), Chaste tree, Small-flowered willow herb, Dodder, Hydrangea, Poplar, Balm of gilead, Passion flower, Chickweed, Linden, Small-leaved lime, Sea buckthorn, Lomatium, English hawthorn, Chinese hawthorn, Buckwheat, Devil's claw, Mayapple, Crocus, Meadow saffron, Ginkgo, Primrose, Cowslip, Chinese licorice, Bayberry, Loosestrife, Cotton, Chamomile, German chamomile, Everlastings, Roman chamomile, Asparagus, Agrimony, Lady's mantle, Wild indigo, Dill, Milk thistle, Black spruce, White spruce, Norway spruce, Stonecrop (small houseleek), Ma Huang, California poppy, Roseroot, English hawthorn, Chinese hawthorn, Black spruce, White spruce, Norway spruce, Sea buckthorn, Gumweed, Bunchberry, St. John's wort, Olive, Black

Flavonoids *(continued)*	snakeroot, Rue, Herb of grace, St. John's wort, Elderberry, Myrtle, Marshmallow, Arnica, Tarragon, Basil thyme, Calamint, Mountain mint, Shepherd's purse, Caraway, Ground ivy, Licorice, St. John's wort, Lavender, Clubmoss, Lemon balm, Balm, Sweet marjoram, Marjoram, Oregano, Parsley, Chinese lantern, Anise, Raspberry, White willow, Pussy willow, Jojoba, Speedwell, European mistletoe, Grape
Flavonol glycosides	Meadow sweet
Folic acid	Sesame
Formic acid	Houseleek, Hens and chicks, Pennyroyal
Formononetin	Astragalus, Huang Qi
Forsythin	Forsythia
Frangulin A and B	Buckthorn, Alder buckthorn, Cascara sagrada
Fumaric acid	Saskatoon, Sage, Clary sage
Furanoid	Sagebrush
Galactose	Dwarf mountain pine, Southern pitch pine, White pine, Lodgepole pine
Galactoside-specific lectin	Mistletoe
Galacturonate	Saskatoon
Gallic acid	Small-flowered willow herb, Gum tree, Eucalyptus, Chinese rhubarb, Rhubarb, Lady's slipper, Birch, Dogwood, Boneset, Raspberry, Mexican mint marigold, Colisfoot
Gallotannin	White peony, Mu Dan Pi, Eyebright
Gentianine	Star thistle
Gentisic	Gum tree, Eucalyptus
Geraniol	Lemon balm, Balm, Ylang-Ylang, Ginger, Wormseed
Germacrene B	English ivy
Ginkgetin	Ginkgo
Ginkgocide A, B, C, J, M	Ginkgo
Ginsenosides	Asian ginseng, Tian Qi, American ginseng
Glabridin	Licorice
Glechomine	Ground ivy
Glucans	Elder
Glucocyanates	Canary creeper, Nasturtium
Glucofrangulin A and B	Buckthorn, Alder buckthorn, Cascara sagrada
Glucoquinone	Stinging nettle
Glucose	Dwarf mountain pine, Southern pitch pine, White pine, Lodgepole pine, Mexican mint marigold

Glucosinolates	Watercress, Radish, Black mustard
Glucuronic acid	Licorice
Glutamic acid	Jojoba
Glutamine	Cow parsley, Cow parsnip
Gluten	Dandelion
Glycoproteins	Narrowleaf echinacea, Pale-flower echinacea, Purple coneflower
Glycorrhizin	Licorice
Glycosides	Lomatium, Chamomile, German chamomile, Basil, Roman chamomile, Tinnevelly senna, Alexandria senna, European peony, Poplar, Balm of gilead, Sesame, Elderberry, White willow, Pussy willow, Buckthorn, Alder buckthorn, Cascara Sagrada, Clover (red), Primrose, Cowslip, Kelp, Yucca, Asparagus, Yucca
Glycyrrhetinic acid	Licorice
Gossypol	Cotton
Grindelic acid	Gumweed
Gum	Dandelion, Lovage, Clove, Mexican mint marigold
Gustanubem 5-hydroxytryp -tamine	Stinging nettle
Hamamelitannin	Witch hazel
Harpagoqside	Devil's claw
Harpagoside	Devil's claw
Hederacoside B	English ivy
Hederacoside C	English ivy
Hederin	English ivy
Helenalin	Elecampane, Arnica
Heliotropin	Meadow sweet
Helonin	False unicorn root
Heraclein	Cow parsley, Cow parsnip
Herniarin	Chamomile, German chamomile
Hesperidin	Hyssop
Hibiscus acid	Hibiscus
Hippuric acid	Cranberry
Histamine	Shepherd's purse, Stinging nettle
Histidine	Radish, Jojoba
Humulene	Hops
Humulone	Hops
Hydrangein	Hydrangea
Hydrastine	Goldenseal
Hydrocotyline	Gotu Kola
Hydrojuglone	Walnut, Black walnut

Hydroquinones	Highbush cranberry
Hydroxybenzoic	Gum tree, Eucalyptus
Hypaconitine	Monkshood
Hypericin	St. John's wort
Hyperoside	Birch, St. John's wort
Indole	Passion flower
Inositol	Juniper, Mountain fly honeysuckle, Dutch honeysuckle, Hemp
Insulins	Blueberry, Bilberry
Intybin	Chicory
Inulin	Mugwort, Burdock, Dandelion, Colisfoot, Elecampane, Arnica, Chicory, Goldenrod
Iodine	English ivy, Watercress, Kelp, Garlic, Tarragon
Iridoids	Blueberry, Bilberry, Chaste tree, Cleavers, Chinese chaste tree, Catnip, Catmint
Iron	Watercress
Iso-ocobullenone	Stinkwood
Isobetanine	Pokeweed
Isoboldine	Sassafras
Isobutylamides	Echinacea
Isoferulic acid	Black cohosh
Isoflavone	Black cohosh, Alfalfa, Wild indigo, Chinese licorice, Soybean, Kudzu, Ge gen
Isofraxin	Siberian ginseng
Isoginkgetin	Ginkgo
Isophthalic acids	English holly, Blue flag
Isopinocamphone	Basil thyme, Calamint, Mountain mint
Isoprebetanine	Pokeweed
Isopulegone	Pennyroyal
Isoquercitrin	Royal jasmine
Isoquinoline alkaloids	Chinese corktree (Huang Bai), Goldenseal, Bloodroot
Isorhamnetin	Russian olive
Isovaltrate	Valerian
Jacoline	Groundsel
Juglandin	Walnut, Black walnut
Juglone	Walnut, Black walnut
Juniperin	Juniper
Kaempferol	Witch hazel
Kava lactones	Kava-Kava

Kawine	Kava-Kava
Kumatakenin	Gumweed, Astragalus, Huang Qi
l-Ephedrine	Ephedrine
Laburnine	Blue cohosh
Lactone	Sweetgrass, Chinese corktree (Huang Bai), Boneset, Ashwagandha, Angelica
Lactucin	Chicory
Laetrile	Japanese apricot
Lanatoside	Foxglove
Lecithin	Soybean, Bistort root, Flowery knotweed (Fu-ti)
Leine	Olive
Leonuride	Motherwort
Levant storax	Gum tree, Storax, Oriental sweet gum
Levulin	Dandelion
Lignans	Sesame, Safflower, Siberian ginseng, Lilac, Mayapple, Schizandra (wu wei zi), curry plant, Arnica, European mistletoe, Sesame
Ligustilide	Angelica
Ligustrin	Lilac
Lilacin	Lilac
Limonene	Celery, Lemon, Bergamot orange, Mandarin, Lime, Bitter orange, Grapefruit, Juniper
Limonic acid	Mountain fly honeysuckle, Dutch honeysuckle, Hyssop, Pennyroyal, Dill, Angelica, Caraway
Linalool	Lemon balm, Balm, Wild ginger, Lavender, Bay laurel, Calamus, Sweet flag, Shi Chang Pu, Ylang-Ylang, Coriander (fruit or seed), Cilantro (leaf), Basil, Savory
Linalyl acetate	Lavender
Linamarin	Flax
Linoleic acid	Psyllium, Sesame, Fenugreek, Flax, Grape, Evening primrose, Sunflower
Linolenic acid	Sunflower, Flax, Evening primrose, Fenugreek, Chickweed
Liposterolic	African prune
Lobelanidine	Lobelia
Lobelidiol	Lobelia
Lobeline	Lobelia
Lupulone	Hops
Luteolin	Chamomile, German chamomile, Agrimon, Olive, Artichoke, Wild artichoke, Olive, Lomatium, Thyme

Lycophilized extract	Houseleek
Lycopodine	Clubmoss
Lysine	Jojoba, Duckweed
Madasiatic acid	Gotu Kola
Madecassic acid	Gotu Kola
Magnoflorine	Blue cohosh
Malic acid	Houseleek, Hens and chicks, Chokecherry, Dog rose, Damask rose, Hibiscus, Japanese apricot, Mountain fly honeysuckle, Dutch honeysuckle, Sage, Clary sage, Colisfoot, Grape
Mallol	Dwarf mountain pine, Southern pitch pine, White pine, Lodgepole pine
Malonic acid	Monkshood
Maltol	Passion flower
Marrubenol	Horsehound
Marrubiin	Horsehound
Massoilactone	Sweetgrass
Maysin	Corn
Menthol	Pennyroyal, Peppermint, Spearmint, Yarrow
Menthone	Peppermint, Spearmint, Basil thyme, Calamint, Mountain mint
Mesaconitine	Monkshood
Methyl ketones	Rue, Herb of grace
Methyl salicylate	Wintergreen, Teaberry, Seneca snakeroot, Coca
Methylamine	Calamus, Sweet flag, Shi Chang Pu
Methylarbutin	Bearberry, Uva-ursi
Methylchavicol	Aniseed
Methylcytisine	Blue cohosh, Black cohosh
Methyleugenol	Sassafras, Basil
Methylsticin	Kava-Kava
Methylxanthines	Green tea
Mn	Watercress
Monoterpenes	Black spruce, White spruce, Norway spruce, Cedarwood, Everlastings, White peony (Bia shao), Tree peony (Mu Dan Pi).
Monotropein	Bearberry, Uva-ursi
Morphine	Poppy
Mucilage	Mullein, Hibiscus, Cinnamon, Camphor, Prickly pear, Feverfew, Couch grass, Daisy, Chestnut, Borage, Black mustard, Marshmallow, Burdock, Licorice, Pansy, Blessed thistle, Plantain, Bay laurel, Houseleek, Hens and chicks, Elecampane,

Mucilage *(continued)*	Comfrey, Slippery elm, Sweet elm, Canary creeper, Nasturtium, Linden, Small-leaved lime, Psyllium, Fireweed, Lemon, Bergamot orange, Mandarin, Lime, Bitter orange, Grapefruit, Mayapple, Calamus, Sweet flag, Shi Chang Pu, Calendula, Marigold, Tinnevelly senna, Alexandria senna, Flax, Linden, Corn
Myoinositol	White peony, Mu Dan Pi
Myrcene	Wormseed
Myricylalchol	Blue flag
Myristic acid	Clubmoss, Hibiscus
Myristicin	Nutmeg, Daffodil, Parsley
Myrosin	Canary creeper, Nasturtium
Myrtocyan	Bilberry
n-Butyldenephthalide	Angelica
n-Methyl anabasine	Stonecrop (small houseleek), Roseroot
n-Tetracosanol	Pygeum
n-trans-Coumaroyltyramine	Puncture vine
n-trans-Feruloyltyramine	Puncture vine
Napelline	Monkshood
Naphthalene glycosides	Tinnevelly senna, Alexandria senna
Neoherculin	Prickly ash, Toothache tree
Neolignan ketone	Stinkwood
Neoline	Monkshood
Neoruscogenin	Butcher's broom
Nicotine	Coca, Horsetail
Nicotinic acid	Fenugreek
Nobilin	Roman chamomile
Norboldine	Sassafras
Nordihydroguaiaretic acid	Chapatral
Nostoclide I	Ground liverwort
Nostoclide II	Ground liverwort
Nupharine	White water lily
Nymphaeine	Basil
Octadecatetraenic acid	Chickweed
Octadecatrienoic acid	Evening primrose
Oleanic acid	Lamb's quarters
Oleasterol	Olive
Oleic acid	Sunflower, Sesame, Fenugreek, Grape, Psyllium, Flax, Gotu Kola
Oleo-resin	Balsam fir, Periwinkle

Oleoresins	Ostrich fern
Oleoropine	Olive
Oligomeric polyphenols	Houseleek
Oligopeptides	Japanese apricot
Oripavine	Iranian poppy, Poppy, Corn poppy
Oxalic acid	Canary creeper, Nasturtium, Saskatoon, Sorrel, Dock, Sage, Clary sage
Oxyberberine	Oregon grape
Oxytocics	Horsebalm, Horsemint, Lily of the valley
p-Coumaric	Ostrich fern
p-Hydroxybenzoic	Ostrich fern
Paclitaxel	Yew
Paeonine	European peony
Paeonol	White peony, Mu Dan Pi
Palmitic acids	Fenugreek, Calamus, Sweet flag, Shi Chang Pu, Sunflower, Psyllium, Grape, White peony, Mu Dan Pi, Hibiscus, Vetiver, Ostrich fern
Panaxosides	Asian ginseng
Papain	Papaya
Pectin	Marshmallow, Mountain ash, Periwinkle, Buckthorn, Alder buckthorn, Cascara sagrada, Raspberry, Black currant, Black gooseberry, Dog rose, Damask rose, English hawthorn, Chinese hawthorn, Siberian ginseng, Raspberry, Mexican mint marigold
Pedunculagin	Myrobalan, emblic
Pentacyclic oxindole alkaloids	Cat's claw
Pentagallotannin	White peony, Mu Dan Pi
Pentagalloyl glucoside	White peony, Bia shao
Pentane	Sagebrush
Pentoses	Slippery elm, Sweet elm, English elm
Penylpropyl cinnamate	Gum tree, Storax
Peregrinine	European peony
Perlolyrin	Codonopsis, Dang Shen, Bellflower
Phelandrine	Tarragon, Angelica
Phenolic acids	Seneca snakeroot
Phenylroparnoids	Siberian ginseng
Phlobaphenes	Alder
Phoratoxin	Mistletoe, American Mistletoe
Phospholipids	Beech, European beech
Phthalides	Ligustrum fruit, Privet, Parsley
Physalien	Chinese lantern, Lycium, Chinese wolfberry, Wolfberry

Physcion	Sorrel, Dock, Buckthorn, Alder buckthorn, Cascara sagrada
Phytoene	Saffron
Phytofluene	Saffron
Phytosterol	Arnica, Hemp, Blue flag, Linden, Small-leaved lime
Picrosalvin	Sage, Clary sage, Roemary
Pinecamphene	Hyssop
Pinenes	Parsley, Juniper, Pennyroyal, Hyssop, Cumin, Wild ginger, Bay laurel, Melaleuca, medicinal tea tree
Piperine	English pepper
Plastoquinones	Chestnut
Podophyllum resin	Mayapple
Polyacetylenes	Narrowleaf echinacea, Pale-flower echinacea, Purple coneflower, Artichoke
Polygalic acid	Goldenrod
Polygalitol	Seneca snakeroot
Polymeric polyphenols	Houseleek
Polypeptides	Shepherd's purse
Polyphenolic acids	Cleavers, Green tea, Beech, European beech, Wild yam, Burdock, Arnica, Clubmoss, Lemon balm, Balm
Polysaccharides	Safflower, Narrowleaf echinacea, Pale-flower echinacea, Purple coneflower, Lungwort, Kelp, Stinging nettle, Boneset, Couch grass, Stevia, Sweet herb of Paraguay, Asian ginseng, Tian Qi, American ginseng, Agrimony, Wild indigo, Caraway, Codonopsis, Dang Shen, Bellflower
Polyynes	Devil's club
Porphyrins	Alfalfa
Potassium	Borage
Proanthocyanidins	Witch hazel, English hawthorn, Chinese hawthorn
Progesteron	Wild yam
Protein	Sesame, English pepper, Mistletoe, Black spruce, White spruce, Norway spruce, Anise, Jojoba, Fenugreek, Soybean, Coriander (fruit or seed), Cilantro (leaf), Almond, Pumpkin, Oat
Protoalnulin	Alder
Protocatechuic	Ostrich fern
Protopine	California poppy
Prunasin	Black cherry, Wild cherry
Prunin	Chokecherry
Pseudoephedrine	Mormon tea
Pseudohypericin	St. John's wort

Psoralen	Cow parsley, Cow parsnip
Psyllic acid	Lycium, Chinese wolfberry, Wolfberry
Puerarin	Kudzu, Ge gen
Pulegone	Pennyroyal
Purine	Green tea
Purpurea glycosides A	Foxglove
Purpurea glycosides B	Foxglove
Putin	Dandelion
Pyrethrins	Pyrethrum
Pyrogallol	Motherwort
Pyrrolizidine	Colisfoot, Groundsel
Pyruvate	Saskatoon
Quercitin	Witch hazel, English hawthorn, Chinese hawthorn, Stinging nettle
Quinate	Saskatoon
Quinic acid	Witch hazel
Rebaudiosides a,b	Stevia, Sweet herb of Paraguay
Resin	Highbush cranberry, Sweet clover, Melilot, Pumpkin, Groundsel, Corn, Yew, Elecampane, Goldenseal, Soapwort, Mayapple, Bay laurel, Juniper, Chapatral, Bindweed, Clubmoss, Borage, Mugwort, Burdock, Barberry, Horse chestnut, Periwinkle, White Baneberry, Red Baneberry, Dwarf mountain pine, Southern pitch pine, White pine, Lodgepole pine, Ground ivy, Alder, Black cohosh, Hops, Horseradish, Calendula, Marigold, Curcuma, Jiang Huang, Turmeric, Siberian ginseng, Lovage, Bayberry, Marjoram, Oregano, Mexican mint marigold, European mistletoe, Prickly ash, Toothache tree
Resmarinic acid	Rosemary
Rhein anthrones	Chinese rhubarb, Tinnevelly senna, Alexandria senna
Rhodioloside	Roseroot
Rosmarinic acid	Sweet marjoram
Ruscogenin	Butcher's broom
Rutaverine	Rue, herb of grace
Rutin	Tarragon, English hawthorn, Chinese hawthorn
Rutoside	Rue, herb of grace
Safrole	Sassafras, Nutmeg, Daffodil

Salicarin	Purple loosestrife
Salicin	White willow, Pussy willow
Salicortin	White willow, Pussy willow
Salicylates	Black cohosh
Salicylic acid	Blue flag, English holly, Pennyroyal, Pansy, Cloudberry, Ylang-Ylang, Lady's mantle, Clover (crimson), Meadow sweet, Mountain fly honeysuckle, Dutch honeysuckle, Goldenrod, Clover (red)
Salvin	Sage, Clary sage
Sanguinarine	Bloodroot
Sapogenin	Wild yam, Soapwort
Saponin	Horsehound, Basil, Ma Huang, Butcher's broom, Puncture vine, Hydrangea, Chinese licorice, Ground ivy, Heal all, Selfheal, Periwinkle, Dogwood, Saw palmetto, Foxglove, Horse chestnut, Wormseed, Calendula, Marigold, Borage, Daisy, Birch, Blue cohosh, Gotu Kola, Thyme, Prickly pear, Loosestrife, Goldenrod, Lemon thyme, Yucca, Soapwort, Black snakeroot, Corn, Mullein, Chinese yam (Shen yao), Wild or Mexican yam, Primrose, Cowslip, Seneca snakeroot, Chickweed, Codonopsis, Dang Shen, Bellflower, False unicorn root, Calamus, Sweet flag, Shi Chang Pu, Pilewort, Siberian ginseng, Licorice, Motherwort, Oat
Sciadopitysin	Ginkgo
Scoparoside	Broom
Scopoletin	Kudzu, Ge gen
Scordinins	Garlic
Scutellarin	Baical skullcap , Virginia skullcap, Skullcap
Sedacrine	Stonecrop (small houseleek), Roseroot
Sedacryptine	Stonecrop (small houseleek), Roseroot
Sedinine	Stonecrop (small houseleek), Roseroot
Selenium	Garlic
Senecionine	Groundsel
Seneciphyline	Groundsel
Sennosides A and B	Senna
Serotonin	Stinging nettle
Sesquiterpene hydrocarbons	Sandalwood
Sesquiterpene lactons	Yarrow
Sesquiterpenes	Arnica, Feverfew, Cedarwood, Periwinkle, Clove, Devils's club, Chicory, Ground ivy, Artichoke
Shogaols	Ginger

Silibinin	Milk thistle
Silicates	Horsetail
Silichristin	Milk thristle
Silicic acid	Borage, Horsetail, Clover (red)
Silymarin	Milk thistle
Sinapine	Black mustard
Sinigrin	Horseradish
Sitosterol	Gotu Kola
Sorbitol	Mountain fly honeysuckle, Dutch honeysuckle
Sparteine	Broom
Sphondin	Cow parsley, Cow parsnip
Stachydrine	Horsehound, Alfalfa
Stachyose	Devil's claw
Starch	Marshmallow
Stearic acid	Flax, Grape
Sterins	Codonopsis, Dang Shen, Bellflower
Sterols	Astragalus, Canary creeper, Nasturtium, Elderberry, Baical skullcap , Virginia skullcap, Skullcap, Wild or Mexican yam, Lycium, Chinese wolfberry, Wolfberry, Hazelnut, Pygeum, Dong quai, Devil's claw, Schizandra (wu wei zi), Stinging nettle, Puncture vine, Tinnevelly senna, Alexandria senna, Saw palmetto, Small-flowered willow herb, Lamb's quarters, Chinese corktree (Huang Bai), soapwort, Yarrow, Black cohosh, Onion, Marjoram, Oregano, Chinese lantern, Kudzu, Ge gen
Steviobioside	Stevia, Sweet herb of Paraguay, Seneca snakeroot
Stevioside	Stevia, Sweet herb of Paraguay
Stigmasterol	Ostrich fern
Stychydrine	Yarrow
Suberins	Beech, European beech, White oak
Sulphoxide	Chives
Sulphur compounds	Garlic, Onion
Tannic acid	Dogwood, Boneset, Licorice, Butcher's broom
Tannins	Bay laurel, Plantain, Purple loosestrife, Walnut, Black walnut, Ma Huang, Dwarf mountain pine, Southern pitch pine, White pine, Lodgepole pine, Sweet clover, Melilot, Geranium, American cranesbill, Baical skullcap , Virginia skullcap, Skullcap , Sea buckthorn, Hyssop, Ground ivy, Fireweed, Motherwort, European peony, Black snakeroot, White willow, Pussy willow, Prickly

Tannins *(continued)*	pear, Elderberry, Gumweed, Basil, Horsehound, Bayberry, Primrose, Cowslip, Black currant, Black gooseberry, Saw palmetto, Wild or Mexican yam, Catnip, Catmint, Loosestrife, Pilewort, Chamomile, German chamomile, Marjoram, Oregano, Pansy Puncturevine, Stinging nettle, Bunchberry, Linden, Small-leaved lime, Thyme, Periwinkle, Corn, Birch, Borage, Chestnut, Agrimony, Lady's mantle, Alder, Angelica, Blessed thistle, Mugwort, Horse chestnut, Feverfew, Daisy, Barberry, Wormwood, Spikenard, Burdock, Arnica, Yarrow, Highbush cranberry, Dogwood, Bindwee, Mullei, Mountain ash, Black spruce, White spruce, Norway spruce, Goldenrod, Hazelnut, Betony, Houseleek, Hens and chicks, Buffalo berry, Witch hazel, Heal all, Selfheal, Blackberry, Eyebright, White oak,.Raspberry, Slippery elm, Sweet elm, Clover (red), English ivy, Speedwell, Comfrey, Strawberry, Blueberry, Bilberry, Cinnamon, Camphor, Clove, Myrtle, Black cherry, Bearberry, Uva-ursi, Hops, Dandelion, Tarragon, English hawthorn, Chinese hawthorn, Meadow sweet, Cleavers, Hydrangea, Lavender, Mountain fly honeysuckle, Dutch honeysuckle, Lemon balm, Balm, Pygeum, Chinese rhubarb, Rhubarb, Sage, Clary sage, Mexican mint marigold, Colisfoot , Grape, Prickly ash, Toothache tree
Taraxasterol	Dandelion
Taraxerin	Dandelion
Taraxerol	Dandelion, Alder, Linden, Small-leaved lime
Tartaric acids	Hibiscus, English hawthorn, Chinese hawthorn, Colisfoot, Grape
Taxol	Yew
Terpenic acid	Lemon balm, Balm, Hyssop
Terpenoids	Stevia, Sweet herb of Paraguay, Ligustrum fruit, Privet, Curry plant, Pennyroyal
Terpineol	Sweet marjoram, Wild ginger, Lovage, Cumin, Bay laurel, Melaleuca, Medicinal tea tree
Terpinine	Coriander, Cilantro, Juniper
Terpins	Sweet marjoram
Terrestriamide	Puncture vine
Tetain	Lamb's quarters
Tetracyclic oxindole alkaloid	Cat's claw
Tetrahydro-cannabinols	Hemp

Tetrahydronaphthalene	Sesame
Thamnolic	Ground liverwort (dog lichen), Lungwort
Thebaine	Iranian poppy, Poppy, Corn poppy
Thriophenes	Mexican mint marigold, Tansy
Thujene	California laurel
Thujone	Mugwort, Sage, Clary sage
Thymol	Marjoram, Oregano, Catnip, Catmint, Thyme, Lemon thyme, Savory
Tiliadine	Linden, Small-leaved lime
Tinnins	White ash, Groundsel, White water lily
Tocopherol	Cloudberry, Boxwood
Toxicodendrol	Poison ivy
Trans-sabinene hydrate	California laurel
Triacylglycerols	Purple loosestrife
Triandrin	White willow, Pussy willow
Tribulusamide A	Puncture vine
Tribulusamide B	Puncture vine
Trigonelline	Hemp, Fenugreek
Triterpene	Black cohosh, Pygeum, Elderberry, Dill, Bayberry, English hawthorn, Chinese hawthorn, English holly, Blue flag, Yarrow, Calendula, Marigold, Gum tree, Storax, Clubmoss, Lemon balm, Balm, Gotu Kola
Triterpenoids	Frankincense, Myrrh
Tyramine	Shepherd's purse, Mistletoe, American mistletoe
Umbelliferone	Chamomile, German chamomile
Umbellulone	California laurel
Uric acid	Horse chestnut
Uronic acid	Basil
Ursolic acid	Royal jasmine, Lavender, Heal all, Cowslip, Savory
Urushiol	Poison ivy
Usnic acid	Lungwort, Ground liverwort (dog lichen)
Valepotriates	Valerian
Valtrate	Valerian
Vanillin	Gum tree, Eucalyptus, Meadow sweet, Linden, Small-leaved lime, Ostrich fern
Vellarin	Gotu Kola
Verbenalin	Dogwood
Vetivenate	Vetiver
Vetivene	Vetiver
Vetivenic acid	Vetiver
Vetivenyl	Vetiver

Vinblastine	Periwinkle
Vincristine	Periwinkle
Violin	Pansy
Viscotoxin	Mistletoe, American Mistletoe
Vitamin A	Pineapple, Lamb's quarters, Dwarf mountain pine, Southern pitch pine, White pine, Lodgepole pine, Radish, Chinese wormwood (Qing Hao), Watercress, Elderberry, Lemon, Bergamot orange, Mandarin, Lime, Bitter orange, Grapefruit, Stinging nettle, Dandelion, Sea buckthorn, Almond, Soybran, Alfalfa, Cranberry
Vitamin B	Radish, Sesame, Blackberry, Heal all, Selfheal, Lemon, Bergamot orange, Mandarin, Lime, Bitter orange, Grapefruit
Vitamin B_1	Buckwheat, Black currant, Black gooseberry, Lycium, Chinese wolfberry, Wolfberry, Cayenne pepper, Sweet pepper, Chilli pepper, Corn, Black spruce, White spruce, Norway spruce, Mugwort, Hemp, Japanese apricot
Vitamin B_2	Buckwheat, Black currant, Black gooseberry, Cayenne pepper, Sweet pepper, Chilli pepper
Vitamin B_{12}	Dong quai, Lycium, Chinese wolfberry, Wolfberry
Vitamin B complex	Sea buckthorn, Oat, Carrot, English hawthorn, Chinese hawthorn, Dog rose, Damask rose, Watercress
Vitamin C	Canary creeper, Nasturtium, Pineapple, Lamb's quarters, Radish, Speedwell, Sea buckthorn, strawberry, Coriander, Cilantro, Raspberry, Blackberry, Dandelion, Heal all, Selfheal, Eyebright, Stinging nettle, Black currant, Black gooseberry, Buckthorn, Alder buckthorn, Cascara sagrada, Sweet marjoram, Chickweed, Schizandra, Mountain ash (wu wei zi), Elderberry, Carrot, Chinese lantern, Lycium, Chinese wolfberry, Wolfberry, Myrobalan, emblic, English hawthorn, Chinese hawthorn, Dog rose, Damask rose, Lemon, Bergamot orange, Mandarin, Lime, Bitter orange, Grapefruit, Cayenne pepper, Sweet pepper, Chilli pepper, Watercress, Cloudberry, Mugwort, Japanese apricot, Alfalfa, Cranberry, Horseradish
Vitamin D	Watercress
Vitamin E	Sesame, Sea buckthorn, Schizandra (wu wei zi), Cayenne pepper, Sweet pepper, Chilli pepper, Watercress, Soybean

Vitamin K	Heal all, Selfheal
Vitamin P	Black currant, Black gooseberry
Vitexin	Purple loosestrife
Viticine	Chaste tree
Withanolides	Ashwagandha, Indian ginseng
Xanthone	Dill, Star thistle
Xylose	Hemp
Yangonin	Kava-Kava

APPENDIX 2: ESSENTIAL OILS AND THEIR SOURCE

Name of Essential Oil	Source
1,2-Benzenedicarboxylic acid	Chinese lantern
1,4-Methylhexadecanoic acid	Southern pitch pine, Lodgepole pine, White pine
1,8-Cineole	Jiang Huang, Ginger, Sagebrush, Ground ivy, Curcuma, California laurel
1,8-Pentadecadiene	Echinacea
1-Hexenol	Dang Shen, Bellflower, Codonopsis
1-Methyoxy-4-(2-propenyl)	Aniseed
1-Octen-3-ol	Ground ivy, Yucca
1-Octen-3-ol,3,5-dimethoxyphenol	Yew
1-p-Menthene	Yucca
1-Pentadecene	Echinacea
1-Phenyl-1,3-butanedione	Skullcap, Baical skullcap, Virginia skullcap
1-Terpinen 4-ol	Juniper
(2E, 4E)-2,4-Decadienal	Japanese parsley, Mitsuba
(2E, 4Z)-2,4-Decadienol	Japanese parsley, Mitsuba
(2E, 4Z)-2,4-Heptadienal	Japanese parsley, Mitsuba
(2E, 6Z)-Nonadienal	Lily of the valley, Pansy
(2E, 6Z)-Nonadienol	Lily of the valley, Pansy
2-Hydroxybenzoic acid methyl ester	Passion flower
2-Hydroxy-4-methoxyacetophenone	Carrot
2-Methyl-butanal	Echinacea
2-Methyl-4-pentenal	Echinacea
2-Methyl-propanal	Echinacea
2-Methyl-tetradeca-5,12-diene	Echinacea
2-Methyl-tetradeca-6,12-diene	Echinacea
2-Nonanone	Rue, Herb of grace
2-Nonylacetate	Rue, Herb of grace
2-Phenylethyl alcohol	Passion flower
2-Propanal	Echinacea
2-Undecanone	Rue, Herb of grace
2-Undecylacetate	Rue, Herb of grace
2,5-trans-p-Methanediol	Mint
3,9-Epoxy-p-menth-1-one	Dill
3,7-Octatrien-3-ol	Elder
3,5-Dimethoxyphenol	Yew
3,4-Benzopyrene	Coffee
3-Carene	Black spruce, White spruce, Norway spruce

3-Hexen-1-ol	Walnut, Black walnut, Yucca, Echinacea
3-Methyl-butanal	Sweetgrass, Echinacea
3-Methyl-butyl octanoate	Sea buckthorn
3-Methyl-butyl benzoate	Sea buckthorn
3-Octanol	Shepherd's purse, Yucca, Ground ivy
4,5-Dimethoxy-6-(2-propenyl)-1,3-benzodioxole	Dill
4-Mercapto-4-methylpentan-2-one	Boxwood
4-Methoxy-2-methyl-2-mercaptobutane	Black currant, Black gooseberry
5-Methylheptan-2,4-dione	St. John's wort
5,9-Octadecadienoic acid	Ginkgo
5,9,12-Octadecatrienoic acid	Ginkgo
6-Methylheptan-2,4-dione	St. John's wort
8,9-Dehydroeoisolongifolene	Cedarwood
14-Methylhexadecanoic (14-MHD)	Lodgepole pine, Ginkgo
Abietatriene	Japanese parsley, Mitsuba
Abrotamine	Qing Hao, Chinese wormwood
Acetaldehyde	Peppermint, Echinacea
Acetic ester	Fu-ti, Bistort root, Flowery knotweed, Cow parsley, Cow parsnip
Acetone	Alfalfa
Acetophenone	Skullcap, Virginia skullcap, Baical skullcap
Acetyl eugenol	Clove
Acetylenes	Horse chestnut
Achilline	Yarrow
Acorin	Calamus, Sweet flag, Shi Chang Pu
Acthusin	Horse chestnut
Acylated sterylglucoside	Ashwagandha
Adenine	Plantains
Aethusanol A	Horse chestnut
Aethusanol B	Horse chestnut
Agropyrene	Couch grass
Alantol	Elecampane
Alcohol farnesol	Hibiscus
Aldehyde	Camphor, Citronella, Cumin, Periwinkle, Lemon grass
Aliphatic alcohols	Olive, Yew
Aliphatic esters	Olive, Sea buckthorn
Aliphatic hydrocarbons	Black walnut, Walnut
Alkane derivative	White oak
Alkylmethoxy pyrazines	Lily of the valley, Pansy
Allicin	Chives, Garlic

Alliin	Garlic
Allo-aromadendrene	Hemp, Groundsel
Allo-ocimene	Angelica
Allphatic hydrocarbons	Yucca
Allyl isothiocyanate	Black mustanrd, Horseradish
Allypropyl	Garlic
Alpha-atalantone	Curcuma, Jiang Huang
Alpha-bergamotol	Passion flower
Alpha-betulenolacetate	Birch
Alpha-bisabolol	Roman chamomile
Alpha-caryophyllene	Popular, Balm of gilead, American aspen
Alpha-copaene	Yucca, Ginger, Gotu Kola, Stinkwood
Alpha-cuprenene	St. John's wort
Alpha-humulene	Gotu Kola, Hops, Clove
Alpha-linoleic acid	Oat
Alpha-muurolene	Sweet herb of Paraguay, Stevia
Alpha-naginatene	Pilewort
Alpha-nepetalactone	Catnip, Catmint
Alpha-phyllandrene	Rose geranium, Curcuma, Wild rose geranium, Lemon Geranium, Jiang Huang, Echinacea
Alpha-pinene	Balsam fir, Stinging nettle, Tarragon, Feverfew, Cilantro, Nutmeg, Forsythia, Coriander, Juniper, Buckwheat, Bay laurel, Hemp, Groundsel, Chinese chaste tree, Chaste tree, Myrtle, Wild rose geranium, Rose geranium, Lemon geranium, Gumweed, Angelica, Gotu Kola, Echinacea, Anise
Alpha-santalols	Sandalwood
Alpha-selinene	Ginger
Alpha-terpinene	Nutmeg, Lime
Alpha-terpineol	Cumin, Buckwheat, Lemon grass, Ground ivy, Bay laurel, Eucalyptus, Gum tree, Yucca
Alpha-terpinyl	Clove
Alpha-thujene	Buckwheat, Sage, Sinkwood
Alpha-tocopherol	Boxwood, Evening primrose, Safflower, Sunflower
Alpha-vetrivone	Vetiver
Alpha-ylangene	Cedarwood
Ambrettolide	Hibiscus
Anabsinthin	Wormwood
Anarin	Strawberry
Anemonin	Buttercup

Anethole	Coriander, Cilantro, Mexican mint marigold, Fennel, California laurel, Bay laurel
Angelic acid esters	Roman chamomile
Anisaldehyde	Anise
Anisic acid	Fennel, Tarragon, Angelica
Anisole	Tarragon
Anthraquinones	Rhubarb, Chinese rhubarb
Apigene	German chamomile, Chamomile
Apigetrin	German chamomile, Chamomile
Apiin	Chamomile, German chamomile
Apiol	Parsley, Celery
Ar-curcumene	Ginger
Ar-turmerone	Curcuma, Jiang Huang
Arachic acid	Papaya
Araliene	Spikenard
Arnidiol	Arnica
Aromadendrene	Eucalyptus, Gum tree, Yucca
Artabsin	Wormwood
Artemisia ketone	Qing Hao, Chinese wormwood
Asarone	Wild ginger, Sassafras, Carrot
Ascaridole	Wormseed, Lamb's quarters
Atlantones	Cedarwood
Aucubin	Plantains
Azelaic acid	Angelica
Azulene	Calamus, Kava-Kava, Sweet flag, Shi Chang Pu, Roman chamomile, Wormwood, Yarrow
Behenic acid	Milk thistle
Benzaldehyde	Chokecherry, Black cherry, Cinnamon bark, Wild cherry, Mountain ash, Japanese apricot
Benzene	Aniseed
Benzenoides	Bay laurel
Benzoic acid	Bia shao, White peony, Japanese apricot
Benzyl acetate	Jasmine, Ylang-Ylang, Yarrow
Benzyl alcohol	Oriental sweet gum, Almond, White oak, Passion flower, Gum tree, Clover (red), Clover (crimson), Jasmine, Daffodil
Benzyl benzoate	Cinnamon leaf, Jasmine
Benzyl cyanide	Canary creeper, nasturtium
Benzyl isothiocyanate	Canary creeper, Nasturtium
Bergapten	Bergamot orange
Bergaterpeme	Cow parsnip, Cow parsley
Beta-amyrin	Pyrethrum, Biashao, White peony

Beta-asarone	Shi Chang Pu, Sweet flag, Calamus
Beta-bisabolene	English pepper, Ginger, Carrot
Beta-borneol	Oregano, Marjoram
Beta-bourbonene	Ground ivy, Chinese wormwood, Qing Hao
Beta-carophyllene	Wild rose geranium, Lemon geranium, Rose geranium, English pepper, White Water lily, Hemp, Hops, Gotu Kola, Stevia, Chaste tree, Chinese chaste tree, Sweet herb of Paraguay, Gotu Kola, Carrot, Mint, Valerian
Beta-cyclopyrethrosin	Pyrethrum
Beta-elemene	Pilewort, Ground ivy, Juniper
Beta-farnesene	Sweet flag, Shi Chang Pu, Calamus, Echinacea
Beta-ionone	Passion flower, Lemon grass
Beta-linoleic acid	Oat
Beta-myrcene	Echinacea
Beta-nepetalactone	Catmint, Catnip
Beta-ocimene	Angelica, Forsythia
Beta-phyllandrene	Hemp, Angelica, Forsythia
Beta-pinene	Roseroot, Bitter orange, English pepper, Hyssop, Forsythia, Ground ivy, Juniper, Melaleuca, Tarragon, Myrtle, Walnut, Lemon, Black walnut, Dwarf mountain pine, Nutmeg, Gumweed, Medicinal tea tree, Echinacea, Hemp, Gotu Kola, Anise
Beta-santalols	Sandalwood
Beta-selinene	Celery, Stevia, Sweet herb of Paraguay, Ginger
Beta-sesquiphyllandrene	Goldenrod, Ginger
Beta-sitosterol	Evening primrose, Hibiscus, Saw palmetto
Beta-terpinene	Nutmeg, Sage
Beta-thujone	Tansy
Beta-vetivone	Vetiver
Betulene	Birch
Betulenol	Birch
Betulol	Birch
Biisobutyl phthalate	Chinese rhubarb, Rhubarb
Bisabolene	Marjoram, Oregano, Carrot
Bisabolol oxide	Chamomile, German chamomile
Bishomophinolenic	Lodgepole pine, Southern pitch pine, White pine
Borneal	Black spruce, White spruce, Norway spruce
Borneol acetate	Rosemary, Kava-Kava, Dwarf mountain pine,

Borneol acetate *(continued)*	Cilantro, Labrador tea, Sandalwood, Lavender, Gumweed, Feverfew, Strawberry, Coriander, Sage, Wild ginger, Angelica, Goldenrod, Garden thyme, Tansy, Curcuma, Jiang Huang, Echinacea
Bornyl acetate	Dwarf mountain pine, Buckwheat, Valerian, Goldenrod, Echinacea
Bulnesol	Devil's club
Butanone	Alfalfa
Butyl acetate	Japanese apricot
Butylaldehyde	Green tea
Butylidene	Lovage, Dong quai
Butylpythalides	Lovage
Butyric acid	Flowery knotweed, Ostrich fern, Fu-ti, Bistort root
Butyrospermol	White peony, Biashao
C_{18} paraffin	Dog rose
Cadenene	Devil's club
Cadinene	Rose geranium, Lemon geranium, Wild rose geranium, Dwarf mountain pine, Yucca, Goldenrod, Juniper
Cadinol	Shi Chang Pu, Calamus, Sweet flag
Calmenen	Calamus
Calamenenol	Calamus, Sweet flag, Shi Chang Pu
Calendulin	Marigold
Campesterol	Hibiscus, Plantains, Saw palmetto
Camphene	Forsythia, Myrtle, Fennel, English pepper, Nutmeg, Citronella, Tarragon, Juniper, Hemp, Camphor, Garden thyme. Yucca, White spruce, Norway spruce, Black spruce, Ginger, Echinacea, Rosemary, Anise
Camphor	Hyssop, Sassafras, Labrador tea, Yarrow, Lavender, Cedarwood, Sage, Rosemary, Shepherd's purse, Tansy, Camphor, Kava-Kava, Feverfew, Angelica, Birch
Canadensis curlone	Goldenrod
Capric acid	Saw palmetto
Caproic acid	Saw palmetto, Cow parsley, Cow parsnip
Capronate	Pineapple
Capronic acid	Japanese apricot, Almond
Caprylic acid	Saw palmetto
Capsaicin	Sweet pepper, Cayenne pepper, Chilli pepper

Carbonic acid	Birch
Carboxylic acids	Seneca snakeroot
Carene	Forsythia
Carlinoxide	Star thistle
Carota-1,4-beta-oxide	Carrot
Carotol	Carrot
Carvacrol	Oregano, Shepherd's purse, Couch grass, Sweet marjoram, Marjoram, Horsemint, Dong quai, Horsebalm, Lemon thyme, Garden thyme, Savory, Plantains, Catnip, Catmint, Angelica
Carveol	Caraway
Carvomenthene	Echinacea
Carvone	Coriander, Gum tree, Cilantro, Couch grass, Eucalyptus, Passion flower, Alexandria senna, Tinnevelly senna, Caraway, Dill, Anise, Carrot
Caryophyllene	Ylang-Ylang, Oregano, Black currant, Black gooseberry, Pilewort, Marjoram, St. John's wort, Walnut, Black walnut, Clove, Celery, Echinacea, Birch, Bayberry
Caryophyllene eposide	Echinacea
Caryophyllene oxide	Melilot, Sweet clover
Cedrene	Cedarwood
Cedrol	Devil's club
Cerene	Rosemary
Chamazulene	Roman chamomile, Tansy, Chamomile, German chamomil, Yarrow
Chavicol	Clove, Highbush cranberry
Cholesterol	Hibiscus, Plantains
Choline	Plantains
Chrysanin	Pyrethrum
Chrysanolide	Pyrethrum
Chrysophanic acid	Rhubarb, Chinese rhubarb
Cineole	Saffron, Bayberry, Labrador tea, Cinnamon bark, Sage, Myrobalan, Emblic, Chinese chaste tree, Chaste tree, Basil, Herb of grace, Camphor, Rue, Myrtle, Pyrethrum, Spearmint, Yarrow, St. John's wort, Bay laurel, Lavender, Juniper, Kava-Kava, Garden thyme, Mugwort, Rosemary, Calamus
Cinnamaldehyde	Cinnamon bark, Cinnamon leaf, Daffodil
Cinnamic acid	Rhubarb, Chinese rhubarb

Cinnamic alcohol	Oriental sweet gum, Gum tree, Storax
Cinnamic aldehyde	Cassia
cis-Anethole	Anise
cis-Asarone	Carrot
cis-Dihydroatlantones	Cedarwood
cis-Jasmone	Jasmine
cis-Ocimene	Ground ivy
cis-trans-Rose oxides	Elder
cis-3-Hexen-1-ol	Shepherd's purse
cis-3-Hexenal	Horsetail
cis-3-Hexenol	Dang Shen, Codonopsis, Bellflower, Elder
cis-6-Octadecatrienoic aicd	Evening primrose
cis-8-Heptadecene	Yucca
cis-9-Nonadecene	Yucca
cis-9-Octadecatrienoic acid	Evening primrose
cis-12-Octadecatrienoic acid	Evening primrose
Citral	Gum tree, Marigold, Lemon grass, Eucalyptus, Gingergrass, Palmorosa, Sweet clover, Lemon balm, Melilot, Balm
Citric acid	Bitter orange, Caraway
Citronellal	Lemon balm, Nettle, Citronella, Melilot, Lemon grass, Balm, Sweet clover, Lime, Eucalyptus, Gum tree, Marigold, Lemon geranium, Wild rose geranium, Palmorosa, Catnip, Rose geranium, Catmint, Gingergrass, Citronella, Yucca, Dog rose, Lemon grass
Citrostadienol	Evening primrose
Coffeasterol	Coffee
Comphor	Mugwort
Coriandrol	Coriander, Cilantro
Coumarines	Sweetgrass, Selfheal, Heal all, Lovage, Melilot, Sweet clover
Creosol	Birch
Cresol	Birch, Green tea
Cryptotaenene	Mitsuba, Japanese parsley
Cuminaldehyde	Gum tree, Eucalyptus, Cinnamon bark
Cyclo-santalal	Sandalwood
Cycloalliin	Onion, Chives
Cycloarienol	White peony, Bia shao, Coffee
Cymbopogonol	Lemon grass
Cymeme	Lamb's quarters, Cumin, Nutmeg, Juniper,

Cymeme *(continued)*	Coriander, Myrtle, Frankincense, Savory, Myrobalan, Cilantro, Horsebalm, Horsemint, Oregano, Medicinal tea tree, Rosemary, Melaleuca, Emblic, Marjoram, Garden thyme, Celery
Cymol	Celery
d-1-Methyl-3-cyclohexanone	Pennyroyal
d-α-Phellandrene	Bayberry
d-Cadeine	Black currant, Black gooseberry
d-Camphor	Selfheal, Heal all
d-Fenchone	Heal all, Selfheal
d-Limonene	Grapefruit, Bergamot orange
Dandelion oil	Dandelion
Daiceme	Carrot
Daucol	Carrot
Decylaldehyde	Coriander, Cilantro
Delta-3-carene	Myrtle
Delta-cadinene	Pilewort, Ylang-Ylang
Delta-hydromatricaria-ester	Mugwort
Delta-sabinene	Juniper
Depentene	Roseroot
Desmethoxyyangenin	Kava-Kava
Di-acids	Pokeweed
Diacetyl	Pineapple
Diacylglycerols	Ashwagandha
Diasarone	Wild ginger
Dibutyl ester	Chinese lantern
Dicarbonic acid	Angelica
Digalactosylglycerol	Ashwagandha
Dihydralantolactone	Elecampane
Dihydroalliin	Onion
Dihydrobutylinene	Lovage
Dihydrocarveol	Caraway
Dihydrocarvone	Caraway, Spearmint, Feverfew
Dihydrochamazulene	Tansy
Dihydroisalantolactone	Elecampane
Dihydrokavin	Kava-Kava
Dilaurocapnin	California laurel
Dillapiole	Ligustrum fruit
Dimethyl azelate	Huang Qi, Astragalus
Dimethyl phenols	Birch
Dimethyl sulfide	Peppermint, Echinacea

Dipentene	Marjoram, Lemon grass, Oregano, Camphor, Frankincense, Dwarf mountain pine, Goldenrod, Hops
Dipropyl disulphide	Onion
Dipropyl trisulphide	Onion
Disulphide	Garlic
Diterpene	Bergamot orange, White oak
Dodeca-2,4-dien-1-yl-isovalerate	Echinacea
Dodecanoid acid	California laurel
Dodecanol	Devil's club, St. John's wort
Dodinene	Devil's club
(E)-2-Hexenal	Japanese apricot
(E)-Anethole	Aniseed
(E)-Beta-ocimene	Walnut, Black walnut
Echinolone	Echinacea
Eicosa-11,14,17-trienoic acid	Loosestrife
Elemene	Celery, Juniper
Elemicin	Pilewort
Elemol	Rue, Herb of grace
Elincin	Nutmeg
Epicyclosantalal	Sandalwood
Epishyobunone	Calamus, Echinacea
Epsilon-bulgarene	Ground ivy
Epsilon-muurolene	Ground ivy
Equinopanacene	Devil's club
Equinopanacol	Devil's club
Ergosterol	Hibiscus
Erucic acid	Calendula, Marigold, Safflower, Flax, Milk thistle
Essential oil	Cascara sagrada, Cloudberry, Watercress, Betony, Woundwort, Chickweed, Buffalo berry, Purging buckthorn, Bindweed, Alder, Buckthorn, Oregon grape, Foxglove, Puncture vine
Esters	Balsam fir, Pyrethrum
Estragole	Hyssop, Tarragon, Mexican mint marigold, Anise, Aniseed
Ethanol	Pineapple, Celery
Ethyl alcohol	Gum tree, Oriental sweet gum, Storax
Ethyl butyrates	Cow parsley, Cow parsnip
Ethyl hexanoate	Sea buckthorn
Ethyl linoleate	Japanese parsley, Mitsuba

Ethyl linolenate	Japanese parsley, Mitsuba
Ethyl palmitate	Japanese parsley, Mitsuba
Eucalyptol	Gum tree, Eucalyptus, Nettle
Eudesmol	Gum tree, Eucalyptus
Eudesmyl acetate	Eucalyptus, Gum tree
Eugenol	Cinnamon leaf, Cinnamon bark, Ylang-Ylang, Passion flower, Myrrh, Dill, Basil, Wild ginger, Nutmeg, Clove, Anise, Cinnamon leaf, Jasmine, Japanese apricot
Famesene	Ylang-Ylang
Faradiol	Arnica
Farnesene	Hops, Chamomile, German chamomile, Roman chamomile, Ginger
Farnesol	Damask rose, Jasmine, Lemon grass, Small-leaved lime, Linden
Fatty acid	Psyllium, White water lily
Fenchene	Rosemary
Fenchone	Fennel
Fenchyl alcohol	Basil
Ferruginol	Japanese parsley, Mitsuba
Ferulic acid	Rhubarb, Chinese rhubarb
Flavanoids	Selfheal, Heal all, Houseleek
Flower essences	Dutch honeysuckle, White spruce, Red baneberry, Norway spruce, Buttercup, Uva-ursi, Mountain ash, Raspberry, Black currant, Sorrel, Dock, Sweet clover, Shepherd's purse, Black spruce, Black gooseberry, Melilot, Dogwood, White baneberry, Alfalfa, Gumweed, Plantains, Couch grass, Honeysuckle, Bindweed, Bearberry, Mountain fly, Aloe, Bunchberry, Saskatoon, Lilac, Coltsfoot, Cow parsnip, Cow parsley
Fluoranthene	Loosestrife
Foradiol	Arnica
Formaldehyde	Wintergreen
Fumaric acid	Celery
Furano-monoterpene	Pilewort
Furanosesquiterpenes	Myrrh
Furfural	Curry plant, Blue flag
Gamma-2-cadinene	Goldenrod
Gamma-atalantone	Jiang Huang, Curcuma

Gamma-decalactone	Japanese apricot
Gamma-elemene	Ground ivy, Toothache tree, Prickly ash
Gamma-gurjunene	Yucca
Gamma-jasmolactone	Japanese apricot
Gamma-linolenic acid	Borage, Evening primrose
Gamma-terpinene	Melaleuca, Gingergrass, Palmorosa, Medicinal tea tree, Grapefruit, Savory, Lime, Lemon, Mandarin, Rose geranium, Wild rose geranium, Lemon geranium
Gamolenic acid	Evening primrose
Gaultheriline	Wintergreen
Gaultherin	Meadow sweet
Geijerene	Rue, Herb of grace
Geranial	Ginger, Ylang-Ylang, Eucalyptus, Horsemint, Lemon grass, Marigold, Curry plant, Wild ginger, Gum tree, Bitter orange, Balm, Dog rose, Wormseed, Shepherd's purse, Horsebalm, Rose geranium, Carrot, Coriander, Lemon balm, Cilantro, Wild rose geranium, Lemon geranium, Palmorosa, Citronella, Gingergrass, Damask rose, Lemon grass, Ginger
Geraniol	Carrot, Catnip, Catmint
Geraniol acetate	Carrot
Geranyl acetate	Gingergrass, Palmorosa, Plantains, Ginger
Geranylisobutyrate	Echinacea
Germacrene	St. John's wort, Ylang-Ylang, Yarrow, Goldenrod, Ground ivy, Pilewort
Germacrene B	Chinese chaste tree, Toothache tree, Chaste tree, Prickly ash
Germacrene D	Gotu Kola, Echinacea
Germacrone	Prickly ash, Toothache tree
Globulol	Eucalyptus, Gum tree
Glycerol	Stinging nettle
Glycolic acid	Caraway
Gossypol	Cotton
Grape seed oil	Grape
Gualacol	Birch
Heerabolene	Myrrh
Henicosane	White water lily
Heptadecanoic acid	Loosestrife
Heptanone	Clove

Heptyl-2-methyl butyrates	Everlastings
Hesperidin	Bitter orange
Hexadecane	Walnut, Black walnut
Hexadecanoic acid	Selfheal, Heal all, Loosestrife
Hexanal	Passion flower, Japanese apricot
Hexanoic acid	Seneca snakeroot
Hexenol	Green tea, Lemon thyme
Hexenyl derivativea and acetals	White oak
Hexyl acetate	Japanese apricot
Hexyl esters	Ostrich fern, Everlastings
Himachalenes	Cedarwood
Humulene	Purple coneflower, Hops, Pale-flower echinacea, Narrowleaf echinacea, Celery
Hydrocarbons	Lodgepole pine, Southern pitch pine, White pine, White water lily, Vetiver, Olive
Hydrocyanic acid	Mountain ash, Wild cherry, Black cherry, Chokecherry
Hymulene	St. John's wort
Indole	Jasmine, Daffodil
Inulin	Burdock, Coltsfoot, Chicory
Ishwarane	St. John's wort
Iso-amyl alcohol	Raspberry
Iso-butanol	Pineapple
Iso-butyric acid	Almond
Iso-butyric ester of phlorol	Arnica
Iso-calamendiol	White water lily
Iso-citric	Celery
Iso-eugenol	Passion flower, Nettle
Iso-lactone	Elecampane
Iso-menthone	Rose geranium, Pennyroyal, Lemon geranium, Wild rose geranium
Iso-phytol	Jasmine
Iso-pimpinellin	Cow parsnip, Cow parsley
Iso-pinocamphone	Hyssop
Iso-pulegol	Gum tree, Eucalyptus
Iso-quercitin	Japanese apricot, Heal all, Selfheal
Iso-ricinoleic acid	Plantains
Iso-salicin	Meadow sweet
Iso-tridecane	St. John's wort
Iso-valeraldehyde	Green tea
Iso-valerate	Valerian
Iso-valeric aldehyde	Curry plant, Peppermint

Jojoba oil	Jojoba
Kavain	Kava-Kava
Ketone	St. John's wort, Vetiver
Ketone carvone	Caraway
L-Borneol	Pyrethrum
L-Camphene	Dwarf mountain pine
L-Camphor	Pyrethrum
L-Carvone	Spearment
Lactate	Pineapple
Lactones	Almond, Chicory, Yarrow
Lanosterol	Coffee
Lauric acid	Bia shao, White peony, Saw palmetto, Cow parsnip, Cow parsley, California laurel
Ledene	Labrador tea
Ledol	Labrador tea
Lepalox	Labrador tea
Ligostilides	Lovage, Dong quai
Ligustilide	Angelica
Limonene	Southern pitch pine, Carrot, White pine, Eucalyptus, Gum tree, Ground ivy, Pennyroyal, Citronella, Lodgepole pine, Yarrow, Horsebalm, Rue, Tarragon, Marigold, Frankincense, Camphor, Black walnut, Herb of grace, Tinnevelly senna, Horsemint, Spearmint, Alexandria senna, Myrtle, Hyssop, Walnut, Cumin, Dill, Rosemary, Hemp, Pyrethrum, Angelica, Shepherd's purse, Peppermint, Fennel, Juniper, Caraway, Tansy, Yucca, English pepper, Lemon, Lime, Mandarin, Celery, Anise, Echinacea, Aniseed, Angelica, Dong quai, Goldenrod
Linalool	Wild ginger, Lemon grass, Curry plant, Clary sage, Cinnamon leaf, Angelica, Jasmine, Myrobalan, Emblic, Lime, Plantains, Forsythia, Sage, Almond, Rose geranium, Wild rose geranium, Lemon geranium, Myrtle, Nutmeg, Gum tree, Bergamot orange, Elder, horsebalm, Lavender, Palmorosa , Savory, Eucalyptus, Shi Chang Pu, Camphor, Japanese

Linalool *(continued)*	apricot, Sweet flag, Gingergrass, Calamus, Yarrow, Strawberry, Melilot, Balm, Bay laurel, Lemon balm, horsemint, Sweet marjoram, Sweet clover, Passion flower, Ylang-Ylang, Yucca, Raspberry, Bitter orange, Spearmint, Basil, Carrot, Anise
Linalyl acetate	Jasmine, Bergamot orange, Lavender, Bitter orange, Clary sage, Carrot
Linoleic acid	Papaya, Motherwort, Pumpkin , Plantains, Grapefruit, White peony, Poppy, Coffee, Flax, Elder, Bia shao, Corn poppy, Iranian poppy, Evening primrose, English ivy, Damask rose, Stinging nettle, Sesame, Saw palmetto, Psyllium, Flax, Strawberry, Motherwort, Plantains, Raspberry, Damask rose, Saw palmetto, Japanese parsley, Mitsuba, Foxglove, Milk thistle, Purple loosestrife, Boxwood, Raspberry, Clover (red), Clover (crimson), Sea buckthorn, Ginseng, Safflower
Linolenic acid	Borage, Echinacea, Mountain ash, Strawberry, Motherwort, Flax, Lycium, Chinese wolfberry, Wolfberry, Plantains, Damask rose, Raspberry, Sage, Elder, Saw palmetto, Milk thistle, Kudzu, Evening primrose, Sea buckthorn, Ginseng
Longiborneal	Black spruce, Norway spruce, White spruce
Longifolene	Black spruce, White spruce, Norway spruce
Longipinene	Cedarwood
Lupeol	White peony, Bia shao
Luteolin	Chamomile, German chamomile
Lycopine	Almond, Japanese apricot
Malic acid	Bitter orange, Mountain ash, Celery
Malonic	Caraway
Massoilactone	Sweetgrass
Maté Absolute	English holly
Melilotic acid	Sweet clover, Melilot
Menthol	Tarragon, Melilot, Sweet clover, Peppermint, Pennyroyal
Menthone	Spearmint, Peppermint, Sassafras
Menthy-2-octane	St. John's wort
Methanol	Pineapple

Methone	Balm, Lemon balm
Methyl acetate	Mint
Methyl anthranilate	Grape
Methyl benzoate	Ylang-Ylang
Methyl branched fatty acid	Ginkgo
Methyl chavicol	Basil, Goldenrod, Tarragon, Fennel, Anise, Clove, Aniseed
Methyl cinnamate	Basil
Methyl ethers	Mint
Methyl eugenol	Shi Chang Pu, Cassia, Calamus, Dog rose, Sweet flag
Methyl heptenone	Lemon grass
Methyl ionone	Lemon grass
Methyl isoeugenol	Wild ginger
Methyl jasmonate	Jasmine
Methyl ketones	Huang Bai, Chinese corktree
Methyl leugenol	Mexican mint marigold
Methyl levulinate	Huang Qi, Astragalus
Methyl linoleate	Toothache tree, Prickly ash
Methyl nonyl ketones	Rue, Herb of grace
Methyl palmitate	Kudzu, Huang Qi, Astragalus, Jasmine
Methyl phenyl esters	Olive
Methyl salicylate	Wormseed, Bia shao, White peony, Seneca snakeroot, Wintergreen, Strawberry, Coca, Birch, Clover (red), Clover (crimson), Garden thyme, Cassia
Methylalliin	Onion
Methylsalicylaldehyde	Cassia
Methystirin + dihyromethysticia	Kava-Kava
Mitsubene	Japanese parsley, Mitsuba
Mono-acids	Pokeweed
Monoacylglycerols	Ashwagandha
Monocarboxylic acid	Labrador tea
Monoenes	Ginseng
Monogalactosylglycerol	Ashwagandha
Monoterpene	Dwarf mountain pine, Hops, English pepper, American ginseng, White oak, Tian Qi, Asian ginseng, Toothache tree, Prickly ash, Geranium, White baneberry, Beet root, Juniper, Balsam fir, Chaste tree, Chinese chaste tree, Calamus, Sweet flag, Shi Chang Pu, Ginseng
Monoterpene alcohols	Green tea

Monoterpene aldehydes	Green tea
Monoterpene hydrocarbons	Groundsel
Monoterpenoid	Bay laurel, Huang Bai, Chinese corktree, Feverfew, California laurel
Monoterpenols	Stevia, Sweet herb of Paraguay
Mucilage	Arnica, Coltsfoot
Murolene	Yucca, Juniper
Mycene	Chinese corktree, Huang Bai, Bitter orange, Goldenrod, Hops, Ground ivy, Frankincense, Mandarin, Tarragon, Forsythia, Lemon grass, Juniper, Cumin, Rosemary, Rue, Herb of grace, Echinacea
Myrcene	Clary sage, Anise
Myristic	White peony, Bia shao, Caraway, Angelica, Celery
Myristicin	Carrot, Parsley, Nutmeg, Sassafras, Celery
Myristoleic acid	White peony, Bia shao, Celery
Myrtenal	Eucalyptus, Gum tree, Myrtle
n-Alkanes	White peony, Bia shao
n-Amyl alcohol	Cotton
n-Butylidene phthafide	Ligustrum fruit, Privet
n-Butyric acid	Pennyroyal
n-Capric acid	Pennyroyal
n-Caprylic acid	Pennyroyal
n-Decane	Shepherd's purse
n-Docosanol	African prune
n-Heptadecane	Yucca
n-Hexanal	Seneca snakeroot, Yucca
n-Nonadecane	Yucca
n-Tetracosanol	African prune
n-Undecylic acid	Sagebrush
Naphthalene	Clove
Narcissine	Daffodil
Neoisopulegol	Gum tree, Eucalyptus
Neomenthol	Mint
Neral	Dog rose, Lemon grass, Horsemint, Horsebalm, Mandarin, Tarragon, Marigold, Damask rose, Lemon grass, Lavender, Ginger
Nerol	Curry plant, Ginger
Nerolidol	Devil's club, Stinkwood
Neryl acetate	Curry plant
Nitribine	Black gooseberry, Black currant

Nonadecane	White water lily
Nonanal	Strawberry, Lime
Nor-alpha-trans-bergamotenone	Sandalwood
Nor-lapachol	Forsythia
Norpinene	Black gooseberry, Black currant
o-Cresol	Seneca snakeroot
Ocimene	Basil, Tarragon, Sage, Echinacea
Octanol	St. John's wort
Octyl alcohol	Cow parsley, Cow parsnip
Octyl esters	Ostrich fern
Olefinic terpenes	Rosemary
Oleic acid	White peony, English ivy, Pumpkin , Psyllium, Coffee, Passion flower, Iranian poppy, Grapefruit, Oat, Bia shao, Flax, Papaya, Olive, Clover (red), Sesame, Stinging nettle, Clover (crimson), Motherwort, Celery, Foxglove, Milk thistle, Ashwagandha, Raspberry, Sea buckthorn, Strawberry, Virginia skullcap, Baical skullcap, Skullcap, dogwood
Oleoresin	Nutmeg, Alder, Buffalo berry, Saw palmetto
Orthocoumaric acid	Melilot, Sweet clover
Oxygenated sesquiterpene	Groundsel, Groundsel, Ginger
p-Anisic acid	Anise
p-Cresyl	Ylang-Ylang
p-Cymene	Wild rose geranium, Lemon geranium, Mandarin, Jiang Huang, Rose geranium, Curcuma, Forsythia, Yucca
p-Hydroxybenzaldehyde	White peony, Bia shao
p-Hydroxycinnamic acid methyl ester	Echinacea
p-Mentha-1,3,8-triene	Celery
Paeonol	Bia shao, White peony, Pilewort
Paliloleic	Celery
Palmitic acid	Psyllium, White peony, African prune, Sagebrush, Passion flower, Bia shao, Hibiscus, Grapefruit, Papaya, Baical skullcap, Elder, Virginia skullcap, Clover (crimson), Clover (red), Skullcap, Hazelnut, Oat, Iranian poppy, Coffee, English ivy, Celery, Japanese parsley, Mitsuba, Saw palmetto, Ashwagandha, Raspberry,

Palmitic acid *(continued)*	Echinacea, Sea buckthorn, Dogwood
Palmitoleic	White peony, Bia shao
Paraffin	Hazelnut
Pelargonic acid	Ostrich fern
Penta-(1,8Z)-diene	Echinacea
Pentadeca-8-en-2-one	Echinacea
Pentadeca-(8Z)-en-2-one	Echinacea
Pentane	Alfalfa, Pineapple, Celery
Persicariol	Bistort root, Flowery knotweed , Fu-ti
Petroselaidic	Celery
Petroselinic	English ivy, Caraway, Celery
Phenethyl	Everlastings
Phenols	Green tea, Birch
Phenyl ethanol	Damask rose
Phenylacetic acid	Jasmine
Phenylacetic aldehyde	Lily of the valley, Pansy
Phenylethyl alcohol	Japanese apricot
Phenylethyl isothiocynate	Horseradish
Phenylpropanes	Calamus, Sweet flag, Shi Chang Pu
Phenylpropionic acid	Chinese rhubarb, Rhubarb
Phenylpropyl alcohol	Oriental sweet gum, Storax, Gum tree
Phosphatidylcholine	Ashwagandha
Phospholipids	Sesame
Phthalide	Dong quai, Celery
Phyellandrene	Cinnamon bark, Wormwood, Aniseed, Balsam fir, Spearmint, Rosemary, Fennel, Parsley, Dill, Frankincense, Dong quai, Camphor, Cumin, Clary sage, Gingergrass, Palmorosa, Caraway, Sassafras, Goldenrod, Shepherd's purse
Phytol	Passion flower, Jasmine
Phytol ester	Jasmine
Pimpinellin	Cow parsley, Cow parsnip
Pinene	Curry plant, St. John's wort, Parsley, Labrador tea, Sassafras, Carrot, Ylang-Ylang, Eucalyptus, Rosemary, White pine, Black spruce, Oregano, Caraway, Dill, Pennyroyal, Peppermint, Lodgepole pine, Clary sage, Rue, White spruce, Gum tree, Cinnamon bark, Camphor, Fennel, Yarrow, Angelica, Nutmeg, Mandarin, Norway spruce, Cumin, Frankincense, Wild ginger, Herb of grace, Wormwood, Marjoram, Southern pitch pine,

Pinene *(continued)*	Lemon thyme, Saffron, Hyssop, Goldenrod, Yucca, Celery
Pinocamphene	Hyssop
Pinocarvone	Gum tree, Eucalyptus, California laurel
Piperetine	English pepper
Piperidine	English pepper
Piperine	English pepper
Piperitone	Lemon balm, Balm, Pennyroyal
Planteose trisaccharides	Plantains
Podophyllin	Mayapple
Polyine	Horse chestnut
Populene	Poplar
Pregeijerene	Rue, Herb of grace
Proazulenes	Roman chamomile
Propanal	Alfalfa, Pineapple
Protocanemonin	Buttercup
Pulegone	Pennyroyal, Mountain mint, Calamint, Basil thyme, Pennyroyal
Pyrobetulin	Birch
Pyrocatechol	Birch
Pyrrolizidine alkaloids	Borage
Pyruvic	Caraway
Quebrachitol	Mugwort
Quercimeritrin	German chamomile, Chamomile
Quercitin	Almond, Daffodil, Meadow sweet
Quinic acid	Black gooseberry, Black currant
Root oil	Sorrel, Dock
Rutin	Chamomile, German chamomile
Sabinene	Curcuma, Jiang Huang, Labrador tea, Rosemary, Black gooseberry, Juniper, Nutmeg, Juniper, English pepper, Ground ivy, Yarrow, Bitter orange, Black currant, Sweet marjoram, Lemon, Sweet marjoram, Tarragon, Chinese chaste tree, Chaste tree, Stinging nettle, Carrot, Anise, California laurel
Safranal	Saffron
Safrole	Cinnamon leaf, Ylang-Ylang, Nutmeg, Camphor, Sassafras, Stinkwood, Anise

Salicin	Meadow sweet
Salicylaldehyde	Meadow sweet, Cassia
Salicylic acid	Meadow sweet
Santalol	Celery
Santalone	Sandalwood
Santene	Rosemary, Sandalwood
Sativene	Cedarwood
Saturated acids	Stinging nettle, Pokeweed
Scutellarin	Baical skullcap, Virginia skullcap, Skullcap
Seaweed Absolute	Kelp
Sedanic acid	Celery
Sedanolide	Celery
Sedanonic acid	Ligustrum fruit
Selinene	Labrador tea, Celery, Mandarin
Sellnadiene	Labrador tea
Sesquiterpene	Chicory, Bayberry, Rosemary, Yarrow, Feverfew, Melaleuca, Medicinal tea tree, Jiang Huang, Curcuma, Geranium, Dong quai, White oak, Curry plant, Angelica, Lomatium, Vetiver, Chaste tree, Chinese chaste tree, Clove, Periwinkle, Burdock, Asian ginseng, Hemp, Calamus, American ginseng, Shi Chang Pu, Tian Qi, Sweet flag, Jasmine, Ginseng
Sesquiterpene alcohol	Stinkwood
Sesquiterpene hydrocarbon	Sandalwood, Groundsel, Ginger
Sesquiterpene lactones	Elecampane, Heal all , Selfheal, Blessed thistle
Sesquiterpenoids	Gotu Kola
Sesquiterphenol	Black spruce, Norway spruce, White spruce
Shyobunone	Calamus
Sitosterol	Mugwort, Plantains
Soybean oil	Soybean
Spathulenol	Juniper
Spiracoside	Meadow sweet
Spiraein	Meadow sweet
Spiroether	Roman Chamomile
Sprondrin	Cow parsnip, Cow parsley
Stearic acid	Coffee, Iranian poppy, White peony, Flax, Grapefruit, Bia shao, Motherwort, Papaya, Sesame, Clover (red), Clover (crimson), Celery, Saw palmetto
Stearopten	Damask rose

Sterylglucoside	Ashwagandha
Stigmasterol	Plantains, Saw palmetto
Styrene	Storax, Oriental sweet gum, Gum tree
Sunflower oil	Sunflower
Tannin	Coltsfoot
Tartaric acids	Celery
Tauremisin	Mugwort
Teresantol	Sandalwood
Terpene	Cloudberry, Lemon thyme, Lamb's quarters, Sea buckthorn, Black gooseberry, Pyrethrum, Black currant, Hemp, Vetiver, Roseroot, Lomatium, Carum
Terpene alcohols	Camphor
Terpene-d-limoneene	Caraway
Terpenic oxides	Hemp
Terpenoid	Lovage
Terpinen-4-ol	Melaleuca, Medicinal tea tree
Terpinene	Frankincense, Bitter orange, Horsebalm, Peppermint, Rosemary, Horsemint, Labrador tea, Juniper, St. John's wort, Garden thyme, Medicinal tea tree, Melaleuca, Coriander, Cilantro, Echinacea, Forsythia
Terpinene-4-ol	Forsythia, Ground ivy
Terpineol	Ylang-Ylang, Nutmeg, Sweet marjoram, Wild ginger, Gumweed
Terpinolene	Yucca, Parsley
Tetracosanol	Mugwort
Tetracyclic triterpenes	Huang Bai, Chinese corktree
Tetrahydrosesquiterpene hydrocarbon	Echinacea
Tetramethozyally benzene	Parsley
Thujene	Juniper, Rosemary, Mandarin
Thujone	Feverfew, Wormwood, Frankincense, Sassafras, Mugwort
Thujyl alcohol	Wormwood
Thymol	Shepherd's purse, Catmint, Catnip, Savory, Marjoram, Oregano, Horsemint, Horsebalm, Garden thyme, Lemon thyme
Thymol-hydroquinone dimethyl ether	Arnica
Tiglic	Roman chamomile
Titerpenoids	Olive
Tocopherol	Raspberry, Sea buckthorn
Torreyol	Devil's club

trans-10-11-Dihydroatlantones	Cedarwood
trans-2-Hexanal	Bellflower, Dang Shen, Codonopsis, Codonopsis, Bellflower, Horsetail
trans-3,7-Dimethyll	Elder
trans-Anethole	Passion flower, Tarragon, Couch grass, Anise
trans-Asarone	Carrot
trans-Beta-farnesene	Angelica, Gotu Kola
trans-Ferutic acid esters	African prune
trans-Ocimene	Echinacea
Tri-acylglycerols	Sesame, Purple loosestrife
Tri-caprin	California laurel
Tri-carboxylic acid	Celery
Tri-consane	White water lily
Tri-cyclene	Rosemary
Tri-cyclo-ekasantalal	Sandalwood
Tri-glycerols	Ashwagandha, California laurel
Tri-laurin	California laurel
Tri-terpenoid acid	Chinese hawthorn, English hawthorn
Turmeron	Jiang Huang, Curcuma
Umbellulone	California laurel, Bay laurel
Undecane	St. John's wort
Urushiol	Poison ivy
Valepotriates	Valerian
Valerianic acid	Hops
Valeric acid	Ylang-Ylang
Vanillin	Storax, Gum tree, Oriental sweet gum, Vetiver, Carrot, Echinacea
Verbenol	Frankincense
Vetivene	Vetiver
Vetiver oil	Asparagus
Vetiverol	Vetiver
Volatile oil	Schizandra, Witch hazel, Pussy willow, Cayenne pepper, Ma Huang, Sweet pepper, Lomatium, Dog rose, Borage, Hazelnut, Wu wei zi, Chilli pepper, European beech, Lovage, Blessed thistle, Cowslip, Dog lichen, Chinese licorice, Soapwort, Small houseleek, Mountain cranberry, Dandelion, Cowberry, Mountain ash, Foxberry, Bilberry, Russian olive, Broom, Licorice, Ground liverwort, Elderberry, Hydrangea, Goldenseal, Cleavers,

Volatile oil *(continued)*	Dodder, Duckweed, Chives, White ash, Radish, Beech, Agrimony, Broom, Scoparium, Primrose, Eyebright, Boneset, Black snakeroot, Lungwort, White willow, Clubmoss, Purple loosestrife, Horehound, Monkshood, Lady's slipper, Fireweed, Ephedra, Daisy, Blueberry, Mullein, Cranberry, Corn, Highbush cranberry, Fenugreek, Stonecrop, Roseroot
Xylenol	Birch
γ-Elemene	Aniseed
Yangonin	Kave-Kava
(Z)-3-Hexenyl acetate	Lily of the valley, Pansy
Zingiberene	Curcuma, Jiang Huang, Ginger
Zizanoic acid	Vetiver

APPENDIX 3: LIST OF COMMON AND SCIENTIFIC NAMES

Common Name	Scientific Name
Absinthe	*Artemisia absinthium* L.
African prune	*Prunus affiricana* L.
Agrimony	*Agrimonia eupatoria* L.
Ague tree	*Sassafras albidum* (Nutt.) Nees.
Ague weed	*Eupatorium perfoliatom* L.
Alder	*Alnus glutinosa* (L.) Gaetn.
Alder	*Alnus incana* (L.) Moench. subsp. *tenufolia*
Alder	*Alnus crispus* (Ait.) Pursh.
Alder buckthorn	*Rhamnus frangula* L.
Alfalfa	*Medicago sativa* L.
All heal	*Prunella vulgaris* L.
Almond	*Prunus dulcis* L.
Aloe	*Aloe vera* (L.) Burm.f.
Althaea	*Althaea officinalis* L.
Amber touch-and-heal	*Hypericum perforatum* L.
American ash	*Fraxinus americana* L.
American cranesbill	*Geranium macrorrhizum* L.
American licorice	*Glycyrrhiza lepidota* (Nutt.) Pursh.
American sanicle	*Sanicula marilandica* L.
American sarsaparilla	*Aralia racemusa* L.
American senna	*Cassia marilandica* L.
American valerian	*Cypripedium calceolus* L.
Angelica	*Angelica archangelica* L.
Angelica tree	*Zanthoxylum americanum* Mill
Anise	*Pimpinella anisum* L.
Anise hyssop	*Agastache foeniculum* L.
Aniseed	*Agastache foeniculum* L.
Aniseed	*Pimpinella anisum* L.
Apricot (Japanese)	*Prunus mume* Siebold & Zucc.
Apricot vine	*Passiflora incarnata* L.
Arnica	*Arnica latifolia* Bong.
Arnica	*Arnica chamissonis* L. subsp. *foliosa*
Arnica	*Arnica montana* L.
Arnica	*A. condifolia* Hook
Arnica	*A. fulgens* Pursh
Arnica	*A. sororia* Greene
Arrowwood	*Rhamnus frangula* L.
Artichoke	*Cynara scolymus* L.

Artichoke (wild)	*Cynara cardunculus* L.
Ashwagandha	*Withania somnifera* Dunal.
Astragalus	*Astragalus americana* Bunge.
Azafran	*Carthamus tinctorius* L.
Bachelor's button	*Chrysanthemum parthenium* (L.) Bernh.
Baical skullcap	*Scutellaria baicalensis* Georgi.
Balm	*Melissa officinalis* L.
Balm of gilead	*Populus nigra* L.
Balm of gilead	*Populus balsamifera* L.
Balm of gilead	*Populus multiflorum* Thunb.
Balsam fir	*Abies balsamea* (L.) Mill.
Balsam poplar	*Populus nigra* L.
Baneberry (red)	*Actaea rubra* (Ait.) Willd.
Baneberry (white)	*Actaea alba* L.
Baplisia	*Baptisia tinctoria* (L.) R. Br.
Barberry	*Berberis vulgaris* L.
Basil	*Ocimum basilicum* L.
Basil thyme	*Calamintha nepeta* (L.) Savi
Basin sagebrush	*Artemisia tridentata* Nutt.
Bastard saffron	*Carthamus tinctorius* L.
Bay	*Laurus nobilis* L.
Bay laurel	*Lanrus nobilis* L.
Bay laurel	*Umbelluslaria california* Nutt.
Bayberry	*Myrica penxylvanica* Lois.
Bearberry	*Arctostaphylos uva-ursi* (L.) Spreng
Bear's foot	*Alchemilla vulgaris* L.
Bear's grape	*Arctostaphylos uva-ursi* (L.) Spreng
Bedstraw	*Galium aparine* L.
Bee bread	*Borago officinalis* L.
Bee bread	*Trifolium pratense* L.
Beech	*Fagus grandifolia* Ehrh.
Beech (European)	*Fagus sylvatica* L.
Beet root	*Beta vulgaris* L.
Bellflower	*Codonopsis pilosula* (Franch.) Nannfeldt.
Bellflower	*Codonopsis tangshen* Oliver
Bergamot orange	*Citrus bergamia* Risso & Poit
Betony	*Stachys officinalis* (L.) Trev.
Bia shao	*Paeonia lactiflora* Pall.
Bilberry	*Vaccinium myrtillus* L.
Bilberry	*Vaccinium oreophilum* Rydb.
Bindweed	*Convolvulus sepium* L. Birch (Birchbark)
Bindweed	*Convolvulus arvensis* L.

Birch *Betula lenta* L.
Bird's foot *Trigonella foenum-graecum* L.
Bistort (Flowery knotweed) *Polygonum multiflorum* Thunb.
Bistort root *Polygonum bistorta* L.
Bitter dock *Rumex obtusifolius* L.
Bitter orange *Citrus aurantium* L.
Bittmore ash *Fraxinus americana* L.
Black alder *Alnus glutinosa* (L.) Gaertn.
Black berry *Rubus fruiticosus* L.
Black cherry *Prunus serotina* J. F. Ehrb.
Black cohosh *Cimicifuga racemosa* (L.) Nutt.
Black currant *Ribes nigrum* L.
Black dogwood *Rhamnus frangula* L.
Black mustard *Brassica nigra* (L.) Kock.
Black poplar *Populus nigra* L.
Black sanicle *Sanicula marilandica* L.
Black snakeroot *Cimicifuga racemosa* (L.) Nutt.
Black snakeroot *Sanicula marilandica* L.
Black walnut *Juglans nigra* L.
Black wort *Sympyhtum officinale* L.
Bladder cherry *Physalis alkekengi* L.
Bladder fucus *Fucus vesiculosus* L.
Bladderwrack *Fucus vesiculosus* L.
Blazing star *Chamaelirium luteum* (L.) A. Gray
Blessed thistle *Cnicus benedictus* L.
Blind nettle *Lamium album* L.
Bloodroot *Sanguinaria canadensis* L.
Bloodwort *Achillea millefolium* L.
Blooming Sally *Episetum arvense* L.
Blue cohosh *Caulophyllum thalietroides* (L.) Michx.
Blue cohosh *Caulophyllum giganteum* (F.) Loconte & W. H. Blackwell
Blue flag *Iris versicolor* L.
Blue ginseng *Caulophyllum thalictroides* (L.) Michx.
Blue lobelia *Lobelia siphilitica* L.
Blue mountain tea *Solidago odora* Act.
Blue pimpernel *Scutellaria lateriflora* L.
Blue sailors *Cichorum intybus* L.
Blue skullcap *Scutellaria lateriflora* L.
Blueberry *Vaccinium myrtilloides* Michx.
Blunt-leaves dock *Rumex obtusifolius* L.
Boneset *Eupatorium perfoliatum* L.
Borage *Borago officinalis* L.
Bottle brush *Equisetum arvense* L.

Bourtree	*Sambucus nigra* L.
Boxberry	*Gaultheria procumbens* L.
Boxtree	*Buxus sempervirens* L.
Boxwood	*Cornus florida* L.
Boxwood	*Buxus sempervirens* L.
Bridewort	*Filipendula ulmaria* (L.) Maxim.
Brigham tea	*Ephedra nevadensis* Wats.
Broad-leaved dock	*Rumex obtusifolius* L.
Broom	*Cytisus scoparius* (L.) Link.
Broom tops	*Cytisus scoparius* (L.) Link
Brueberry	*Vaccinium mrytilloides* Ait.
Buckeye	*Aesculus hippocastanum* L.
Buckthorn	*Rhamnus cathartica* L.
Buckwheat	*Fagopyrum tataricum* L.
Buffalo berry	*Shepherdia canadensis* L.
Buffalo herb	*Medicago sativa* L.
Bugbane	*Cimicifuga racemosa* (L.) Nutt.
Bunchberry	*Cornus canadensis* L.
Bunny's ears	*Verbascum thapsus* L.
Burdock	*Arctium lappa* L.
Butcher's broom	*Ruscus aculeatus* L.
Buttercup	*Ranunculus occidentalis* Nutt.
Calamint	*Calamintha ascendens* L.
Calamintha	*Calamintha officinalis* L.
Calamus	*Acorus calamus* L. var. Americanus Wolff.
Calendula	*Calendula officinalis* L.
California bay	*Umbellularia californica* Nutt.
California laurel	*Umberlluslaria california* Nutt.
California poppy	*Eschscholtzia california* Cham.
Caltrops	*Centaurea calcitrapa* L.
Camphor	*Cinnamomum camphora* (L.) Nees & Ebern.
Canada snakeroot	*Asarum canadense* L.
Canary creeper	*Phataris canariensis* L.
Caraway	*Carum caroi* L.
Carrot	*Daucus carota* L.
Cascara buckthorn	*Rhamnus purshiana* DC
Cascara sagrade	*Rhamnus purshianus* L.
Caseweed	*Capsella bursa-pastoris* (L.) Medic.
Catchweed	*Galium aparine* L.
Catmint	*Nepeta cataria* L.
Catnip	*Nepeta cataria* L.
Cat's claw	*Uncaria tomentosa* (Willd.) DC.

Cat's claw	*Uncaria guianensis* (Aubl.) Gmel.
Cat's foot	*Glechoma hederacea* L.
Cedarwood	*Cedrus libani* A. Rich subsp. *atlantica*
Celery	*Apium graveolens* L.
Chamomile	*Matricaria chamomilla* L.
Chapatral	*Larrea tridentata* (Sesse. & Moc. ex DC.) Coville.
Chaste tree	*Vitex agnus-castus* L.
Checkber berry	*Gaultheria procumbens* L.
Cherry birch	*Betula lenta* L.
Chestnut	*Castanea sativa* Mill.
Chickweed	*Stellaria media* (L.) Vill.
Chicory	*Cichorium intybus* L.
China rose	*Hibiscus rosa-sinenesis* L.
Chinese chaste	*Vitex negundo* L.
Chinese corktree	*Phellodendron chinensis* Schneid.
Chinese hibiscus	*Hibiscus rosa-sinenesis* L.
Chinese lantern	*Physalis alkekengi* L.
Chinese licorice	*Glycyrrhizas uralensis* Fisch ex DC
Chinese rhubarb	*Rheum palmatum* L.
Chinese wolfberry	*Lycium chinense* Mill.
Chittem bark	*Rhamnus purshiana* DC
Chives	*Allium schoenoprasum* L.
Chokecherry	*Prunus virginiana* L.
Choublac	*Hibiscus rosa-sincensis* L.
Church steeples	*Agrimonia eupatoria* L.
Cilantro	*Coriandrum sativum* L.
Cinnamon	*Cinnamomum cassia* (Nees) Nees & Eberm.
Cinnamon	*Cinnamomum verum* J. Presl.
Cinnamon wood	*Sassafras albidum* (Nutt.) Nees.
Citronella	*Cymbopogon nardus* L.
Citronella	*Cymbopogon winterianus* Jowitt
Clary	*Salvia sclarea* L.
Clary sage	*Salvia sclarea* L.
Clear eye	*Salvia sclarea* L.
Cleavers	*Galium aparine* L.
Cloudberry	*Rubus chamaemorus* L.
Clove	*Syzygium aromaticum* (L.) Merr. & Perry
Clover (crimson)	*Trifolium incarnatum* L.
Clover (red)	*Trifolium pratense* L.
Clover broom	*Baptisia tinctoria* (L.) R. Br.
Clubmoss	*Lycopodium clavatum* L.
Coca	*Erythroxylum coca* Lam
Cockle buttons	*Arctium lappa* L.

Cocklebur	*Agrimonia eupatoria* L.
Codonopsis	*Codonopsis tangshen* Oliver
Codonopsis	*Codonopsis pilosula* (Franch.) Nannfeldt.
Coffee	*Coffea arabica* L.
Coffeweed	*Cichorum intybus* L.
Colchicum	*Colchicum autumnale* L.
Colicroot	*Asarum canadense* L.
Colicroot	*Dioscorea villosa* L.
Coltsfoot	*Tussilago farfara* L.
Comfrey	*Symhytum officinale* L.
Common bugloss	*Borago officinalis* L
Coriander	*Coriandrum sativum* L.
Corn	*Zea mays* L.
Corn poppy	*Papaver rhoens* L.
Corn rose	*Papaver rhoeas* L.
Cotton	*Gossypium hirsutum* L.
Couch grass	*Agropyron repens* (L.) Beauvois
Couch grass	*Elymus repens* L.
Coughwort	*Tussilago farfara* L.
Countryman's treacle	*Ruta graveolens* L.
Cow clover	*Trifolium pratense* L.
Cow parsley	*Heracleum maximum* Bartr.
Cow parsnip	*Heracleum lanatum* Michx.
Cow parsnip	*Heracleum sphondylium* L.
Cowberry	*Vaccinium vitis-idaea* L.
Cowslip	*Primula veris* L.
Crampbark	*Viburnum opulus* L.
Cranberry	*Vaccinium macrocarpon* L.
Creeping charlie	*Glechoma hederacea* L.
Creeping thyme	*Thymus serpyllum* L.
Crimson clover	*Trifolium incarnatum* L.
Crocus	*Colchicum autumnale* L.
Crosswort	*Eupatorium perfoliatom* L.
Crowberry	*Arctostaphylos uva-usi* (L.) Spreng
Cumin	*Cuminum cyminum* L.
Curcuma	*Curcuma domestica* L.
Curcuma	*Curcuma longa* L.
Curcuma	*Curcuma xanthorrhiza* L.
Curry plant	*Helichrysum angustifolium* (Lam) DC.
Daffodil	*Narcissus pseudonarcissus* L.
Daisy	*Bellis perennis* L.
Damask rose	*Rosa damascena* Mill.

Dandelion	*Taraxacum officinale* G. H.Weber ex Wigg.
Dang Shen	*Codonopsis Pilosula* (Franch.) Nannfeldt.
Dang Shen	*Codonopsis tangshen* Oliver
Dead nettle	*Lamium album* L.
Deadmen's bells	*Digitalis purpurea* L.
Desert tea	*Ephedra nevadensis* Wats.
Devil's apple	*Podophyllum peltatum* L.
Devil's bit	*Chamaelirium luteum* (L.) A. Gray
Devil's bones	*Dioscorea villosa* L.
Devil's claw	*Harpagophytum procumbens* DC.
Devil's club	*Echinopanax horridus* (Sm) Decne & Planch. ex. H. A. T. Harms
Devil's club	*Oplopanax horridus* L.
Dill	*Anethum graveolens* L.
Dock	*Rumex obtusifolia*
Dodder	*Cuscuta chinesis* Lam.
Dodder	*Cuscuta epithymum* Murr.
Dog brier	*Rosa canina* L.
Dog grass	*Agropyron repens* (L.) Boauvcis.
Dog lichen	*Peltigera canina* L.
Dog rose	*Rosa canina* L.
Dog tree	*Cornus florida* L.
Dogwood	*Cornus florida* L.
Dong quai	*Angelica sinensis* (Oliv.) Diels.
Dragon's mugwort	*Artemisia dracunculus* L.
Duckweed	*Lemna minor* L.
Dutch honeysuckle	*Lonicera caprifolium* L.
Dwarf juniper	*Juniperus communis* L.
Dwarf mountain pine	*Pinus mugo* Turra var. Pumilio
Echinacea (pale-flower)	*Echinacea pallida* (Nutt.) Nutt.
Echinacea (purple coneflower)	*Echinacea pururea* (L.) Moench.
Echinacea (narrowleaf)	*Echinacea angustifolia* DC.
Elder	*Sambucus racemosa* L.
Elderberry	*Sambucus nigra* L.
Elderberry	*Sambucus canadensis* L.
Elecampane	*Inula helenium* L.
Emblic	*Phyllanthus emblica* L.
English alder	*Alnus glutinosa* (L.) Gaertn.
English chamomile	*Anthemis nobilis* L.
English holly	*Ilex aquifolium* L.
English ivy	*Hedera helix* L.
Ephedra	*Ephedra distachya* L.

Eucalyptus	*Eucalyptus globulus* Labill.
European alder	*Alnus glutinosa* (L.) Gaertn.
European barberry	*Berberis vulgaris* L.
European buckthorn	*Rhamnus cathartica* L.
European chestnut	*Castanea sativa* Mill.
European elder	*Sambucus nigra* L.
European holly	*Ilex aquifolium* L.
European peony	*Paeonia officinalis* L.
Evening primrose	*Oenothera biennis* L.
Evening star	*Oenothera biennis* L.
Everlastings	*Anaphalis margaritacea* (L.) Bench & Hook
Everlastings	*Antennaria margaritacea* Schultz Bip.
Eyebright	*Euphrasia officinalis* L.
Eyeroot	*Hydrastis canadensis* L.
Fairy cup	*Primula veris* L.
Fairy wand	*Chamaelirium luteum* (L.) A. Gray
False coltsfoot	*Asarum canadense* L.
False unicorn root	*Chamaelirium luteum* (L.) A. Gray
Featherfew	*Chrysanthemum parthenium* (L.) Bernh.
Felon herb	*Artemisa vulgaris* L.
Fennel	*Foeniculum vulgare* Mill.
Fenugreek	*Trigonella foenum-gracecum* L.
Fever tree	*Eucalyptus globulus* Labill.
Feverfew	*Tanacetum parthenium* (L.) Schultz-Bip.
Feverfew	*Chrysanthemum parthenium* (L.) Berhn.
Feverfew	*Tanacetum arthenium* (L.) Schultz.
Feverwort	*Eupatorium perfoliatom* L.
Field balm	*Glechoma hederacea* L.
Field poppy	*Papaver rhoeas* L.
Fir balsam	*Abies balsames* (L.) Mill
Fir pine	*Abies balsames* (L.) Mill
Fireweed	*Epilobium angustifolium* L.
Fireweed	*Chamaenerion angustifolium* (L.) Scop.
Flanneleaf	*Verbascum thapsus* L.
Flax	*Linum usitatissimum* L.
Flowering dogwood	*Cornus florida* L.
Flowery knotweed	*Polygonum multiflorum* Thunb
Forsythia	*Forsythia suspensa* (Thunb.) Vahl.
Fox berry	*Vaccinium vitis-idaea* L.
Fox berry	*Arctostaphylos uva-usi* (L.) Spreng
Foxberry	*Vaccinium vitis-idaea* L.
Foxglove	*Digitalis purpurea* L.

Frankincense	*Boswellia carteri* Birdwood
Frankincense	*Boswellia sacra* Roxb. ex Colebr.
French tarragon	*Artemisia dracunculus* L.
Fu Ti	*Polygonum multiflorum* Thunb
Garden rue	*Ruta graveolens* L.
Garden sage	*Salvia officinalis* L.
Garden sorrel	*Rumex acetosa* L.
Garlic	*Allium sativum* L.
Genista	*Cytisus scoparius* (L.) Link
Geranium	*Robertium macrorrhizum* Pic.
Geranium	*Geranium macrorrhizum* L.
German chamomile	*Matricaria recutita* L.
Ginger	*Zingiber officinale* Roscoe
Gingergrass	*Cymbopogon martinii* (Roxb) Wats.
Ginkgo	*Ginkgo biloba* L.
Ginseng (American)	*Panax quinquefolium* L.
Ginseng (Asian)	*Panax ginseng* C. A. Meyer.
Ginseng (Tian Qi)	*Panax notoginseng* (Burk.) F. H. Chen
Glossy buckthorn	*Rhamnus frangula* L.
Glue gum	*Eucalyptus globulus* Labill.
Goatweed	*Hypericum perforatum* L.
Golden bough	*Phoradendron flavescens* (Pursh) Nutt.
Golden loosestrife	*Lysimachia vulgaris* L.
Golden rose	*Viburnum opulus* L.
Goldenrod	*Solidago virgaurea* L.
Goldenseal	*Hydrastis canadensis* L.
Gooseberry (Black)	*Ribes lacustre* (Pers.) Poir.
Goosegrass	*Galium aparine* L.
Gotu Kola	*Centella asiatica* (L.) Urb.
Grape	*Vitis vinifera* L.
Grape	*Vitis labrusca* L.
Grapefruit	*Citrus paradisi* Macfad.
Great burdock	*Arctium lappa* L.
Greek hayseed	*Trigonella foenum-graecum* L.
Green ginger	*Artemisia absinthium* L.
Green sauce	*Rumex acetosa* L.
Green tea	*Camellia sinensis* (L.) Kunize
Groats	*Avena sativa* L.
Ground cherry	*Physalis alkekengi* L.
Ground glutton	*Senecio vulgaris* L.
Ground ivy	*Glechoma hederacea* L.
Ground juniper	*Juniperus communis* L.

Ground liverwort	*Peltigera canina* L.
Ground pine	*Lycopodium clavatum* L.
Ground raspberry	*Hydrastis canadensis* L.
Ground swallow	*Senecio vulgaris* L.
Groundsel	*Senecio vulgris* L.
Gum tree	*Eucalyptus citriodora* Hool.
Gumweed	*Grindelia robusta* Nutt.
Gumweed	*Grindelia squarrosa* (Pursh) Donal.
Gypsyweed	*Veronica officinalis* L.
Haw	*Crataegus monogyna* Jacq.
Hawthorn (Chinese)	*Crataegus pinnatifida* Bge.
Hawthorn (English)	*Crataegus monogyna* Jacq.
Hazelnut	*Corylus rostrata* Marsh.
Hazelnut	*Corylus avellana* L.
Hazelnut	*Corylus cornuta* Marsh.
Heal all	*Prunella vulgaris* L.
Healing herb	*Sympyhtum officinale* L.
Heartsease	*Viola tricolor* L.
Hemp	*Cannabis sativa* L.
Hens and chicks	*Sempervivum tectorum* L.
Herb of grace	*Ruta graveolens* L.
Hibiscus	*Hibiscus sabdariffa* L.
Hibiscus (Chinese)	*Hibiscus rosa-sinensis* L.
Highbush cranberry	*Viburnum opulus* L.
Hoarhound	*Marrubium vulgare* L.
Hog apple	*Podophyllum peltatum* L.
Holly	*Ilex aquifolium* L.
Holy thistle	*Cnicus benedictus* L.
Honeysuckle (Dutch)	*Lonicera caprifolium* L.
Honeysuckle (Mountain fly)	*Lonicera caerulea* L.
Hoodwort	*Scutellaria lateriflora* L.
Hops	*Humulus lupulus* L.
Horehound	*Marrubium vulgare* L.
Horse chestnut	*Aesculus hippocastanum* L.
Horsebalm	*Monarda odoratissima* Benth.
Horsefly weed	*Baptisia tinctoria* (L.) R. Br.
Horseheal	*Inula helenium* L.
Horsehound	*Cochlearia amorucia* L.
Horsehound	*Marrubium vulgare* L.
Horsemint	*Monarda punctata* T. J. Howell
Horseradish	*Armoracia rusticana* Gaetn, Mey & Scherb.
Horsetail	*Equisetum arvense* L.

Houseleek	*Sempervivum tectorum* L.
Huang Bai	*Phellodendron chinensis* Schneid
Huang Qi	*Astragalus sinensis* L.
Huang Qin	*Scutellaria baicalensis* Georgi.
Hungarian chamomile	*Matricaria chamomilla* L.
Hydrangea	*Hydrangea arborescens* L.
Hyssop	*Hyssopus officinalis* L.
Indian dye	*Hydrastis canadensis* L.
Indian elm	*Ulmus rubra* Muhl.
Indian ginger	*Asarum canadense* L.
Indian ginseng	*Withania somnifera* Dunal.
Indian paint	*Sanguinaria canadensis* L.
Iranian poppy	*Papaver bracteatum* L.
Irish broom	*Cytisus scoparius* (L.) Link
Japanese parsley	*Cryptotaenia japonica* Hassk.
Jasmine	*Jasminum grandiforum* L.
Jerusalem oak	*Chenopodium ambrosioides* L.
Jiang Huang	*Curcuma domestica* L.
Jiang Huang	*Curcuma longa* L.
Jiang Huang	*Curcuma xanthorrhiza* L.
Jojoba	*Simmondsia chinensis* (Link) C. Schneid
Juniper	*Juniperus communis* L.
Kava-Kava	*Piper methysticum* G. Frost
Kava-Kava	*Macropiper excelsum* G. Frost
Kelp	*Fucus vesiculosus* L.
Kelp	*Laminaria digitata* (Huddss.) Lamk.
Kelp	*Laminaria saccharina* (L.) Lamk.
Kelpware	*Fucus vesiculosus* L.
Key flower	*Primula veris* L.
King's clover	*Melilotus officinalis* (L.) Lamk.
Kudzu	*Pueraria lobata* (Willd) Ohwi
Labrador tea	*Ledum latifolium* Jacq.
Ladder to heaven	*Convallaria majalis* L.
Ladies delight	*Viola tricolor* L.
Lady's mantle	*Alchomilla volgaris* L.
Lady's slipper	*Cypripedium calceolus* L.
Lamb's quarter	*Chenopodium album* L.
Laurel	*Laurus nobilis* L.
Lavender	*Lavandula angustifolia* Mill.

Lavender	*Lavandula spica* L.
Lavender	*Lavandula officinalis* Chaix
Lemon	*Citrus limon* (L.) Burm.
Lemon balm	*Melissa officinalis* L.
Lemon geranium	*Pelargunium graveolens* L'Her. ex. Aiton.
Lemon grass	*Cymbopogon citratus* (DC. ex Nees) Stapf.
Lemon thyme	*Thymus citriodorus* (Pers.) Schreb.
Lent lily	*Narcissus pseudonarcissus* L.
Leontopodium	*Alchemilla vulgaris* L.
Leopard's bane	*Arnica montana* L.
Lesser periwinkle	*Vinca minor* L.
Lian qiao	*Forsythia suspensa* (Thunb.) Vahl.
Licorice	*Glycyrrhiza glabra* L.
Ligustrum fruit	*Ligustrum lucidium* W. T. Aiton.
Lilac	*Syringa vulgaris* L.
Lily cenvalle	*Convallaria majalis* L.
Lily of the valley	*Convallaria majalis* L.
Lily of the valley	*Myosotis scorpioides* L.
Lime	*Citrus aurantifolia* Swingle
Linden	*Tillia cordata* Mill.
Lingberry	*Vaccinium vitis-idaea* L.
Linseed	*Linum usitatissimum* L.
Lion's ear	*Leonurus cardiaca* L.
Lion's tail	*Leonurus cardiaca* L.
Liverlily	*Iris versicolor* L.
Lobelia	*Lobelia inflata* L.
Lodgepole pine	*Pinus contorta* Dougl. ex. Loud.
Lomatium	*Lomatium dissectum* (Nutt.) Math. & Const.
Loosestrife	*Lysimachia vulgaris* L.
Lovage	*Levisticum officinale* W. Koch.
Love vine	*Cuscuta epitlymum* Murr.
Low speed well	*Veronica officinalis* L.
Lucerne	*Medicago sativa* L.
Lungwort	*Lobaria pulmonaria* L.
Lycium	*Lycium barbarum* L.
Ma Huang	*Ephedra sinica* Stapf.
Mad weed	*Scutellaria lateriflora* L.
Maiden hair tree	*Ginkgo biloba* L.
Mandarin	*Citrus reticulata* Blanco
Mandrake	*Podophyllum peltatum* L.
Marian thistle	*Silybum marianum* (L.) Gaertn.
Marigold	*Calendula officinalis* L.

Marjoram	*Origanum vulgare* L. subsp. *hirtum* (Link.) Ietswaart.
Marrubium	*Marrubium vulgare* L.
Marsh parsley	*Apium graveolens* L.
Marshmallow	*Althea officinalis* L.
May lily	*Convallaria majalis* L.
Mayapple	*Podophyllum peltatum* L.
Maybush	*Crataegus monogyna* Jacq.
Maypop	*Passiflora incarnata* L.
Meadow eyebright	*Euphrasia officinalis* L.
Meadow safforn	*Colchicum autumnale* L.
Meadow sage	*Salvia officinalis* L.
Meadow sweet	*Filipendula ulmaria* (L.) Maxim.
Mealberry	*Arctostaphylos uva-usi* (L.) Spreng
Medicinal tea tree	*Melaleuca alternifolia* (Maid. & Bet.) Cheel.
Melaleuca	*Melaleuca alternifolia* (Maid. & Bet.) Cheel.
Melilot	*Melilotus arvensis* L.
Melissa	*Melissa officinalis* L.
Mexican mint marigold	*Tagetes lucida* Cav.
Mexican tea	*Chenopodium ambrosioides* L.
Milfoil	*Achillea millefolium* L.
Milk thistle	*Silybum marianum* (L.) Gaertn.
Milkwort	*Polygala senega* L.
Milsuba	*Cryptotaenia japonica* Hassk.
Mistletoe	*Viscum album* L.
Mistletoe	*Phoradendron leucarpum* (Raf.) Rev. & M. C. Johnst.
Mistletoe (American)	*Phoradendron flavescens* (Push.) Nutt.
Mitsuba	*Cryptotaenia japonica* Hassk.
Moccasin flower	*Cypripedium calceolus* L.
Monkshood	*Aconitum napellus* L.
Moose elm	*Ulmus rubra* Muhl.
Mormon tea	*Ephedra nevadensis* Wats
Mortification root	*Althaea officinalis* L.
Motherwort	*Leonurus cardiaca* L.
Mother's heart	*Capsella bursa-pastoris* (L.) Medic.
Mountain ash	*Sorbus aucuparia* L.
Mountain cranberry	*Viburnum vitis-idaea* L.
Mountain fly honeysuckle	*Lonicera caerulea* L.
Mountain holly	*Ilex aquifolium* L.
Mountain mahogany	*Betula lenta* L.
Mountain mint	*Calamintha officinalis* L.
Mountain tea	*Gaultheria procumbens* L.
Mountain tobacco	*Arnica montana* L.
Mu Dan Pi	*Paeonia suffruiticosa* Andr.

Mugwort	*Artemisia vulgaris* L.
Mullein	*Verbascum thapsus* L.
Myrobalan	*Phyllanthus emblica* L.
Myrrh	*Commiphora molmol* Engl. ex. Tschirch
Myrrh	*Commiphora myrrha* (Nees) Engl.
Myrtle	*Myrtus communis* L.
Naked ladies	*Colchicum autumnale* L.
Nasturtium	*Nasturtium officinale* L.
Nasturtium	*Tropaeolum majus* L.
Neroli oil	*Citrus bergamia* Risso & Poit.
Nettle	*Lamium album* L.
New England pine	*Pinus strobus* L.
Night willow herb	*Oenothera biennis* L.
Northern pine	*Pinus strobus* L.
Northern prickly ash	*Zanthoxylum americanum* Mill
Nutmeg	*Myristica fragrans* Houtt.
Oak (white)	*Quercus alba* L.
Oat	*Avena sativa* L.
Old man	*Rosmarinus officinalis* L.
Olive	*Olea europaea* L.
Onion	*Allium cepa* L.
Oregon grape	*Mahonia aquifolium* (Pursh) Nutt.
Oregano	*Origanum vulgare* L.
Oregano	*Origanum vulgare* L. subsp. *hirtum* (Link)
Oriental sweet gum	*Liquidambar orientalis* L.
Ostrich fern	*Matteucia struthiopteris* (L.) Tod.
Pacific myrtle	*Umbellularia california* Nutt.
Palmarosa	*Cymbopogon martinii* (Roxb) Wats.
Pansy	*Viola tricolor* L.
Papaya	*Carica papaya* L.
Pariwinkle	*Catharanthus roseus* (L.) G. Don
Parsley	*Petroselinum crispum* (Mill.) Nym.
Passion flower	*Passiflora incarnata* L.
Passion vine	*Passiflora incarnata* L.
Patience dock	*Polygonum bistorta* L.
Pennyroyal mint	*Hedeoma pulegiodes* (L.) Pers.
Pepper (Cayenne)	*Capsicum annuum* L. var. annuum
Pepper (Chilli)	*Capasicum annum* L. var. acuminatum
Pepper (English)	*Piper nigrum* L.
Pepper (Sweet)	*Capasicum annum* L. var. grossum

Peppermint	*Mentha x piperita* L.
Periwinkle	*Vinca minor* L.
Pewterwort	*Equisetum arvense* L.
Philanthropos	*Agrimonia eupatoria* L.
Pilewort	*Collinsonia canadensis* L.
Pinckly ash	*Zanthoxylum americanum* Mill.
Pineapple	*Ananas comosus* (L.) Merr.
Plantains	*Plantago major* L.
Plantains	*Plantago lanceolata* L.
Poison flag	*Iris versicolor* L.
Poison ivy	*Rhus radicans* L.
Poison ivy	*Rhus toxicodendron* L.
Pokeweed	*Phytolacca americana* L.
Poorman's treacle	*Allium sativum* L.
Poplar	*Populus balsamifera* L.
Poplar	*Populus candicans* L.
Popotillo	*Ephedra nevadensis* Wats.
Poppy	*Papaver rhoens* L.
Prickly pear	*Opuntia fragilis* (Nutt.) Haw.
Priest's crown	*Taraxacum officinale* Weber.
Prim	*Ligustrum vulgare* L.
Primrose	*Primula vulgaris* Huds.
Privet	*Ligustium vulgare* L.
Psyllium	*Plantago asiatica* L.
Psyllium	*Plantago psyllium* L.
Pumpkin	*Cucurbita pepo* L.
Puncture vine	*Tribulus terrestris* L.
Purging buckthorn	*Rhamnus catharticus* L.
Purple fire top	*Episetum arvense* L.
Purple loosestrife	*Lythrum salicaria* L.
Purple medic	*Medicago sativa* L.
Pussy willow	*Salix discolour* Muhlenb.
Pygeum	*Pygeum africanum* L.
Pyrethrum	*Tanacetum cinerarifolium* L.
Pyrethrum	*Chrysanthemum cinerarifolium* (Trevir.) Vis.
Qing Hao	*Artemisia annua* L.
Quackgrass	*Agropyron repens* (L.) Boauvcis.
Queen Anne's lace	*Daucus carota* L.
Queen of the meadow	*Filipendula ulmaria* (L.) Maxim
Quickbean	*Sorbus aucuparia* L.
Radish	*Raphanus sativus* L.

Rainbow weed	*Lythrum salicaria* L.
Raspberry	*Rubus idaeus* L.
Rattle snake root	*Polygala senega* L.
Red clover	*Trifolium pratense* L.
Red poppy	*Papaver rhoens* L.
Red puccoon	*Sanguinaria canadensis* L.
Red-veined dock	*Rumex obtusifolius* L.
Redroot	*Sanguinaria canadensis* L.
Rhubarb	*Rheum officinal* Baill.
Rhubarb	*Rheum tanguticum* L.
Roman chamomile	*Chamaemelum nobile* (L.) All.
Root of the holy ghost	*Angelica archangelica* L.
Rose (Damask)	*Rosa damascena* L.
Rose (Dog)	*Rosa canina* L.
Rose geranium	*Pelargunium capitatum* (L.) L'Her
Roselle	*Hibiscus sabdariffa* L.
Rosemary	*Rosmarinus officinalis* L.
Roseroot	*Rhus toxicodendron* L.
Roseroot	*Rhodiola rosea* L.
Roseroot	*Sedum rosea* (L.) Scop. subsp. *integrifolium*
Rowan tree	*Sorbus aucuparia* L.
Royal jasmine	*Jasminum auriculatum* Vahl.
Royal jasmine	*Jasminum grandiflorum* L.
Rue	*Ruta graveolens* L.
Russian olive	*Elaeagnus angustifolia* L.
Safflower	*Carthamus tinctorius* L.
Saffron	*Crocus sativus* L.
Sage	*Salvia officinalis* L.
Sagebrush	*Artemisia tridentata* Nutt.
Sandalwood	*Santalum album* L.
Sanicle	*Sanicula marilandica* L.
Saskatoon	*Amelanchier alnifolia* Nutt.
Sassafras	*Sassafras albidum* (Nutt.) Nees
Satin flower	*Stellaria media* (L.) Cyrillo
Savory	*Satureja hortensis* L.
Saw palmetto	*Serenoa repens* (Bartr.) Small.
Scented fern	*Tanacetum vulgare* L.
Schizandra	*Schisandra chinensis* (Turcz.) Baill.
Scouring rush	*Equisetum arvense* L.
Sea buckthorn	*Hippophae rhamnoides* L.
Sea parsley	*Levisticum officinales* W. Koch.
Selfheal	*Prunella vulgars* L.

Seneca snakeroot	*Polygala senega* L.
Senna (Alexandra)	*Cassia angustifolia* Vahl.
Senna (Alexandra)	*Senna alexandrina* L.
Senna (Tinnevelly)	*Cassia senns* L.
Senna (Tinnevelly)	*Cassia marilandica* L.
Sesame	*Sesamum orientale* L.
Sesame	*Sesamum indicum* L.
Shamrock	*Oxalis acetosella* L.
Sheep sorrel	*Rumex acetosella* L.
Shen yao	*Dioscorea oppositae* Thunb.
Shepherd's purse	*Capsella bursa-pastoris* (L.) Medik.
Shi chang Pu	*Acorus tatarinowii* L.
Siberian ginseng	*Eleuthrococcus senticosus* (Rupr. ex Maxim) Maxim.
Silver pine	*Abies balsames* (L.) Mill
Skullcap	*Scutellaria baicalensis* Georgi.
Skullcap	*Scutellaria galericulata* L.
Slack birch	*Betula lenta* L.
Slippery elm	*Ulmus rubra* Muhl.
Small age	*Apium graveolens* L.
Small houseleek	*Sedum acre* L.
Small-flowered willow herb	*Epilobium parviflorum* Schreb.
Small-leaved lime	*Tillia cordata* Mill.
Smelling stick	*Sassafras albidum* (Nutt.) Nees.
Snakeroot	*Polygala senega* L.
Snakeweed	*Polygonum bistorta* L.
Snapping hazel	*Hamamelis virginiana* L.
Snowball tree	*Viburnum opulus* L.
Snowflake	*Lamium album* L.
Soapwort	*Saponaria officinalis* L.
Sorrel	*Rumex acetosella* L.
Sorrel	*Oxalis acetosella* L.
Southern pitch pine	*Pinus palustris* Mill.
Soybean	*Glycine max* (L.) Merrill.
Spearmint	*Mentha spicata* L.
Speedwell	*Veronica officinalis* L.
Spiked loosestrife	*Lythrum salicaria* L.
Spikenard	*Aralia racemosa* L.
Spikenard	*Aralia nudicaulis* L.
Spotted thistle	*Cnicus benedictus* L.
Spruce (black)	*Picea mariana* (Mill) Black
Spruce (Norway)	*Picea ables* (L.) Karst.
Spruce (white)	*Picea glauca* (Moench) Voss.
Squawroot	*Cimicifuga racemosa* (L.) Nutt.

St. John's wort	*Hypericum perforatum* L.
Stag's horn	*Lycopodium clavatum* L.
Star flower	*Borago officinalis* L
Star thistle	*Centaurea calcitrapa* L.
Star weed	*Stellaria media* (L.) Cyrillo
Stave oak	*Quercus alba* L.
Stellaria	*Alchemilla vulgaris* L.
Stevia	*Stevia rebaudiana* (Bertoni) Bertoni
Stickwort	*Agrimonia eupatoria* L.
Stinging nettle	*Urtica dioica* L.
Stinging nettle	*Urtica urens* L.
Stinking weed	*Chenopodium ambrosioides* L.
Stinking Willie	*Tanacetum vulgare* L.
Stinkwood	*Ocotea bullata* (Birth) E. May
Stone oak	*Quercus alba* L.
Stonecrop	*Sedum acre* L.
Storax	*Liquidambar styraciflua* L.
Strawberry	*Fragaria vesca* L.
Strawberry tomato	*Physalis alkekengi* L.
Sugar wrack	*Laminaria digitata* (Huds.) Lamk.
Sunflower	*Helianthus annuus* L.
Suterberry	*Zanthoxylum americanum* Mill
Sweet Annie	*Artemisia annua* L.
Sweet balm	*Abies balsames* (L.) Mill
Sweet bay	*Laurus nobilis* L.
Sweet birch	*Betula lenta* L.
Sweet chestnut	*Castanea sativa* Mill.
Sweet clover	*Melilotus officinalis* (L.) Lamk.
Sweet cumin	*Pimpinella anisum* L.
Sweet elm	*Ulmus rubra* Muhl.
Sweet fennel	*Foeniculum vulgare* Mill.
Sweet flag	*Acorus calamus* L. var. Americanus Wolff.
Sweet goldenrod	*Solidago odora* Act.
Sweet grass	*Hierochloe odorata* (L.) Beauv.
Sweet herb of Paraguay	*Stevia rebaudiana* (Bertoni) Bertoni
Sweet Lucerne	*Melilotus officinalis* (L.) Lamk.
Sweet marjoram	*Origanum majorana* L.
Sweet weed	*Althaea officinalis* L.
Sweet wrack	*Laminaria saccharina* (L.) Lamk.
Tangleweed	*Laminaria digitata* (Huds.) Lamk.
Tansy	*Tanacetum vulgare* L.
Tarragon	*Artemisia dracunculus* L.

Teaberry	*Gaultheria procumbens* L.
Telltime	*Taraxacum officinale* Weber.
Tetterwort	*Sanguinaria canadensis* L.
Throwwort	*Leonurus cardiaca* L.
Thyme (Garden)	*Thymus vulgaris* L.
Thyme (Lemon)	*Thymus citriodorus* (Pers.) Schreb.
Toothache tree	*Zanthoxylum americanum* Mill.
Tree peony	*Paeonia suffruiticosa* Andr.
Trifoil	*Trifolium pratense* L.
True wood sorrel	*Oxalis acetosella* L.
Turmeric	*Curcuma longa* L.
Uva-ursi	*Arctostaphylos uva-ursi* (L.) Spreng
Valerian	*Valeriana officinalis* L.
Vetiver	*Vetiveria zizanioides* L. (Nush.)
Virginia dogwood	*Cornus florida* L.
Virginia skullcap	*Scutellaria lateriflora* L.
Walnuts	*Juglans regia* L.
Water lily	*Nymphaea alba* L.
Watercress	*Nasturtium officinale* L.
White ash	*Fraxinus americana* L.
White ash	*Fraxinus excelsior* L.
White peony	*Paeonia lactiflora* Pall.
White pine	*Pinus strobus* L.
White pine	*Pinus albicaulis* Engelm.
White water lily	*Nymphaea alba* L.
White willow	*Salix alba* L.
Wild celery	*Apium graveolens* L.
Wild chamomile	*Curcuma longa* L.
Wild chamomile	*Chrysanthemum parthenium* (L.) Bernh.
Wild cherry	*Prunus serotina* J. F. Ehrb.
Wild fennel	*Foeniculum vulgare* Mill.
Wild ginger	*Asarum canadensis* L.
Wild indigo	*Baptisia tinctora* (L.) R.Br. Ex Ait.f.
Wild lemon	*Podophyllum peltatum* L.
Wild licorice	*Glycyrrhiza lepidota* (Nutt.) Pursh.
Wild marjoram	*Origanum vulgare* L.
Wild rose geranium	*Pelargunium capitatum* (L.) L'Her
Wild senna	*Cassia marilandica* L.
Wild strawberry	*Fragaria vesca* L.
Wild sunflower	*Inula helenium* L.

Wild thyme	*Thymus serpyllum* L.
Willow (white)	*Salix alba* L.
Willow herb	*Episetum arvense* L.
Willow herb	*Lysimachia vulgaris* L.
Winter bloom	*Hamamelis virginiana* L.
Winter green	*Gaultheria procumbens* L.
Winter savory	*Satureja montana* L.
Witch hazel	*Hamamelis virginana* L.
Witch's bells	*Digitalis purpurea* L.
Wolfberry	*Lycium pallidum* L.
Wolfberry (Chinese)	*Lycium chinesis* Mill.
Wood sorrel	*Oxalis acetosella* L.
Woodland strawberry	*Fragaria vesca* L.
Worm wood	*Artemisia tridentata* Nutt.
Wormseed	*Chenopodium ambrosioides* L.
Wormwood (Chinese)	*Artemisia annua* L.
Wormwood	*Artemisia absinthium* L.
Wu Wei Zi	*Schisandra chinesis* (Turcz.) Ball
Yam (Mexican)	*Dioscorea villosa* L.
Yam (Chinese)	*Dioscorea oppositae* Thunb.
Yarrow	*Achillea millefolium* L.
Yellow ginseng	*Caulophyllum thalictroides* (L.) Michx.
Yellow Indian shoe	*Cypripedium calceolus* L.
Yellow melilot	*Melilotus officinalis* (L.) Lamk.
Yellow puccoon	*Hydrastis canadensis* L.
Yellow rochet	*Lysimachia vulgaris* L.
Yew	*Taxus brevifolia* Nutt.
Yew	*Taxus x media* Rehd.
Ylang-Ylang	*Cananga odorata* (Lam) J. D. Hook & T. Thompson
Yu Chin	*Curcuma longa* L.
Yucca	*Yucca aloifolia* L.

INDEX

1-Hexenol, 122, 431
1-Homostarchydrine, 30, 407
1-Methyoxy-4-(2-propenyl), 112, 431
1-Octen-3-ol, 129, 150, 153, 431
1-Octen-3-ol,3,5-dimethoxyphenol, 150, 431
1-P-menthene, 153, 431
1-Pentadecene, 126, 431
1-Phenyl-1,3-butanedione, 147, 431
1-Terpinen 4-ol, 132, 431
1,2-Benzenedicarboxylic acid, 140, 431
1,4-Methylhexadecanoic acid, 141, 431
1,8-Cineole, 48, 116, 124, 129, 145, 151, 407, 431
1,8-Dihydroxy-anthracene derivatives, 5, 407
1,8-Pentadecadiene, 126, 431
11-α,13-Dihydrohelenalin, 7, 407
2-Hydroxybenzoic acid methyl ester, 139, 431
2-Hydroxy-4-methoxyacetophenone, 125, 431
2-Methyl-butanal, 126, 431
2-Methyl-4-pentenal, 126, 431
2-Methyl-propanal, 126, 431
2-Methyl-tetradeca-5,12-diene, 126, 431
2-Methyl-tetradeca-6,12-diene, 126, 431
2-Nonanone, 145, 431
2-Nonylacetate, 145, 431
2-Phenylethyl alcohol, 139, 431
2-Propanal, 126, 431
2-Undecanone, 145, 431
2-Undecylacetate, 145, 431
2-Vinyl-4H-1,3-dithin, 4, 407
2-3,4-Dihydroxyphenyl-ethanol, 25, 407
2,5-Trans-p-methanediol, 136, 431
2-β-Glucuronosyl, 22, 407
(2E, 4E)-2,4-Decadienal, 123, 431
(2E, 4Z)-2,4-Decadienal, 123, 431
(2E, 4Z)-2,4-Heptadienal, 123, 431
(2E, 6Z)-Nonadienal, 122, 152, 431
(2E, 6Z)-Nonadienol, 122, 152, 431
3-Carene, 140, 431
3-Hexen-1-ol, 126, 132, 153, 432
3-Methyl-butanal, 131, 432
3-Methyl-butyl benzoate, 131, 432
3-Methyl-butyl octanoate, 131, 432
3-n-Pentadecylcatechol, 40, 407
3-O-β-D-Glucuronide, 19, 407
3-Octanol, 118, 153, 432
3,4-Benzopyrene, 122, 431
3,5-Dimethoxyphenol, 150, 431
3,7-Octatrien-3-ol, 146, 431
3,9-Epoxy-p-menth-1-ene, 114, 431
4-Mercapto-4-methylpentan-2-one, 64, 117, 146, 165, 432
4-Methoxy-2-methyl-2-mercaptobutane, 144, 432
4'-O-Methylpyridoxin, 21, 52, 109, 407
4,5-Dimethoxy-6-(2-propenyl)-1,3-benzodioxole, 114, 432
5-Methylheptan-2,4-dione, 131, 432
5,9-Octadecadienoic acid, 129, 432
5,9,12-Octadecatrienoic acid, 129, 432
6-Methylheptan-2,4-dione, 131, 432
8,9-Dehydroeoisolongifolene, 119, 432
α-Acids, 24, 407
β-Acids, 24, 407
(E)-2-Hexenal, 143, 440
(E)-Anethole, 112, 440
(E)-Beta-ocimene, 132, 440
(Z)-3-Hexenyl acetate, 122, 152, 454

Aabies balsames, 456, 462, 471, 472
Abietatriene, 123, 432
Abrotamine, 7, 115, 407, 432
Absinthe, 455
Absinthol, 7, 407
Acacetin, 22, 407
Acetaldehyde, 126, 136, 432
Acetate, 14, 26, 36, 112, 118, 121, 122, 124,

Acetate *(continued)*, 125, 127, 128, 130, 132, 133, 136, 140, 141, 143, 146, 148, 149, 151-153, 407, 410, 418, 434-436, 441-443, 445-447, 454
Acetic acid, 10, 133
Acetic ester, 131, 432
Acetone, 136, 432
Acetopenone glucoside, 49, 407
Acetophenone, 147, 432
Acetyl eugenol, 110, 149, 164
Acetylcholine, 10, 50, 407
Acetylenes, 112, 432
Acetylenic compounds, 34, 407
Acetylinic, 36, 407
Achillea millefolium, 3, 57, 68, 112, 158, 168, 228, 258, 301, 364, 457, 467, 474
Achilleine, 3, 407
Achilline, 112, 432
Aconitine, 3, 63, 68, 221, 407, 409
Aconitum napellus, 3, 68, 112, 168, 228, 258, 301, 364, 467
Acoric acid, 3, 407
Acorin, 112, 432
Acorus calamus, 3, 53, 68, 112, 164, 168, 214, 228, 258, 293, 301, 364, 458, 472
Acorus tatarinowii, 471
Acrylic acid, 5, 407
Actaea alba, 3, 68, 112, 168, 228, 258, 301, 364, 456
Actaea rubra, 456
Actein, 407
Acthusin, 112, 432
Acylated sterylglucoside, 152, 432
Acylphloroglucinal derivatives, 23, 407
Adenine, 141, 432
Aesculin, 407
Aesculus hippocastanum, 3, 68, 112, 228, 258, 364, 458, 464
Aethusanol, 112, 432
African prune, 38, 52, 98, 142, 154, 200, 249, 283, 344, 387, 418, 447, 448, 453, 455
Agastache foeniculum, 4, 68, 112, 161, 168, 219, 228, 258, 296, 302, 364, 455
Agglutinin, 48, 407
Aglycone, 23, 407
Agnuside, 50, 407
Agrimonia eupatoria, 4, 228, 455, 459, 460, 469, 472
Agrimony, 4, 69, 112, 169, 228, 258, 302, 364, 412, 414, 422, 426, 454, 455
Agropyrene, 112, 432
Agropyron repens, 4, 18, 83, 127, 184, 228, 302, 321, 460, 461, 469
Ague tree, 455
Ague weed, 455
Ajoene, 4, 407
Alanine, 11, 407
Alantol, 132, 432
Albiflorin, 33, 407
Albitocin, 31, 407
Albuminoides, 18, 407
Alchemilla vulgaris, 4, 69, 113, 169, 228, 258, 303, 364, 456, 466, 472
Alcohol farnesol, 131, 432
Aldehydes, 50, 111, 118, 124, 128, 138, 152, 407, 447
Alder, 5, 39, 69, 99, 113, 144, 169, 201, 229, 250, 259, 284, 303, 347, 365, 388-390, 408, 410, 414-416, 421-423, 426, 428, 440, 448, 455, 457, 461, 462
Alder buckthorn, 39, 99, 144, 201, 250, 284, 347, 388, 408, 414-416, 421, 422, 428, 455
Alfalfa, 30, 91, 136, 194, 244, 277, 335, 382, 407, 408, 412, 417, 422, 425, 428, 432, 436, 441, 449, 450, 455
Aliphatic alcohols, 138, 432
Aliphatic hydrocarbons, 132, 432
Alkaloides, 408
Alkane derivative, 143, 432
Alkenyl, 14, 408
Alkenyl glycoside, 14, 408

Alkylamides, 18, 408
Alkylmethoxy pyrazines, 122, 152, 432
Alkyl-phthalides, 27, 408
All heal, 455
Allantoin, 42, 45, 408
Allicin, 4, 113, 408, 432
Alliin, 4, 113, 408, 433
Allium cepa, 4, 69, 113, 169, 228, 259, 303, 364, 468
Allium sativum, 463, 469
Allium schoenoprasum, 459
Allo-aromadendrene, 147, 433
Allo-ocimene, 114, 433
Allphatic hydrocarbons, 433
Allyl isothiocyanate, 117, 433
Allyl phenylethyl isothiocyanate, 115
Allypropyl, 113, 433
Almond, 38, 98, 142, 200, 249, 283, 345, 387, 408, 413, 414, 422, 428, 434, 436, 443-445, 450, 455
Alnus crispus, 5, 69, 113, 169, 229, 259, 303, 365, 455
Alnus glutinosa, 455, 457, 461, 462
Alnus incana subsp. *tenufolia*, 455
Aloe, 5, 11, 54, 61, 70, 113, 170, 215, 219, 229, 259, 303, 365, 407-409, 412, 414, 441, 455
Aloe-emodin, 408
Aloe vera, 5, 70, 113, 170, 229, 259, 303, 365, 455
Aloeresin, 5, 408
Aloin, 5, 408
Alpha, 24, 128, 132, 158
Alpha-atalantone, 124, 433
Alpha-bergamotol, 139, 433
Alpha-betulenolacetate, 117, 433
Alpha-bisabolol, 29, 120, 408, 433
Alpha-caryophyllene, 142, 433
Alpha-copaene, 120, 153, 433
Alpha-cuprenene, 433
Alpha-humulene, 120, 131, 149
Alpha-linoleic acid, 116, 433
Alpha-muurolene, 149, 433
Alpha-naginatene, 122, 433
Alpha-nepetalactone, 137, 433
Alpha-peltatin, 37, 408
Alpha-phyllandrene, 124, 126, 139, 433
Alpha-phytosterol, 25
Alpha-pinene, 15, 26, 36, 112, 114, 115, 118, 120, 121, 123, 126, 128, 130, 132, 133, 136, 137, 139, 144, 147, 151, 152, 407, 432
Alpha-santalols, 146, 432
Alpha-selinene, 153, 432
Alpha-terpinene, 121, 136, 137, 139, 432
Alpha-terpineol, 16, 26, 30, 124, 127, 128, 129, 133, 137 153, 433
Alpha-terpinyl, 149, 432
Alpha-thujene, 128, 433
Alpha-tocopherol, 9, 60, 117, 119, 130, 138, 433
Alpha-vetrivone, 433
Alpha-ylangene, 119, 433
Althaea, 5, 70, 113, 170, 229, 259, 303, 365, 455, 467, 472
Althaea officinalis, 5, 70, 113, 170, 229, 259, 303, 365, 455, 467, 472
Amber touch-and-heal, 455
Ambrettolide, 131, 433
Amelanchier alnifolia, 5, 70, 113, 170, 229, 259, 304, 365, 470
Amidendiol, 7, 408
Amines, 10, 408
Amines choline, 10
Amygdalin, 16, 408
Amyrin, 11, 120, 138, 408, 434
Anabsinthin, 115, 433
Anagyrine, 12, 408
Anahygrine, 51, 408
Ananas comosus, 5, 70, 113, 170, 229, 259, 304, 365, 469
Anaphalis margaritacea, 5, 6, 70, 71, 113, 114, 170, 229, 259, 304, 365, 462
Anarin, 128, 433
Anemonin, 15, 143, 408, 433
Anethole, 20, 36, 112, 115, 128, 139, 140,

Anethole *(continued)*, 149, 151, 408, 434, 438, 440, 453
Anethum graveolens, 6, 70, 114, 170, 229, 260, 304, 365, 461
Angelic acid esters, 12, 120, 434
Angelica, 6, 66, 70, 114, 170, 223, 229, 260, 304, 365, 408, 410, 412, 414, 418, 420, 421, 426, 433-437, 439, 444, 447, 449, 451, 453, 455, 461, 470
Angelica archangelica, 6, 70, 114, 170, 229, 260, 304, 365, 455, 470
Angelica sinensis, 66, 223, 461
Angelica tree, 455
Angelicide, 6, 408
Anisaldehyde, 140, 408, 434
Anise, 36, 96, 140, 157, 198, 216, 248, 282, 342, 386, 407, 408, 413-415, 422, 433-438, 440, 441, 444-448, 450, 453, 455
Anise hyssop, 455
Aniseed, 4, 68, 112, 168, 228, 258, 282, 302, 364, 408, 419, 431, 434, 440, 444, 446, 449, 454, 455
Anisic acid, 128, 140, 434, 448
Anisole, 115, 434
Antennaria margaritacea, 462
Anthemidin, 29, 408
Anthemis nobilis, 6, 71, 114, 171, 229, 260, 305, 366, 461
Anthocyanidin, 33
Anthocyanin, 5, 24, 36
Anthocyanosides, 40, 48, 49, 408
Anthraquinone, 39, 408
Anthraquinone glycosides, 5, 39, 408
Anti-protease, 409
Apigene, 135, 434
Apigenin, 46
Apigernin, 40, 409
Apigetrin, 135, 434
Apiin, 6, 35, 135, 409, 434
Apiole, 35, 139, 409
Apium graveolens, 6, 54, 71, 114, 171, 215, 230, 260, 305, 366, 459, 467, 471, 473
Apricot (Japanese), 455
Apricot vine, 455
Ar-curcumene, 153, 434
Ar-turmerone, 124, 434
Arabinose, 9, 409
Arachic acid, 119, 409, 434
Aralia nudicaulis, 471
Aralia racemusa, 455
Araliene, 114, 434
Arborinine, 41, 409
Arbutin, 7, 27, 49, 110, 223, 409
Arctic acid, 7
Arctiin, 7, 409
Arctiol, 7, 409
Arctium lappa, 7, 71, 115, 171, 230, 260, 305, 366, 458, 459, 463
Arctostaphylos uva-ursi, 7, 60, 71, 115, 171, 230, 260, 305, 366, 456, 460, 462, 467, 473
Aretylcholine, 48, 409
Arginine, 27, 39, 44, 409
Arislolochic acid, 8, 409
Armoracia rusticana, 7, 14, 71, 79, 115, 122, 171, 180, 230, 235, 260, 266, 306, 315, 366, 371, 464
Arnica, 7, 58, 59, 65, 72, 115, 160, 172, 230, 261, 293, 306, 366, 407-409, 411, 414-418, 422, 424, 426, 434, 441, 443, 447, 452, 455, 466, 467
Arnica chamissonis subsp. *foliosa*, 455
Arnica condifolia, 7, 72, 115, 172, 230, 261, 306, 366, 455
Arnica fulgens, 7, 72, 115, 172, 230, 261, 306, 366, 455
Arnica latifolia, 7, 72, 115, 172, 230, 261, 306, 366, 455
Arnica montana, 58, 293, 455, 466, 467
Arnica sororia, 7, 72, 115, 172, 230, 261, 306, 366, 455
Arnicin, 7, 409
Arnidiol, 7, 115, 434

Aromadendrene, 118, 127, 153, 433, 434
Arrowwood, 455
Artemisia absinthium, 7, 72, 115, 172, 230, 261, 306, 367, 455, 463, 474
Artemisia annua, 469, 472, 474
Artemisia dracunculus, 461, 463, 472
Artemisia ketone, 115, 149, 434
Artemisia tridentata, 456, 470, 474
Artemisia vulgaris, 462, 471
Artemisinin, 7, 409
Artichoke, 17, 81, 125, 182, 237, 268, 319, 374, 410, 411, 413, 418, 422, 424, 455, 456
Artichoke (wild), 456
Asarone, 3, 68, 112, 116, 125, 147, 409, 434, 435, 438, 453
Asarum canadense, 458, 460, 462, 465
Ascaridole, 13, 120, 409, 434
Ash, 20, 21, 44, 51, 85, 103, 108, 129, 148, 153, 186, 207, 213, 239, 253, 257, 271, 288, 292, 325, 354, 362, 376, 392, 395, 401, 409, 412, 414, 420, 421, 423, 426-428, 434, 441-443, 445, 446, 453-455, 457, 467-469, 473
Ash (American), 455
Ashwagandha, 51, 63, 108, 152, 212, 256, 292, 297, 362, 395, 408, 410, 418, 429, 432, 439, 446, 448, 449, 452, 453, 456
Asiatic acid, 409
Asiaticoside, 409
Asparagine, 5, 7, 8, 409
Aspirin, 98, 409
Astragalin, 30, 33, 409
Astragalus, 8, 73, 116, 161, 173, 219, 231, 261, 307, 367, 409, 411, 415, 418, 425, 439, 446, 456, 465
Astragalus americana, 8, 73, 116, 173, 231, 261, 307, 367, 456
Astragalus sinensis, 465
Astrogen, 22, 409
Atlantol, 409
Atlantones, 119, 434
Aucubigenin, 409
Aucubin, 36, 38, 50, 141, 409, 434
Aucuboside, 49, 409
Authraquinone, 11, 409
Avena sativa, 8, 73, 116, 173, 231, 261, 307, 367, 463, 468
Azafran, 456
Azelaic acid, 434
Azulene, 12, 23, 29, 112, 120, 141, 409, 434

Bachelor's button, 456
Baical skullcap, 43, 102, 147, 205, 252, 287, 352, 390, 409, 411, 414, 424, 425, 431, 432, 448, 451, 456
Baldrianic acid, 409
Balm, 30, 37, 92, 97, 136, 142, 194, 200, 244, 249, 277, 283, 336, 344, 382, 387, 412, 414-416, 418, 422, 426, 427, 433, 438, 442, 445, 446, 450, 456, 462, 466, 472
Balm of gilead, 37, 200, 249, 387, 414, 416, 433, 456
Balm of gilead fir, 456
Balsam fir, 3, 68, 112, 168, 228, 258, 301, 364, 420, 433, 440, 446, 449, 456
Balsam poplar, 456
Baneberry (red), 456
Baneberry (white), 456
Baptisia, 8, 73, 116, 173, 231, 261, 307, 367, 456, 459, 464, 473
Baptisia tinctoria, 456, 459, 464
Barbaloin, 5, 409
Barberry, 9, 73, 116, 174, 231, 262, 308, 367, 410, 423, 426, 456, 462
Basil, 10, 32, 74, 93, 117, 137, 175, 195, 232, 245, 263, 279, 309, 338, 368, 383, 415-420, 424, 426, 427, 437, 441, 445, 446, 448, 450, 456
Basil thyme, 10, 74, 117, 175, 232, 263, 309, 368, 415, 417, 419, 450, 456
Basin sagebrush, 456
Bastard saffron, 456

Bay, 26, 48, 56, 89, 106, 133, 151, 158, 191, 210, 242, 255, 275, 290, 332, 359, 380, 394, 407, 412, 418, 419, 422, 423, 425, 426, 433, 434, 437, 445, 447, 453, 456, 458, 472
Bay laurel, 26, 48, 89, 106, 133, 151, 191, 210, 242, 255, 275, 290, 332, 359, 380, 394, 407, 412, 418, 419, 422, 423, 425, 426, 433, 434, 437, 445, 447, 453, 456
Bayberry, 31, 92, 137, 195, 245, 278, 337, 383, 414, 423, 426, 427, 437, 439, 451, 456
Bear's foot, 456
Bear's grape, 456
Bearberry, 7, 60, 71, 115, 171, 230, 260, 305, 366, 409, 419, 426, 441, 456
Bedstraw, 456
Bee bread, 456
Beech, 20, 62, 84, 128, 162, 185, 239, 271, 323, 376, 413, 421, 422, 425, 453, 454, 456
Beech (European), 456
Beet root, 9, 73, 117, 174, 231, 262, 308, 367, 410, 446, 456
Behenic acid, 434
Bellflower, 14, 79, 122, 180, 235, 266, 371, 408, 421, 422, 424, 425, 431, 438, 453, 456
Bellis perennis, 8, 73, 116, 173, 231, 262, 308, 367, 460
Benzaldehyde, 38, 121, 143, 148, 409, 434
Benzene, 112, 114, 139, 434, 452
Benzenoides, 133, 434
Benzoic acid, 10, 14, 33, 40, 43, 138, 143, 409, 410, 434
Benzoly-aconitine, 409
Benzophenanthridine alkaloids, 42, 410
Benzoquinone, 410
Benzoylecgonine, 19, 410
Benzyl acetate, 118, 132, 434
Benzyl alcohol, 134, 137, 139, 142, 150, 434
Benzyl bezoate, 121, 434
Benzyl cyanide, 434
Benzyl isothiocyanate, 150, 434
Berbamine, 9, 410
Berberastine, 25, 410
Berberine, 9, 25, 29, 410
Berberis vulgaris, 174, 231, 262, 308, 367, 456, 462
Berberry, 456
Berberubine, 410
Bergamot orange, 14, 78, 121, 179, 235, 266, 315, 371, 410, 412, 418, 420, 428, 434, 439, 440, 444, 445, 456
Bergapten, 6, 121, 410, 434
Bergaterpeme, 131, 434
Bergegin, 410
Beta vulgaris, 9, 73, 117, 174, 231, 262, 308, 367, 456
Beta-amyrin, 120, 138, 434
Beta-asarone, 68, 112, 435
Beta-bisabolene, 125, 141, 153, 435
Beta-borneol, 138, 435
Beta-boswellic acid, 9, 410
Beta-bourbonene, 115, 129, 435
Beta-caryophyllene, 118, 120, 125, 131, 136, 149, 151-152
Beta-cyclopyrethrosin, 120, 435
Beta-elemene, 122, 129, 132, 410, 435
Beta-farnesene, 112, 120, 126, 435
Beta-ionone, 124, 139, 435
Beta-linoleic acid, 116, 435
Beta-myrcene, 126, 435
Beta-nepetalactone, 137
Beta-ocimene, 114, 435
Beta-peltatin, 37, 410
Beta-phenylethylamine, 35, 410
Beta-phyllandrene, 114, 118, 435
Beta-pinene, 30, 36, 115, 120, 121, 126, 128-130, 132, 136, 137, 140, 141, 144, 435
Beta-santalols, 146, 435
Beta-selinene, 114, 149, 153, 435
Beta-sesquiphyllandrene, 148, 435

Beta-sitosterol, 5, 13, 15, 18, 27, 28, 30, 38, 39, 43, 44, 48, 131, 138, 148, 410, 435
Beta-terpinene, 137, 435
Beta-thujone, 146, 149, 435
Beta-vetivone, 152, 435
Betaine, 9, 28, 45, 410
Betonicial, 29
Betonicine, 45, 410
Betony, 45, 104, 148, 207, 253, 288, 354, 392, 410, 412, 426, 440, 456
Betula lenta, 9, 73, 117, 174, 231, 262, 308, 367, 457, 459, 467, 471, 472
Betulene, 117, 435
Betulenol, 117, 435
Betulinic, 38, 410
Betulol, 117, 435
Bia shao, 33, 94, 196, 246, 339, 384, 409, 419, 421, 456
Biisobutyl phthalate, 144, 435
Bilberry, 49, 55, 106, 151, 211, 255, 291, 360, 394, 408, 409, 417, 420, 426, 453, 456
Bilobalide, 21, 410
Bilobetin, 21, 410
Bindweed, 15, 80, 123, 180, 235, 267, 316, 372, 412, 423, 440, 441, 456
Bioflavonoids, 14, 410
Birch, 9, 73, 117, 174, 231, 245, 262, 279, 308, 338, 367, 383, 415, 417, 424, 426, 433, 435-439, 442, 446, 449, 450, 454, 456, 457, 459, 471, 472
Bird's foot, 457
Bisabolene, 125, 141, 153, 435
Bisabolol oxide, 135, 435
Bishomophinolenic acid, 36, 410
Bistort, 37, 97, 142, 200, 248, 283, 344, 386, 408, 418, 432, 436, 449, 457
Bistort root, 37, 97, 142, 200, 248, 283, 344, 386, 408, 418, 432, 436, 449, 457
Bitter dock, 457
Bitter orange, 14, 78, 121, 179, 235, 266, 315, 410, 412, 418, 420, 428, 435, 438, 442, 443, 445, 447, 450, 452, 457
Bittmore ash, 457
Black alder, 457
Black berry, 457
Black cherry, 38, 98, 143, 200, 249, 283, 345, 387, 409, 412, 414, 422, 426, 434, 443, 457
Black cohosh, 13, 77, 121, 179, 234, 265, 314, 371, 407, 412, 417, 419, 423-425, 427, 457
Black currant, 40, 100, 144, 388, 408, 409, 412, 421, 426, 428, 429, 432, 437, 439, 441, 447, 448, 450, 452, 457
Black dogwood, 457
Black mustard, 9, 74, 117, 174, 262, 309, 368, 416, 419, 425, 457
Black poplar, 457
Black sanicle, 457
Black snakeroot, 42, 102, 146, 286, 351, 390, 408, 411, 424, 425, 454, 457
Black walnut, 26, 88, 132, 190, 242, 274, 331, 379, 416, 417, 425, 432, 435, 437, 440, 443, 444, 457
Black wort, 457
Bladder cherry, 457
Bladder fucus, 457
Bladder wrack, 376, 457
Blazing star, 457
Blessed thistle, 11, 14, 75, 79, 118, 122, 176, 179, 233, 235, 263, 264, 266, 311, 315, 369, 371, 414, 419, 426, 451, 453, 457
Blind nettle, 457
Bloodroot, 42, 102, 146, 203, 251, 286, 351, 390, 410, 417, 424, 457
Bloodwort, 457
Blooming sally, 457
Blue cohosh, 12, 76, 119, 177, 233, 264, 296, 312, 370, 408, 411, 418, 419, 424, 457
Blue flag, 25, 88, 132, 190, 242, 274, 330, 379, 417, 422, 424, 427, 441, 457

Blue ginseng, 457
Blue lobelia, 28, 90, 134, 192, 243, 276, 333, 381, 408, 457
Blue mountain tea, 457
Blue pimpernel, 457
Blue sailors, 457
Blue skullcap, 457
Blueberry, 49, 61, 106, 151, 211, 220, 255, 291, 360, 394, 408, 409, 417, 426, 454, 457
Blunt-leaves dock, 457
Boldine, 42, 410
Boneset, 20, 84, 128, 185, 239, 270, 323, 376, 414, 415, 418, 422, 425, 454, 457
Borage, 9, 60, 73, 117, 161, 174, 231, 262, 309, 368, 409, 411, 419, 422-426, 442, 445, 450, 453, 457
Borago officinalis, 9, 73, 117, 174, 231, 262, 309, 368, 456, 457, 460, 472
Borneol acetate, 36, 435
Bornyl acetate, 128, 141, 148, 151, 410, 436
Boswellia carteri, 463
Boswellia sacra, 74, 117, 174, 231, 262, 309, 368, 463
Bottle brush, 457
Bourtree, 458
Boxberry, 458
Boxtree, 458
Boxwood, 9, 74, 117, 232, 262, 309, 368, 408, 427, 432, 433, 445, 458
Brassica nigra, 9, 74, 117, 174, 231, 262, 309, 368, 457
Brefeldin, 6, 410
Bridewort, 458
Brigham tea, 458
Bromelain, 5, 410
Broom, 17, 41, 42, 64, 82, 100, 102, 116, 125, 145, 146, 165, 182, 203, 204, 237, 251, 252, 268, 285, 286, 319, 342, 350, 351, 359, 360, 362, 374, 389, 390, 405, 414, 420, 423-425, 453, 454, 458, 459, 465
Buckeye, 458
Buckthorn, 24, 39, 59, 61, 87, 99, 131, 144, 155, 161, 164, 189, 201, 214, 219, 241, 250, 273,
284, 295, 329, 347, 378, 379, 388, 400, 401, 408, 411, 414-416, 421, 422, 425,
428, 432, 440, 445, 448, 449, 452, 455, 458, 462, 463, 469, 470
Buckwheat, 20, 84, 128, 185, 239, 270, 323, 376, 412, 414, 428, 433, 436, 458
Buffalo berry, 44, 103, 148, 206, 253, 288, 353, 391, 426, 440, 448, 458
Buffalo herb, 458
Bugbane, 458
Bulnesol, 138, 436
Bunchberry, 15, 80, 123, 181, 236, 267, 317, 372, 412, 414, 426, 441, 458
Bunny's ears, 458
Burdock, 7, 71, 115, 171, 230, 260, 305, 366, 409, 411, 413, 417, 419, 422, 423, 426, 443, 451, 458, 463
Bursine, 10, 410
Butane, 148, 410
Butanolide, 148, 410
Butanone, 136, 436
Butcher's broom, 41, 100, 145, 203, 251, 285, 350, 389, 420, 423- 425, 458
Buttercup, 39, 99, 143, 201, 249, 284, 347, 388, 408, 433, 441, 450, 458
Butyl acetate, 143, 436
Butyl aldehyde, 118, 436
Butyl idene, 134, 436
Butyl phthalidine, 27, 410
Butyl pythalides, 134, 436
Butyric acid, 130, 135, 436, 443, 447
Butyrospermol, 138, 436
Buxus sempervirens, 9, 64, 74, 117, 165, 174, 232, 262, 309, 368, 458

C18 paraffin, 144, 436
Cadenene, 138, 436
Cadinene, 118, 122, 132, 139, 141, 148, 153, 436, 439, 441

Cadinol, 112, 436
Caffeic acid, 4, 11, 17, 18, 22, 23, 30, 33, 47, 50, 53, 410
Caffeine, 10, 14, 25, 180, 208, 411
Calamenenol, 112, 436
Calamint, 10, 74, 117, 175, 232, 263, 309, 368, 415, 417, 419, 450, 458
Calamintha, 10, 74, 117, 175, 232, 263, 309, 368, 456, 458, 467
Calamintha ascendens, 458
Calamintha nepeta, 10, 74, 117, 175, 232, 263, 309, 368, 456
Calamintha officinalis, 467
Calamus, 3, 53, 68, 112, 164, 168, 214, 228, 258, 293, 301, 364, 397, 407, 409, 418-421, 424, 432, 434-437, 440, 445, 446, 449, 451, 458, 472
Calcium oxalate, 411
Calcyosm, 8, 411
Calendula, 10, 74, 117, 175, 232, 263, 309, 368, 411, 420, 423, 424, 427, 440, 458, 466
Calendula officinalis, 10, 74, 117, 175, 232, 263, 309, 368, 458, 466
Calendulin, 117, 436
California bay, 56, 158, 458
California laurel, 48, 106, 151, 210, 255, 290, 359, 394, 407, 412, 414, 427, 431, 434, 440, 447, 450, 453, 458
California poppy, 19, 84, 127, 184, 238, 270, 322, 375, 411, 413, 414, 422, 458
Calmenen, 436
Caltrops, 458
Camellia sinensis, 10, 58, 74, 118, 175, 218, 232, 263, 310, 368, 463
Campesterol, 30, 48, 141, 148, 411, 436
Camphene, 14, 115, 118, 121, 124, 126, 128, 132, 137, 140, 141, 145, 150, 153, 411, 436, 444
Camphor, 3, 11, 13, 14, 26, 40-42, 78, 112, 118-121, 132, 133, 141, 142, 145-147, 149, 179, 234, 266, 314, 371, 411, 412, 419, 426, 432, 436, 437, 439, 440, 444, 449, 450, 452, 458
Canada snakeroot, 458
Canadensis curlone, 148, 436
Canadine, 25, 411
Cananga odorata, 10, 74, 118, 175, 232, 263, 310, 368, 474
Canary creeper, 35, 47, 95, 105, 140, 150, 197, 209, 247, 255, 281, 290, 341, 358, 385, 393, 415, 420, 421, 425, 428, 434, 458
Cannabis sativa, 10, 75, 118, 175, 232, 263, 310, 368, 464
Capric acid, 436, 447
Capronate, 113, 436
Capronic acid, 142, 143, 436
Caprylic acid, 130, 436, 447
Capsaicin, 11, 118, 411, 436
Capsella bursa-pastoris, 10, 75, 118, 175, 232, 263, 310, 368, 458, 467, 471
Capsicum annuum var. Acuminatum, 11, 75, 118, 176, 232, 263, 311, 369, 468
Capsicum annuum var. Grossum, 11, 75, 118, 176, 232, 263, 311, 369, 468
Caraway, 11, 56, 75, 119, 157, 158, 176, 216, 233, 264, 311, 369, 400, 411, 415, 418, 422, 437-439, 442, 444, 445, 447, 449, 450, 452, 458
Carbohydrates, 22, 165, 411
Carbonic acid, 117, 437
Carbosylic acid, 411
Carene, 128, 140, 431, 437, 439
Caria ester, 411
Carica papaya, 11, 75, 119, 176, 233, 264, 311, 369, 468
Carlinoxide, 437
Carota-1,4-beta-oxide, 437
Carotenes, 17, 33, 36, 40, 411
Carotol, 125, 437
Carrot, 17, 82, 125, 183, 237, 268, 320, 365, 366, 374, 385, 411, 428, 431, 434, 435, 437-439, 442, 444, 445, 447, 449, 450, 453, 458

Carthamin, 119, 176, 411
Carthamone, 11, 411
Carthamus tinctorius, 11, 56, 75, 119, 176, 216, 233, 264, 311, 369, 456, 470
Carum caroi, 233, 264, 311, 369, 458
Carvacrol, 32, 42, 46, 112, 114, 118, 136-138, 141, 147, 150, 411, 437
Carveol, 119, 437
Carvomenthene, 126, 437
Carvone, 6, 11, 30, 112, 114, 119, 123, 125, 127, 136, 139, 140, 411, 437, 444
Caryophyllene, 35, 114, 117, 118, 120, 122, 125, 126, 131, 132, 136, 138, 144, 149, 151, 411, 433, 437
Caryophyllene eposide, 126, 437
Caryophyllene oxide, 136, 437
Cascara buckthorn, 458
Cascara sagrade, 201, 284, 347, 388, 458
Caseweed, 458
Cassia angustifolia, 43, 103, 147, 205, 287, 353, 391, 471
Cassia marilandica, 455, 471, 473
Casticin, 50, 411
Cat's claw, 48, 61, 106, 151, 210, 255, 290, 296, 359, 394, 408, 421, 426, 458-459
Cat's foot, 459
Catalpol, 43, 411
Catchweed, 458
Catechins, 16, 23, 27, 411
Catharanthus roseus, 12, 76, 110, 119, 177, 223, 233, 264, 312, 369, 468
Catmint, 32, 93, 137, 195, 245, 279, 337, 383, 411, 417, 426, 427, 433, 435, 437, 438, 442, 452, 458
Catnip, 32, 93, 137, 195, 245, 279, 337, 383, 411, 417, 426, 427, 433, 435, 437, 438, 442, 452, 458
Caulophylline, 12, 411
Caulophyllum giganteum, 457
Caulophyllum thalietroides, 76, 233, 264, 370, 457
Caulosaponin, 12, 411
Cedarwood, 12, 76, 119, 177, 233, 264, 312, 370, 409, 419, 424, 432-434, 436-438, 443, 445, 451, 453, 459
Cedrene, 119, 437
Cedrol, 138, 437
Cedrus libani subsp. *atlantica*, 12, 54, 76, 119, 156, 177, 233, 264, 312, 370, 459
Celery, 6, 54, 71, 114, 171, 181, 215, 230, 260, 305, 320, 333, 366, 403, 409-412, 418, 434, 435, 437, 439-441, 443-445, 447-453, 459, 473
Centapicrin, 12, 411
Centaurea calcitrapa, 12, 76, 119, 177, 233, 264, 313, 370, 458, 472
Centella asiatica, 12, 58, 63, 76, 120, 159, 163, 166, 177, 218, 221, 234, 265, 313, 370, 463
Cerene, 145, 437
Chamaelirin, 12, 411
Chamaelirium luteum, 12, 23, 76, 87, 120, 130, 177, 188, 234, 265, 273, 313, 328, 370, 378, 457, 461, 462
Chamaemelum nobile, 6, 12, 71, 76, 114, 120, 171, 178, 229, 234, 260, 265, 305, 313, 366, 370, 470
Chamaenerion angustifolium, 462
Chamazulene, 3, 112, 135, 149, 411, 437
Chamomile, 6, 12, 29, 71, 76, 91, 114, 120, 135, 171, 178, 193, 229, 234, 244, 260, 265, 277, 305, 313, 335, 366, 370, 382, 408, 409, 412-414, 416, 418, 420, 426, 427, 433-435, 437, 441, 445, 450-452, 459, 461, 463, 465, 470, 473
Chapatral, 26, 89, 133, 191, 242, 275, 332, 380, 420, 423, 459
Chaste tree, 50, 107, 152, 212, 256, 292, 361, 395, 407, 409, 411, 412, 414, 417, 429, 433, 435, 437, 442, 446, 450, 451, 459
Chavicol, 8, 32, 112, 115, 128, 137, 140, 148, 149, 161, 219, 296, 437, 446

Checkber berry, 459
Chelidonine, 19, 411
Chenopodium album, 13, 55, 77, 120, 178, 215, 234, 265, 370, 465
Chenopodium ambrosioides, 465, 467, 472, 474
Cherry birch, 459
Chestnut, 3, 11, 68, 76, 112, 119, 168, 177, 228, 233, 258, 264, 302, 312, 364, 369, 407, 412, 419, 422-424, 426, 427, 432, 450, 459, 462, 464, 472
Chickweed, 45, 104, 149, 207, 253, 288, 355, 392, 412, 414, 418, 420, 424, 428, 440, 459
Chicory, 13, 77, 120, 179, 234, 265, 314, 371, 408, 417, 418, 424, 443, 444, 451, 459
Chittem bark, 459
Chives, 4, 69, 113, 169, 229, 259, 303, 365, 408, 425, 432, 438, 454, 459
Chlorogenic acid, 11, 17, 25, 32, 45, 411
Choke cherry, 38, 98, 143, 200, 249, 283, 346, 387, 419, 422, 434, 443, 459
Cholesterol, 4, 9, 17, 20, 29, 47, 48, 131, 141, 176, 411, 437
Choline, 10, 12, 17, 20, 22, 27, 46-48, 50, 141, 412, 437
Choublac, 459
Chrysanin, 120, 437
Chrysanolide, 120, 437
Chrysanthemum cinerarifolium, 39, 46, 77, 98, 104, 143, 149, 208, 249, 254, 284, 289, 346, 356, 387, 392, 469
Chrysanthemum parthenium, 46, 104, 149, 254, 289, 356, 392, 456, 462, 473
Chrysophanic acid, 5, 37, 144, 412, 437
Church steeples, 459
Chymopapain, 11, 412
Cichorium intybus, 13, 77, 120, 179, 234, 265, 314, 371, 459
Cilantro, 15, 80, 123, 181, 236, 267, 316, 372, 408, 418, 422, 426, 428, 433, 434, 436-439, 442, 452, 459
Cimicifugin, 13, 412
Cimicifuga racemosa, 13, 77, 121, 179, 234, 265, 314, 371, 457, 458, 471
Cimin aldehyde, 16, 412
Cincole, 8, 123, 412
Cineole, 3, 20, 26, 30, 40, 41, 48, 116, 121, 124, 129, 131-133, 136, 137, 141, 145, 146, 150-153, 407, 412, 431, 437
Cinerins, 13, 412
Cinnamic acid, 14, 23, 27, 28, 39, 144, 412, 437
Cinnamic alcohol, 134, 438
Cinnamic aldehyde, 121, 438
Cinnamomum camphora, 458
Cinnamomum cassia, 459
Cinnamomum verum, 14, 78, 121, 179, 234, 266, 314, 371, 459
Cinnamon, 14, 78, 121, 179, 234, 266, 314, 371, 412, 419, 426, 434, 437, 438, 441, 444, 449, 450, 459
Cinnamon wood, 459
Cinnamyl cinnamate, 27, 28, 412
Cinnamylcoaine, 19, 412
Cinnamolaurine, 42, 412
cis-3-Hexen-1-ol, 118, 438
cis-3-Hexenal, 127, 438
cis-3-Hexenol, 438
cis-8-Heptadecene, 153, 438
cis-9-Nonadecene, 153, 438
cis-Aconitate, 5, 412
cis-Anethole, 140, 438
cis-Asarone, 125, 438
cis-Dihydroatlantones, 119, 438
cis-Jasmone, 132, 438
cis-Ocimene, 129, 438
cis-Sabinene hydrate, 48, 412
cis-trans-Rose oxides, 139, 438
Citral, 17, 30, 117, 124, 127, 136, 412, 438
Citrate, 5, 412
Citric acid, 3, 5, 11, 28, 38, 40, 54, 121, 438
Citronella, 17, 81, 124, 158, 182, 237, 268, 319, 374, 399, 412, 432, 436, 438,

Citronella *(continued)*, 442, 444, 459
Citrostadienol, 138, 438
Citrus aurantifolia, 466
Citrus aurantium, 457
Citrus bergamia, 456, 468
Citrus limon, 14, 78, 121, 179, 235, 266, 315, 371, 466
Citrus paradisi, 463
Citrus reticulata, 466
Clamintha officinalis, 458
Clary, 41, 101, 146, 203, 251, 286, 350, 389, 410-412, 415, 419, 421, 422, 424, 426, 427, 444, 445, 447, 449, 459
Clary sage, 41, 101, 146, 203, 251, 286, 350, 389, 410-412, 415, 419, 421, 422, 424, 426, 427, 444, 445, 447, 449, 459
Clear eye, 459
Cleavers, 21, 85, 129, 186, 239, 271, 325, 376, 417, 422, 426, 453, 459
Cloudberry, 40, 100, 145, 160, 202, 250, 285, 349, 389, 409, 424, 427, 428, 440, 452, 459
Clove, 20, 45, 66, 84, 104, 128, 149, 166, 184, 208, 239, 253, 270, 289, 322, 356, 375, 392, 414, 416, 424, 426, 432, 433, 437, 441, 442, 446, 447, 451, 459
Clover (crimson), 47, 105, 150, 209, 254, 290, 358, 393, 414, 424, 434, 445, 446, 448, 451, 459
Clover (red), 47, 105, 150, 209, 254, 290, 358, 393, 411, 412, 416, 424-426, 434, 445, 446, 448, 451, 459
Clover broom, 116, 459
Clubmoss, 29, 91, 135, 193, 243, 276, 334, 381, 413, 415, 419, 420, 422, 423, 427, 454, 459
Cnicus benedictus, 11, 14, 75, 79, 118, 122, 176, 179, 233, 235, 263, 264, 266, 311, 315, 369, 371, 457, 464, 471
Coca, 19, 84, 127, 184, 238, 270, 322, 375, 410, 412, 413, 419, 420, 446, 459
Cocaine, 19, 184, 412
Cochlearia amorucia, 14, 79, 122, 180, 235, 266, 315, 371, 464
Cockle buttons, 459
Cocklebur, 460
Codeine, 34, 197, 412
Codonopsis, 14, 59, 79, 122, 160, 180, 235, 266, 295, 315, 371, 408, 421, 422, 424, 425, 431, 438, 453, 456, 460, 461
Codonopsis pilosula, 14, 79, 122, 180, 235, 266, 315, 371, 456, 460, 461
Codonopsis tangshen, 456, 460, 461
Coffea arabica, 14, 79, 122, 180, 235, 266, 315, 372, 460
Coffeasterol, 122, 438
Coffee, 14, 52, 53, 79, 122, 154, 158, 179, 180, 208, 235, 266, 315, 372, 411, 431, 438, 444, 445, 448, 451, 460
Coffeweed, 460
Colchicine, 14, 412
Colchicum autumnale, 14, 180, 235, 266, 316, 372, 460, 467, 468
Colicroot, 460
Collinsonia canadensis, 15, 39, 79, 122, 180, 235, 266, 316, 372, 469
Coltsfoot, 150, 441, 443, 447, 452, 460, 462
Columbamine, 412
Comfrey, 45, 57, 104, 149, 207, 218, 253, 289, 355, 392, 408, 420, 426, 460
Commiphora molmol, 15, 79, 122, 180, 267, 316, 372, 468
Commiphora myrrha, 468
Common bugloss, 460
Comphor, 438
Convallaria majalis, 15, 31, 80, 122, 136, 180, 194, 235, 245, 267, 278, 316, 337, 372, 382, 465-467
Convolvuline, 15, 412
Convolvulus arvensis, 15, 80, 123, 180, 235, 267, 316, 372, 456
Convolvulus sepium, 456

Coriander, 15, 80, 123, 181, 236, 267, 316, 372, 408, 418, 422, 426, 428, 433, 434, 436-439, 442, 452, 460
Coriandrol, 123, 438
Coriandrum sativum, 15, 80, 123, 181, 236, 267, 316, 372, 459, 460
Corn, 34, 51, 95, 108, 139, 153, 197, 213, 246, 257, 280, 292, 340, 363, 373, 378, 384, 396, 401, 414, 419-421, 423, 424, 426-428, 445, 454, 460
Corn poppy, 34, 95, 139, 197, 246, 280, 340, 384, 421, 427, 445, 460
Corn rose, 460
Cornic acid, 15, 412
Cornine, 15, 412
Cornus canadensis, 15, 80, 123, 181, 236, 267, 317, 372, 458
Cornus florida, 458, 461, 462, 473
Corosolic acid, 52, 123
Corylus avellana, 464
Corylus cornuta, 15, 80, 123, 181, 236, 267, 317, 373, 464
Corylus rostrata, 464
Cotton, 22, 86, 130, 187, 240, 272, 327, 370, 371, 377, 393, 414, 416, 442, 447, 460
Couch grass, 4, 18, 69, 83, 112, 127, 169, 184, 228, 238, 258, 269, 302, 321, 364, 375, 419, 422, 432, 437, 441, 453, 460
Coughwort, 460
Coumaric, 20, 30, 47, 412, 421
Coumarines, 142, 438
Countryman's treacle, 460
Cow clover, 460
Cow parsley, 24, 87, 131, 188, 241, 273, 328, 378, 416, 423, 425, 432, 434, 436, 440, 441, 443, 444, 448, 449, 451, 460
Cow parsnip, 24, 87, 131, 188, 241, 273, 328, 378, 416, 423, 425, 432, 434, 436, 440, 441, 443, 444, 448, 449, 451, 460
Cowberry, 48, 106, 151, 211, 255, 291, 359, 394, 408, 453, 460
Cowslip, 37, 98, 142, 200, 249, 283, 344, 387, 414, 416, 424, 426, 427, 453, 460
Crampbark, 395, 460
Cranberry, 48, 49, 61, 62, 106, 107, 151, 152, 210, 211, 220, 255, 256, 291, 294, 359, 361, 394, 395, 408, 412, 416, 417, 423, 426, 428, 437, 453, 454, 460, 464, 467
Cranesbill (American), 21, 240, 271, 326, 377, 425, 455
Crataegus acid, 16
Crataegus monogyna, 16, 80, 123, 181, 236, 267, 317, 373, 464, 467
Crataegus oxyacantha, 16
Crataegus pinnatifida, 52, 154, 464
Creeping charlie, 460
Creeping thyme, 460
Creosol, 36, 117, 413, 438
Creosote, 20, 185, 413
Crimson clover, 460
Crocine glycosides, 16, 413
Crocus, 14, 16, 79, 80, 122, 123, 181, 235, 236, 266, 267, 317, 372, 373, 412, 414, 460, 470
Crocus sativus, 16, 80, 123, 181, 236, 267, 317, 373, 470
Crosswort, 460
Crowberry, 460
Cryptopine, 19, 413
Cryptotaenene, 123, 438
Cryptotaenia japonica, 16, 80, 123, 181, 236, 267, 318, 373, 465, 467
Cucurbita pepo, 16, 81, 123, 181, 236, 267, 318, 373, 469
Cucurbitin, 16, 413
Cumin, 16, 81, 124, 182, 236, 268, 318, 373, 400, 412, 422, 426, 432, 433, 438, 444, 447, 449, 460, 472
Cuminaldehyde, 121, 127, 438
Cuminum cyminum, 16, 81, 124, 182, 236,

Cuminum cyminum (continued), 16, 81, 124, 182, 236, 268, 318, 373, 460
Curcuma, 16, 81, 124, 165, 182, 236, 268, 318, 373, 413, 423, 431, 433, 434, 436, 441, 448, 450, 451, 453, 454, 460, 465, 473, 474
Curcuma domestica, 81, 124, 182, 236, 268, 318, 373, 460, 465
Curcuma longa, 460, 465, 473, 474
Curcuma xanthorrhiza, 460, 465
Curcumin, 42, 413
Curcuminoids, 16, 413
Curry plant, 23, 87, 130, 188, 240, 273, 328, 378, 407, 409, 412, 418, 426, 441-444, 447, 449, 451, 460
Cuscuta chinesis, 461
Cuscuta epithymum, 16, 81, 124, 182, 237, 268, 373, 461
Cutins, 39, 413
Cyanide, 38, 413, 434
Cyanogenic glycoside, 24, 38, 413
Cycloalliin, 113, 438
Cycloarienol, 138, 438
Cyclolignans, 148, 413
Cyclosantalal, 146, 438
Cymbopogon citratus, 17, 81, 124, 182, 237, 268, 319, 374, 466
Cymbopogon martinii, 463, 468
Cymbopogon nardus, 459
Cymbopogon winterianus, 459
Cymbopogonol, 124, 438
Cymeme, 120, 438
Cymene, 26, 30, 42, 114, 117, 121, 123, 124, 128, 132, 136-140, 145, 147, 150, 153, 413, 448
Cymol, 114, 439
Cynara cardunculus, 54, 215, 456
Cynara scolymus, 17, 81, 125, 182, 237, 268, 319, 374, 455
Cynarin, 17, 413
Cypripedin, 17, 413
Cypripedium calceolus, 17, 81, 125, 182, 237, 268, 319, 374, 455, 465, 467, 474
Cytisus scoparius, 17, 42, 82, 125, 182, 237, 252, 268, 286, 319, 351, 374, 390, 458, 463, 465

d-1-Methyl-3-cyclohexanone, 130, 439
d-α-Phyllandrene, 137, 439
d-Cadeine, 144, 439
d-Camphor, 14, 142, 439
d-Fenchone, 142, 439
d-Limonene, 14, 121, 122, 139
d-Pseudoephedrine, 19, 413
Daffodil, 31, 93, 137, 195, 245, 278, 337, 383, 408, 420, 423, 434, 437, 443, 447, 450, 460
Daiceme, 439
Daidzein, 38, 413
Daisy, 8, 73, 116, 231, 262, 308, 367, 419, 424, 426, 454, 460
Damask rose, 40, 100, 144, 202, 250, 285, 349, 389, 411, 412, 419, 421, 428, 441, 442, 445, 447, 449, 451, 460
Dammaranedienol, 413
Dandelion, 46, 59, 105, 149, 160, 208, 218, 254, 289, 295, 357, 393, 412, 416-418, 423, 426, 428, 439, 453, 461
Dandelion oil, 149, 439
Dang shen, 14, 79, 122, 180, 235, 266, 315, 371, 408, 421, 422, 424, 425, 431, 438, 453, 461
Daucol, 125, 439
Daucus carota, 17, 82, 125, 183, 237, 268, 320, 374, 458, 469
Dead nettle, 461
Deadmen's bells, 461
Deanolic acid, 413
Decylaldehyde, 123, 439
Dehydrofukinone, 7, 413
Delta-3-carene, 137, 439
Delta-cadinene, 118, 122, 439
Delta-sabinene, 132, 139
Dendrolasin, 42, 413
Depentene, 439

Desert tea, 461
Desmethoxyyangenin, 439
Devil's bit, 12, 461
Devil's bones, 461
Devil's claw, 23, 86, 130, 187, 240, 272, 327, 377, 414, 416, 425, 461
Devil's club, 18, 33, 83, 94, 126, 138, 183, 196, 238, 246, 269, 279, 321, 339, 374, 384, 409, 422, 436, 437, 440, 447, 452, 461
Di-acids, 140, 439
Diacetyl, 439
Diacylglycerols, 439
Dianthrone glucosides, 11, 413
Diasarone, 116, 439
Dibutyl ester, 140, 439
Dicarbonic acid, 114, 439
Dicoumarol, 413
Digalactosyl glycerol, 152, 439
Digitalis, 17, 82, 125, 155, 183, 237, 268, 320, 374, 411, 412, 461, 462, 474
Digitalis purpurea, 17, 82, 125, 183, 237, 268, 320, 374, 461, 462, 474
Digitoxin, 17, 413
Digoxin, 17, 413
Dihydralantolactone, 132, 439
Dihydro-beta-agacofuran, 42, 413
Dihydroalliin, 113, 439
Dihydrobutylinene, 134, 439
Dihydrocarveol, 119, 439
Dihydrocarvone, 119, 120, 136, 439
Dihydrochamazulene, 149, 439
Dihydroisalantolactone, 439
Dihydrokavin, 439
Dihydrokawain, 36, 413
Dihydrolycopodine, 29, 413
Dihydromethysticin, 36, 413
Dilaurocapnin, 439
Dill, 6, 70, 114, 170, 229, 260, 304, 365, 411, 412, 414, 418, 427, 429, 431, 432, 437, 441, 444, 449, 461
Dillapiole, 134, 439
Dimeric indole alkaloids, 50, 413
Dimethyl azelate, 116, 439
Dimethyl ether, 115, 128, 452
Dimethyl phenols, 117, 439
Dimethyl sulfide, 126, 136, 439
Dioscorea oppositae, 18, 82, 125, 183, 237, 269, 320, 374, 471, 474
Dioscorea villosa, 460, 461, 474
Diosgenin, 18, 56, 216, 413
Diosmin, 40
Dipentene, 14, 117, 121, 124, 131, 138, 141, 148, 440
Dipropyl disulphide, 113, 440
Dipropyl trisulphide, 440
Disulphide, 113, 440
Diterpene acids, 6, 413
Diterpenes, 29, 143
Dock, 41, 100, 145, 203, 251, 285, 350, 389, 408, 421, 422, 441, 450, 457, 458, 461, 468, 470
Docosane, 150
Dodder, 16, 81, 124, 182, 237, 268, 318, 373, 402, 410, 412, 414, 454, 461
Dodeca-2,4-dien-1-yl-isovalerate, 126, 440
Dodecanoid acid, 151, 440
Dodecanol, 440
Dodinene, 138, 440
Dog brier, 461
Dog grass, 461
Dog lichen, 34, 95, 139, 197, 247, 280, 340, 385, 427, 453, 461
Dog rose, 40, 100, 144, 202, 250, 285, 349, 389, 411, 412, 419, 421, 428, 436, 438, 442, 446, 447, 453, 461
Dog tree, 461
Dogwood, 15, 58, 80, 123, 181, 218, 236, 267, 317, 372, 415, 424-427, 441, 448, 449, 457, 461, 462, 473
Dong quai, 6, 70, 114, 170, 229, 260, 304, 365, 412, 425, 428, 436, 437, 444, 449, 451, 461
Dragon's mugwort, 461
Duckweed, 27, 89, 133, 191, 242, 275, 332, 380, 409, 419, 454, 461

Dutch honeysuckle, 28, 90, 134, 192, 243, 381, 412, 417-419, 424-426, 441, 461
Dwarf juniper, 461
Dwarf mountain pine, 36, 199, 248, 282, 342, 386, 408, 410, 415, 419, 423, 425, 428, 435, 436, 444, 446, 461

Ecdysteroids, 13, 55, 215, 413
Ecgonine, 19, 413
Echinacea (narrowleaf), 18, 64, 83, 126, 183, 237, 269, 321, 374, 461
Echinacea (pale-flower), 18, 64, 83, 126, 183, 237, 269, 321, 374, 461
Echinacea (purple coneflower), 18, 64, 83, 126, 183, 237, 269, 321, 374, 461
Echinacea angustifolia, 18, 64, 83, 126, 183, 237, 269, 321, 374, 461
Echinacea pallida, 18, 64, 83, 126, 183, 237, 269, 321, 374, 461
Echinacea pururea, 18, 64, 83, 126, 183, 237, 269, 321, 374, 461
Echinolone, 126, 440
Echinopanax horridus, 18, 83, 126, 183, 238, 269, 321, 374, 461
Egin, 7, 414
Eicosa-11,14,17-trienoic acid, 135, 440
Elaeagnus angustifolia, 18, 83, 126, 183, 238, 269, 321, 375, 470
Elder, 42, 101, 146, 203, 251, 286, 351, 390, 409, 415, 431, 438, 444, 445, 448, 453, 461, 462
Elderberry, 41, 101, 146, 203, 251, 286, 351, 389, 408, 413-416, 425-428, 453, 461
Elecampane, 25, 88, 132, 190, 241, 274, 330, 379, 413, 416, 417, 419, 423, 432, 439, 443, 451, 461
Elemene, 112, 114, 122, 129, 132, 153, 410, 435, 440, 442, 454
Elemicin, 122, 440
Elemol, 145, 152, 440
Eleutherosides, 18, 414
Eleuthrococcus senticosus, 471
Elincin, 137, 440
Elixen, 23, 414
Elymus repens, 18, 83, 127, 184, 238, 269, 321, 375, 460
Emblic, 35, 96, 140, 198, 247, 281, 341, 385, 414, 421, 428, 437, 439, 444, 461
Emblicanins, 35, 414
Emodin, 5, 11, 39, 408, 414
Emulsin, 16, 414
English alder, 461
English chamomile, 461
English holly, 25, 88, 132, 190, 241, 274, 330, 379, 411, 417, 424, 427, 445, 461
English ivy, 23, 86, 130, 188, 240, 272, 328, 378, 407, 410, 414-417, 426, 445, 448, 449, 461
Enicin, 14, 414
Ephedra, 19, 61, 83, 127, 184, 219, 238, 269, 322, 375, 408, 413, 414, 454, 458, 461, 466, 467, 469
Ephedra distachya, 19, 83, 127, 184, 238, 269, 322, 375, 461
Ephedra nevadensis, 458, 461, 467, 469
Ephedra sinica, 466
Ephedrine, 19, 61, 219, 414, 418
Epicyclosantalal, 146, 440
Epilobium angustifolium, 12, 19, 76, 83, 120, 127, 178, 234, 238, 265, 270, 313, 370, 375, 405, 462
Epilobium parviflorum, 19, 83, 127, 184, 238, 270, 322, 375, 471
Episetum arvense, 457, 469, 474
Epishyobunone, 112, 440
Epsilon-bulgarene, 129, 440
Epsilon-muurolene, 129, 440
Equinopanacene, 440
Equinopanacol, 138, 440
Equisetum arvense, 19, 84, 127, 184, 238, 270, 322, 375, 457, 464, 469, 470
Equisitine, 19, 414

Ergosterol, 131, 440
Erucic acid, 119, 148, 440
Erythroxylum coca, 19, 84, 127, 184, 238, 270, 322, 375, 459
Eschscholtzia california, 19, 84, 238, 322, 375, 458
Esters, 10, 12, 19, 34, 49, 111, 120, 131, 134, 135, 138, 202, 432, 434, 440, 443, 446, 448, 453
Estragole, 8, 72, 112, 115, 140, 149, 414, 440
Estrogens, 39, 414
Ethanol, 113, 144, 165, 407, 440, 449
Ethyl alcohol, 440
Ethyl butyrates, 131, 440
Ethyl hexanoate, 131, 440
Ethyl linolenate, 123, 441
Ethyl palmitate, 123, 441
Eucalyptol, 20, 48, 127, 133, 414, 441
Eucalyptus, 20, 84, 127, 158, 184, 238, 270, 322, 375, 398, 399, 410, 415, 417, 427, 433, 434, 437, 438, 441-444, 447, 449, 450, 462-464
Eucalyptus citriodora, 20, 84, 127, 184, 238, 270, 322, 375, 464
Eucalyptus globulus, 462, 463
Eucosine, 150
Eudesmic acid, 414
Eudesmol, 127, 152, 441
Eudesmyl acetate, 127, 441
Eugenol, 45, 86, 110, 112, 114, 118, 121, 122, 125, 130, 132, 137, 139, 140, 143, 144, 149, 150, 164, 414, 432, 441, 443, 446
Eugenol acetate, 121
Eupatorin, 20, 414
Eupatorium perfoliatom, 455, 460, 462
Euphrasia officinalis, 20, 84, 128, 185, 239, 270, 323, 376, 462, 467
European alder, 462
European barberry, 462
European buckthorn, 462
European chestnut, 462
European elder, 462
European holly, 462
European mistletoe, 50, 107, 152, 411, 415, 418, 423
European peony, 33, 94, 138, 196, 246, 280, 339, 384, 408, 416, 421, 425, 462
Evening primrose, 32, 60, 94, 138, 158, 196, 217, 221, 245, 279, 338, 383, 418, 420, 433, 435, 438, 442, 445, 462
Evening star, 462
Everlastings, 5, 6, 70, 71, 113, 114, 170, 229, 259, 260, 304, 365, 407, 414, 419, 443, 449, 462
Eyebright, 20, 84, 128, 185, 239, 270, 323, 376, 409, 411, 415, 426, 428, 454, 462, 467
Eyeroot, 462

Fagopyrum tataricum, 84, 128, 163, 185, 239, 270, 323, 376, 458
Fagus grandifolia, 20, 84, 128, 185, 271, 323, 376, 456
Fagus sylvatica, 162, 456
Fairy cup, 462
Fairy wand, 12, 462
False coltsfoot, 462
False unicorn root, 12, 76, 87, 120, 130, 177, 188, 234, 241, 265, 273, 313, 328, 370, 378, 411, 416, 424, 462
Famesene, 118, 441
Faradiol, 441
Farnesene, 112, 120, 126, 131, 135, 153, 435, 441, 453
Farnesol, 124, 131, 132, 144, 150, 432, 441
Fatty acid, 16, 22, 54, 110, 129, 154, 157, 159, 441, 446
Featherfew, 462
Felon herb, 462
Fenchene, 145, 441
Fenchyl alcohol, 137, 441
Fennel, 20, 84, 128, 185, 239, 271, 324, 376, 408, 414, 434, 436, 441, 444, 446, 449, 462, 472, 473

Fenugreek, 47, 53, 58, 105, 150, 155, 160, 209, 254, 290, 358, 393, 412, 413, 418, 420-422, 427, 454, 462
Ferric oxide, 15, 414
Ferruginol, 441
Ferulic acid, 6, 144, 414, 441
Fever tree, 462
Feverfew, 13, 46, 57, 77, 104, 120, 149, 178, 208, 234, 254, 265, 289, 295, 314, 356, 370, 392, 411, 419, 424, 426, 433, 436, 439, 447, 451, 452, 462
Feverwort, 462
Fiber, 22, 175, 188, 192, 414
Field balm, 462
Field poppy, 462
Filipendula ulmaria, 20, 44, 84, 103, 128, 148, 185, 207, 239, 253, 271, 323, 354, 376, 392, 458, 467, 469
Fir balsam, 462
Fir pine, 462
Fireweed, 19, 83, 127, 184, 238, 270, 322, 375, 405, 407, 420, 425, 454, 462
Flanneleaf, 462
Flavanoids, 17, 32, 142, 441
Flavonol glycosides, 20, 55, 415
Flax, 27, 90, 134, 192, 243, 276, 333, 380, 418, 420, 425, 440, 445, 448, 451, 462
Flower essence, 126, 145, 168, 170, 192, 194, 196, 204, 205, 211
Flowering dogwood, 462
Flowery knotweed, 37, 97, 142, 200, 248, 283, 344, 386, 408, 418, 432, 436, 449, 457, 462
Fluoranthene, 441
Foeniculum vulgare, 20, 84, 128, 185, 239, 271, 324, 376, 462, 472, 473
Folic acid, 415
Foradiol, 115, 441
Formaldehyde, 129, 398, 441
Formic acid, 43, 151, 415
Formononetin, 8, 415
Forsythia, 20, 58, 85, 128, 161, 185, 239, 271, 324, 376, 415, 433, 435-437, 444, 447, 448, 452, 462, 466
Forsythia suspensa, 20, 85, 128, 185, 239, 271, 324, 376, 462, 466
Forsythin, 20, 415
Fox berry, 462
Fox glove, 17, 82, 125, 155, 183, 237, 268, 294, 320, 374, 411-413, 418, 423, 424, 440, 445, 448, 462
Fragaria vesca, 21, 85, 128, 185, 239, 271, 324, 376, 472-474
Frangulin, 39, 415
Frankincense, 9, 74, 117, 174, 231, 262, 309, 368, 410, 427, 439, 440, 444, 447, 449, 452, 453, 463
Fraxinus americana, 21, 85, 129, 186, 239, 271, 325, 376, 455, 457, 473
Fraxinus excelsior, 473
French tarragon, 463
Fu ti, 463
Fucus vesiculosus, 21, 26, 60, 85, 89, 129, 133, 161, 186, 190, 239, 242, 271, 325, 331, 376, 379, 457, 465
Fumaric acid, 5, 114, 415, 441
Furano-monoterpene, 122, 441
Furanoid, 8, 415
Furanosesquiterpenes, 441
Furfural, 441

Galactose, 36, 415
Galactoside-specific lectin, 50, 415
Galacturonate, 5, 415
Galium aparine, 21, 85, 129, 186, 239, 271, 325, 376, 456, 458, 459, 463
Gallic acid, 15, 19, 23, 39, 46, 415
Gallotannin, 33, 415
Gamma-2-cadinene, 148, 441
Gamma-atlantone, 124, 441
Gamma-decalactone, 143, 442
Gamma-elemene, 129, 153, 442
Gamma-gurjunene, 153, 442
Gamma-jasmolactone, 143, 442

Gamma-linolenic acid, 32, 60, 94, 117, 138, 442
Gamma-terpinene, 121, 122, 124, 128, 136, 139, 147, 149, 442
Gamolenic acid, 138, 442
Garden rue, 463
Garden sage, 463
Garden sorrel, 463
Garlic, 4, 69, 113, 169, 229, 259, 303, 365, 407, 408, 417, 424, 425, 432, 433, 440, 463
Gaultheria procumbens, 21, 85, 129, 186, 239, 271, 325, 377, 458, 459, 467, 473, 474
Gaultheriline, 442
Gaultherin, 128, 442
Geijerene, 442
Genista, 463
Gentianine, 12, 415
Gentisic, 20, 415
Geranial, 153, 442
Geraniol, 10, 30, 34, 40, 116-118, 120, 121, 123-125, 127, 130, 133, 136, 139, 144, 153, 415, 442
Geraniol acetate, 125, 442
Geranium, 21, 34, 40, 54, 85, 95, 100, 129, 139, 144, 156, 186, 197, 202, 215, 240, 247, 250, 271, 280, 285, 326, 340, 348, 377, 385, 388, 425, 433, 435, 436, 438, 442-444, 446, 448, 451, 455, 463, 466, 470, 473
Geranium macrorrhizum, 21, 40, 85, 129, 144, 202, 271, 285, 377, 388, 455, 463
Geranyl acetate, 118, 124, 141, 153, 442
Geranylisobutyrate, 126, 442
Germacrene, 23, 112, 118, 120, 122, 126, 129, 131, 148, 152, 153, 156, 415, 442
German chamomile, 29, 91, 135, 193, 244, 277, 335, 382, 408, 409, 412-414, 416, 418, 426, 427, 434, 435, 441, 445, 450, 463
Ginger, 8, 17, 51, 53, 62, 72, 108, 116, 153, 155, 163, 173, 213, 221, 231, 257, 261, 292, 307, 363, 367, 396, 409, 410, 415, 418, 422, 424, 426, 431, 433-436, 439, 441, 442, 444, 446-449, 451, 452, 454, 463, 465, 473
Gingergrass, 81, 124, 182, 237, 268, 319, 374, 412, 438, 442, 445, 449, 463
Ginkgetin, 21, 415
Ginkgo, 21, 52, 66, 85, 109, 129, 154, 159, 166, 186, 214, 240, 272, 326, 377, 407, 410, 411, 414, 415, 417, 424, 432, 446, 463, 466
Ginkgo biloba, 21, 52, 85, 109, 129, 154, 159, 186, 214, 240, 272, 326, 377, 463, 466
Ginkgocide, 21, 415
Ginseng (American), 463
Ginseng (Asian), 463
Ginseng (tian qi), 463
Ginsenosides, 34, 415
Glabridin, 22, 415
Glechoma hederacea, 21, 85, 129, 186, 240, 272, 326, 377, 459, 460, 462, 463
Glechomine, 21, 415
Globulol, 127, 442
Glossy buckthorn, 463
Glucans, 42, 415
Glucocyanates, 47, 415
Glucofrangulin, 39, 415
Glucoquinone, 415
Glucose, 25, 29, 36, 46, 415
Glucosinolates, 39, 416
Glucuronic acid, 416
Glue gum, 463
Glutamic acid, 44, 416
Glutamine, 24, 416
Gluten, 46, 416
Glycerol, 151, 442
Glycine max, 22, 86, 129, 187, 240, 272, 326, 377, 471
Glycoproteins, 18, 416
Glycorrhizin, 22, 416

Glycosides, 8, 11-13, 15-17, 20, 29, 32-34, 37, 39, 41, 42, 45, 47, 54, 55, 57, 66, 156, 215, 408, 413, 415, 416, 420, 423
Glycyrrhetinic acid, 22, 416
Glycyrrhiza glabra, 22, 57, 86, 109, 130, 187, 217, 240, 272, 326, 377, 466
Glycyrrhiza lepidota, 455, 473
Glycyrrhiza uralensis, 22, 86, 130, 187, 217, 240, 272, 326, 377, 466
Goatweed, 463
Golden bough, 463
Golden loosestrife, 463
Golden rose, 463
Goldenrod, 44, 103, 148, 206, 253, 288, 354, 391, 411, 417, 422, 424, 426, 435, 436, 440-442, 444, 446, 447, 449, 450, 463, 472
Goldenseal, 25, 53, 87, 131, 189, 214, 217, 274, 293, 294, 330, 379, 400, 410, 411, 416, 417, 423, 463
Gooseberry (black), 463
Goosegrass, 463
Gossypium hirsutum, 22, 86, 130, 187, 240, 272, 327, 377, 460
Gossypol, 22, 416, 442
Gotu kola, 12, 76, 120, 177, 234, 265, 313, 370, 409, 416, 419, 420, 424, 425, 427, 433, 435, 442, 451, 453, 463
Grape, 9, 29, 51, 73, 91, 107, 116, 135, 152, 173, 193, 212, 231, 244, 256, 262, 277, 292, 308, 335, 362, 367, 381, 395, 408, 410, 412, 415, 418-421, 425, 426, 440, 442, 446, 456, 463, 468
Grape seed oil, 152, 442
Grapefruit, 14, 79, 122, 179, 266, 315, 371, 410, 412, 418, 420, 428, 439, 442, 445, 448, 451, 463
Great burdock, 463
Greek hayseed, 463
Green ginger, 463
Green sauce, 463
Green tea, 10, 74, 118, 175, 232, 263, 310, 368, 411, 419, 422, 423, 436, 438, 443, 446, 447, 449, 463
Grindelia robusta, 22, 86, 130, 187, 240, 272, 327, 377, 464
Grindelia squarrosa, 464
Grindelic acid, 22, 416
Groats, 463
Ground cherry, 463
Ground glutton, 463
Ground ivy, 21, 85, 129, 186, 240, 272, 326, 377, 415, 423-425, 431-433, 435, 438, 440, 442, 444, 447, 450, 452, 463
Ground juniper, 463
Ground liverwort, 34, 95, 139, 197, 247, 280, 340, 385, 420, 427, 453, 464
Ground pine, 464
Ground raspberry, 464
Ground swallow, 464
Groundsel, 43, 103, 147, 205, 252, 287, 352, 391, 417, 423, 424, 427, 433, 447, 448, 451, 464
Gualacol, 117, 442
Gum, 9, 15, 20, 27, 28, 33, 46, 84, 90, 117, 122, 127, 134, 174, 180, 184, 192, 238, 243, 261, 262, 270, 276, 322, 329, 333, 375, 380, 410, 412, 415-418, 421, 427, 433, 434, 437, 438, 440-444, 447, 449, 450, 452, 453, 463, 464, 468
Gum tree, 20, 27, 84, 90, 127, 134, 184, 192, 238, 243, 270, 276, 322, 333, 375, 380, 410, 412, 415, 417, 418, 421, 427, 433, 434, 437, 438, 440-444, 447, 449, 450, 452, 453, 464
Gumweed, 22, 86, 130, 187, 240, 272, 327, 377, 407, 414, 416, 418, 426, 433, 435, 436, 441, 452, 464
Gustanubem 5-hydroxytryptamine, 416
Gypsyweed, 464

Hamamelin, 130

Hamamelis virginiana, 23, 55, 66, 86, 130, 187, 240, 272, 327, 377, 471, 474
Hamamelitannin, 23, 416
Harpagophytum procumbens, 86, 461
Harpagoside, 23, 416
Havonoids, 35
Haw, 33, 94, 138, 196, 246, 279, 339, 384, 464, 469
Hawthorn (Chinese), 464
Hawthorn (English), 464
Hazelnut, 15, 80, 123, 181, 236, 267, 317, 373, 414, 425, 426, 448, 449, 453, 464
Heal all, 38, 98, 142, 200, 249, 283, 344, 387, 409, 410, 413, 424, 426-429, 438, 439, 441, 443, 451, 464
Healing herb, 464
Heartsease, 464
Hedeoma pulegiodes, 23, 86, 130, 188, 240, 272, 328, 377, 468
Hedera helix, 23, 86, 130, 188, 240, 272, 328, 378, 461
Hederacoside, 23, 416
Hederin, 23, 416
Heerabolene, 122, 442
Helenalin, 7, 25, 416
Helianthus annuus, 23, 86, 130, 188, 240, 273, 328, 378, 472
Helichrysum angustifolium, 87, 240, 273, 378, 460
Heliotropin, 20, 416
Helonin, 416
Hemp, 10, 75, 118, 175, 232, 263, 310, 368, 411, 412, 417, 422, 426-429, 433, 435, 436, 444, 451, 452, 464
Henicosane, 137, 442
Hens and chicks, 43, 147, 205, 252, 287, 352, 391, 415, 419, 426, 464
Heptadecanoic acid, 135, 442
Heptanone, 149, 442
Heptyl-2-methyl bútyrates, 113, 443
Heraclein, 24, 416
Heracleum lanatum, 460
Heracleum maximum, 24, 87, 131, 188, 241, 273, 328, 378, 460
Heracleum sphondylium, 460
Herb of grace, 41, 101, 145, 203, 251, 286, 350, 389, 409, 412, 415, 419, 423, 431, 437, 440, 442, 444, 446, 447, 449, 450, 464
Herniarin, 29, 416
Hesperidin, 25, 121, 416, 443
Hexadecanoic acid, 135, 142, 443
Hexanal, 122, 139, 142, 143, 443, 447, 453
Hexanoic acid, 142, 443
Hexenol, 118, 122, 150, 152, 431, 438, 443
Hexenyl derivatives and acetals, 143, 443
Hexyl, 113, 135, 143, 443
Hexyl acetate, 143, 443
Hibiscus, 24, 87, 131, 188, 241, 273, 329, 378, 408, 412, 416, 419-421, 426, 432, 433, 435-437, 440, 448, 459, 464, 470
Hibiscus (Chinese), 464
Hibiscus acid, 24, 416
Hibiscus rosa-sinensis, 459, 464
Hibiscus sabdariffa, 24, 87, 131, 188, 241, 273, 329, 378, 464, 470
Hierochloe odorata, 24, 87, 131, 165, 189, 241, 273, 329, 378, 472
Highbush cranberry, 49, 107, 152, 211, 256, 291, 361, 395, 412, 417, 423, 426, 437, 454, 464
Himachalenes, 119, 443
Hippophae rhamnoides, 24, 59, 61, 87, 131, 161, 189, 219, 241, 273, 295, 329, 378, 400, 401, 470
Hippuric acid, 48, 416
Histamine, 10, 48, 416
Histidine, 39, 44, 416
Hoarhound, 464
Hog apple, 464
Holly, 25, 88, 132, 190, 241, 274, 330, 368, 378-381, 411, 417, 424, 427, 445, 461, 462, 464, 467
Holy thistle, 464

Honeysuckle (Dutch), 464
Honeysuckle (mountain fly), 28, 90, 134, 192, 243, 276, 334, 381, 412, 417-419, 424-426, 441, 464-467
Hoodwort, 464
Hops, 87, 131, 189, 217, 241, 273, 294, 329, 378, 400, 433, 435, 441, 443, 446, 447, 453, 464
Horehound, 91, 135, 193, 454, 464
Horse chestnut, 3, 68, 112, 168, 228, 258, 302, 364, 407, 412, 423, 424, 426, 427, 432, 450, 464
Horsebalm, 31, 92, 136, 194, 245, 278, 336, 382, 407, 421, 437, 439, 442, 444, 447, 452, 464
Horsefly weed, 464
Horseheal, 464
Horsehound, 29, 244, 277, 335, 381, 410, 419, 424-426, 464
Horsemint, 31, 92, 136, 194, 245, 278, 336, 382, 407, 421, 437, 439, 442, 444, 445, 447, 452, 464
Horseradish, 7, 14, 71, 79, 115, 122, 171, 180, 230, 235, 260, 266, 306, 315, 366, 371, 409, 423, 425, 428, 433, 449, 464
Horsetail, 19, 84, 127, 184, 238, 269, 270, 322, 375, 408, 414, 420, 425, 438, 453, 464
Houseleek, 43, 102, 103, 147, 205, 252, 287, 352, 391, 414, 415, 419-422, 424, 426, 441, 453, 465, 471
Huang bai, 35, 95, 140, 197, 247, 281, 341, 385, 417, 418, 425, 446, 447, 452, 464
Huang qi, 73, 116, 173, 231, 261, 307, 367, 439, 446, 465
Huang qin, 465
Humulene, 24, 114, 120, 126, 131, 149, 416, 433, 443
Humulus lupulus, 24, 57, 87, 131, 189, 217, 241, 273, 329, 378, 464
Hungarian chamomile, 465
Hydrangea, 24, 87, 131, 169, 189, 241, 273, 330, 379, 413, 414, 416, 424, 426, 453, 465
Hydrangea arborescens, 24, 87, 131, 169, 241, 273, 330, 379, 465
Hydrangein, 24, 416
Hydrastine, 25, 416
Hydrastis canadensis, 25, 87, 131, 189, 241, 274, 330, 379, 462-465, 474
Hydrocarbons, 35, 42, 53, 132, 137, 138, 141, 147, 152, 153, 424, 432, 433, 443, 447
Hydrocotyline, 12, 416
Hydrocyanic acid, 143, 148, 443
Hydrojuglone, 416
Hydroquinones, 49, 417
Hydroxybenzoic, 20, 30, 139, 417, 421, 431
Hymulene, 131, 443
Hypaconitine, 3, 68, 417
Hypericin, 25, 417
Hypericum perforatum, 25, 87, 131, 189, 241, 274, 330, 379, 455, 463, 472
Hyperoside, 9, 25, 417
Hyssop, 25, 88, 132, 189, 241, 274, 330, 379, 416, 418, 422, 425, 426, 435, 436, 440, 443, 444, 450, 455, 465
Hyssopus officinalis, 25, 88, 132, 189, 241, 274, 330, 379, 465

Ilex aquifolium, 25, 190, 241, 274, 330, 379, 461, 462, 464, 467
Indian dye, 465
Indian elm, 465
Indian ginger, 465
Indian ginseng, 51, 108, 152, 212, 256, 292, 362, 395, 408, 410, 429, 465
Indian paint, 465
Indole, 34, 50, 132, 413, 417, 443
Inositol, 10, 26, 28, 417
Insulins, 49, 417
Intybin, 13, 417
Inula helenium, 25, 88, 132, 190, 241, 274, 330, 379, 461, 464, 473

Inulin, 7, 8, 13, 25, 44, 46, 47, 115, 120, 150, 417, 443
Iodine, 4, 8, 21, 23, 123, 417
Iranian poppy, 34, 95, 139, 197, 246, 280, 340, 384, 421, 427, 445, 448, 451, 465
Iridoids, 21, 32, 49, 50, 417
Iris versicolor, 25, 88, 132, 190, 242, 274, 330, 379, 457, 466, 469
Irish broom, 465
Iron, 13, 27, 417
Ishwarane, 131, 443
Iso-amyl alcohol, 145, 443
Iso-butanol, 113, 443
Iso-butyric acid, 142, 443
Iso-butyric ester of phlorol, 115, 443
Iso-calamendiol, 137, 443
Iso-citric, 114, 443
Iso-eugenol, 133, 139, 443
Iso-lactone, 132, 443
Iso-menthone, 136, 139, 443
Iso-ocobullenone, 417, 443
Iso-phytol, 132, 443
Iso-pimpinellin, 131, 443
Iso-pinocamphone, 10, 132, 417, 443
Iso-pulegol, 127, 443
Iso-quercitin, 142-143, 443
Iso-ricinoleic acid, 141, 443
Iso-salicin, 128, 443
Iso-tridecane, 131, 443
Iso-valeraldehyde, 118, 443
Iso-valerate, 126, 151, 443
Iso-valeric aldehyde, 130, 136, 443
Isobetanine, 35, 417
Isoboldine, 42, 417
Isobutylamides, 417
Isocaproic, 5
Isoferulic acid, 13, 417
Isoflavones, 8, 13, 30
Isofraxin, 18, 417
Isoginkgetin, 21, 417
Isomenthone, 30, 136, 139
Isophthalic acids, 417
Isopinocamphone, 132, 417
Isoprebetanine, 417
Isopulegone, 31, 417
Isoquercitrin, 25, 417
Isoquinoline alkaloids, 25, 35, 42, 417
Isorhamnetin, 18, 417
Isovaltrate, 417

Jacoline, 43, 417
Japanese apricot, 38, 54, 98, 143, 166, 200, 223, 249, 283, 345, 387, 411-413, 418, 419, 421, 428, 434, 436, 440-443, 445, 449
Japanese parsley, 16, 80, 123, 181, 236, 267, 318, 373, 431, 432, 438, 440, 441, 445, 446, 448, 465
Jasmine, 25, 58, 88, 132, 159, 190, 218, 242, 274, 331, 379, 407, 417, 427, 434, 438, 441, 443-446, 449, 451, 465, 470
Jasminum auriculatum, 470
Jasminum grandiflorum, 25, 53, 88, 132, 242, 274, 331, 379, 470
Jerusalem oak, 465
Jiang huang, 16, 81, 124, 236, 268, 318, 413, 423, 431, 433, 434, 436, 441, 448, 450, 451, 453, 454, 465
Jojoba, 44, 103, 148, 206, 253, 288, 353, 391, 409, 415, 416, 419, 422, 444, 465
Jojoba oil, 148, 444
Juglandin, 26, 417
Jug*lans nigra*, 156, 457
Juglone, 26, 88, 417
Juniper, 26, 88, 132, 190, 242, 274, 331, 379, 410, 413, 417, 418, 422, 423, 426, 431, 433, 435-440, 444, 446, 447, 450-452, 461, 463, 465
Juniperin, 26, 417
Juniperus communis, 26, 88, 132, 190, 242, 274, 331, 379, 461, 463, 465

Kaempferol, 417

Kava kava, 436, 439, 444, 446, 465
Kava lactones, 36, 417
Kavain, 141, 444
Kawine, 36, 418
Kelp, 21, 26, 65, 85, 89, 129, 133, 165, 186, 190, 222, 239, 242, 271, 274, 325, 331, 376, 379, 416, 417, 422, 451, 465
Kelpware, 465
Ketone carvone, 444
Ketones, 41, 131, 145, 419, 444, 446
Key flower, 465
King's clover, 465
Kudzu, 38, 98, 143, 201, 249, 283, 346, 387, 413, 417, 423-425, 445, 446, 465
Kumatakenin, 22, 418

L-borneol, 120, 444
L-camphene, 141, 444
L-camphor, 120, 444
L-carvone, 136, 444
L-ephedrine, 19, 418
Labrador tea, 27, 89, 133, 191, 242, 275, 332, 380, 409, 436, 437, 444, 446, 449-452, 465
Laburnine, 12, 418
Lactate, 113, 444
Lactones, 3, 7, 20, 36, 51, 59, 112, 122, 132, 142, 160, 417, 444, 451
Lactucin, 13, 418
Ladder to heaven, 465
Ladies delight, 465
Lady's mantle, 4, 69, 113, 169, 228, 258, 303, 364, 414, 424, 426, 465
Lady's slipper, 17, 81, 125, 182, 237, 413, 415, 454, 465
Laetrile, 38, 418
Lamb's quarter, 13, 77, 120, 178, 234, 265, 313, 370, 413, 420, 425, 426, 428, 434, 438, 452, 465
Laminaria digitata, 26, 89, 133, 190, 242, 274, 331, 379, 465, 472
Laminaria saccharina, 465, 472
Lamium album, 26, 89, 133, 191, 242, 275, 332, 380, 457, 461, 468, 471
Lanatoside, 17, 418
Lanosterol, 122, 444
Lanrus nobilis, 456
Lantern (Chinese), 35, 96, 140, 198, 247, 281, 341, 385, 408, 415, 421, 425, 428, 431, 439, 459
Larrea tridentata, 26, 89, 133, 191, 242, 275, 332, 380, 459
Laurel, 26, 48, 89, 106, 133, 151, 191, 210, 242, 255, 275, 290, 332, 359, 380, 394, 407, 412, 414, 418, 419, 422, 423, 425-427, 431, 433, 434, 437, 440, 445, 447, 450, 453, 456, 458, 465
Lauric acid, 31, 62, 131, 151, 162, 444
Laurus nobilis, 26, 89, 133, 191, 242, 275, 332, 380, 456, 465, 472
Lavandula angustifolia, 26, 89, 133, 191, 242, 275, 332, 380, 465
Lavandula officinalis, 466
Lavandula spica, 26, 89, 133, 191, 332, 466
Lavender, 26, 89, 133, 163, 191, 221, 242, 275, 332, 380, 410, 411, 413, 415, 418, 426, 427, 436, 437, 444, 445, 447,.465, 466
Lecithin, 22, 418
Ledene, 133, 444
Ledol, 133, 444
Ledum latifolium, 27, 89, 133, 191, 242, 275, 332, 380, 465
Leine, 418
Lemna minor, 27, 89, 133, 191, 242, 275, 332, 380, 461
Lemon, 14, 17, 30, 34, 46, 78, 81, 92, 95, 105, 121, 124, 136, 139, 150, 179, 182, 194, 195, 197, 209, 235, 237, 244, 247, 254, 266, 268, 277, 280, 289, 315, 319, 336, 340, 357, 371, 374, 382, 385, 393, 410-412, 415, 418, 420, 422, 424, 426-428, 432,

Lemon *(continued)*, 433, 435-438, 440-448, 450, 452, 466, 473
Lemon balm, 30, 92, 136, 194, 244, 277, 336, 382, 412, 415, 418, 422, 426, 427, 438, 442, 445, 446, 450, 466
Lemon geranium, 34, 95, 139, 197, 247, 280, 340, 385, 433, 435, 436, 438, 442-444, 448, 466
Lemon grass, 17, 81, 124, 182, 237, 268, 319, 374, 412, 432, 433, 435, 438, 440-442, 444, 446, 447, 466
Lemon thyme, 46, 105, 150, 209, 254, 289, 357, 393, 411, 424, 427, 437, 443, 450, 452, 466
Lent lily, 466
Leontopodium, 466
Leonuride, 27, 418
Leonurin, 27
Leonurus cardiaca, 27, 89, 133, 191, 242, 275, 332, 380, 466, 467, 473
Leopard's bane, 466
Lepalox, 444
Lesser periwinkle, 466
Levant storax, 27, 28, 418
Levisticum officinale, 27, 89, 134, 158, 192, 243, 275, 333, 380, 400, 403, 466
Levulin, 46, 418
Lian qiao, 20, 185, 239, 271, 324, 376, 466
Licorice (American), 455
Licorice (Chinese), 22, 86, 130, 187, 240, 272, 326, 377, 414, 417, 424, 453, 459
Lignans, 7, 11, 18, 37, 43, 45, 50, 148, 418
Ligostilides, 134, 444
Ligustrin, 45, 418
Ligustrum fruit, 27, 90, 134, 192, 243, 275, 333, 380, 421, 426, 439, 447, 451, 466
Ligustrum lucidium, 90, 192, 243, 275, 333, 380, 466
Ligustrum vulgare, 469
Lilac, 45, 54, 104, 149, 156, 208, 215, 253, 289, 355, 380, 392, 418, 441, 466
Lilacin, 45, 418
Lily cenvalle, 466
Lily of the valley, 15, 31, 80, 92, 122, 136, 180, 194, 235, 245, 267, 278, 316, 337, 372, 382, 407, 421, 431, 432, 449, 454, 466
Lime, 14, 47, 78, 105, 121, 150, 179, 209, 235, 254, 266, 290, 315, 357, 371, 393, 410-412, 414, 418, 420, 422, 426-428, 433, 438, 441, 442, 444, 448, 466, 471
Limonene, 6, 11, 14, 23, 25, 26, 30, 112, 114, 115, 117-122, 124, 126-129, 132, 136, 137, 140, 141, 144, 145, 148, 149, 153, 418, 439, 444
Limonic acid, 418
Linalool, 3, 8, 10, 11, 14, 15, 26, 30, 32, 42, 112, 114, 116, 118, 121, 123-125, 127, 128, 130, 132, 133, 136-139, 141-143, 145, 146, 153, 418, 444
Linalyl acetate, 14, 26, 121, 125, 132, 133, 146, 418, 445
Linamarin, 418
Linden, 47, 105, 150, 209, 212, 254, 290, 357, 393, 411, 414, 420, 422, 426, 427, 441, 466
Lingberry, 466
Linoleic acid, 4, 11, 13, 27, 31, 32, 116, 119, 123, 125, 133, 135, 138, 141, 144-146, 151, 418, 433, 435, 445
Linolenic acid, 27, 32, 45, 60, 94, 117, 126, 128, 135, 138, 141, 145, 150, 418, 442, 445
Linseed, 27, 134, 466
Linum usitatissimum, 27, 90, 134, 192, 243, 276, 333, 380, 462, 466
Lion's ear, 466
Lion's tail, 466
Liposterolic, 38, 44, 418
Liquidambar orientalis, 468
Liquidambar styraciflua, 27, 90, 134, 243, 276, 380, 472
Liverlily, 466

Lobaria pulmonaria, 466
Lobelanidine, 28, 418
Lobelia, 28, 90, 134, 192, 243, 276, 333, 381, 408, 411, 418, 457, 466
Lobelia inflata, 28, 90, 134, 192, 243, 276, 333, 381, 466
Lobelia siphilitica, 28, 90, 134, 192, 243, 276, 333, 381, 457
Lobelidiol, 28, 418
Lobeline, 28, 418
Lodgepole pine, 36, 96, 141, 199, 248, 282, 342, 386, 410, 415, 419, 423, 425, 428, 431, 432, 435, 443, 444, 449, 466
Lomatium, 28, 65, 90, 134, 192, 222, 243, 276, 334, 381, 414, 416, 418, 451-453, 466
Lomatium dissectum, 28, 65, 90, 134, 192, 222, 243, 276, 334, 381, 466
Longiborneal, 140, 445
Longifolene, 140, 445
Longipinene, 119, 445
Lonicera caerulea, 28, 90, 134, 192, 243, 276, 334, 381, 464
Lonicera caprifolium, 461, 464
Loosestrife, 29, 91, 135, 193, 244, 276, 277, 334, 381, 410, 414, 424-427, 429, 440-443, 445, 453, 454, 463, 466, 469, 471
Lovage, 27, 89, 134, 158, 192, 243, 275, 333, 380, 400, 408, 410, 413, 416, 423, 426, 436, 438, 439, 444, 452, 453, 466
Love vine, 466
Low speed well, 466
Lucerne, 466, 472
Lungwort, 28, 90, 134, 192, 243, 276, 333, 381, 408, 422, 427, 454, 466
Lupeol, 138, 445
Lupulone, 24, 418
Luteolin, 4, 17, 29, 32, 135, 418, 445
Lycium, 28, 91, 135, 154, 193, 243, 276, 298, 334, 381, 410-412, 421, 423, 425, 428, 445, 459, 466, 474
Lycium barbarum, 28, 91, 135, 193, 243, 276, 298, 334, 381, 466
Lycium chinense, 459
Lycium pallidum, 474
Lycophilized extract, 43, 419
Lycopine, 142, 143, 445
Lycopodine, 29, 419
Lycopodium clavatum, 29, 91, 135, 193, 243, 276, 334, 381, 459, 464, 472
Lysimachia vulgaris, 29, 91, 135, 193, 244, 276, 334, 381, 463, 466, 474
Lysine, 27, 44, 419
Lythrum salicaria, 29, 91, 135, 193, 244, 277, 334, 381, 469-471

Ma huang, 19, 83, 127, 184, 238, 269, 322, 375, 408, 414, 424, 425, 453, 466
Macropiper excelsum, 29, 91, 135, 193, 244, 277, 335, 381, 465
Mad weed, 466
Madasiatic acid, 419
Madecassic acid, 419
Magnoflorine, 29, 419
Mahonia aquifolium, 9, 29, 60, 73, 91, 116, 135, 173, 193, 231, 244, 262, 277, 308, 335, 367, 381, 468
Maiden hair tree, 466
Malic acid, 11, 38, 51, 121, 148, 419, 445
Mallol, 36, 419
Malonic, 3, 114, 419, 445
Maltol, 34, 419
Mandrake, 466
Marian thistle, 466
Marigold, 10, 46, 74, 104, 117, 149, 175, 208, 232, 254, 263, 289, 309, 356, 368, 392, 411, 412, 415, 416, 420, 421, 423, 424, 426, 427, 434, 436, 438, 440, 442, 444, 446, 447, 466, 467
Marjoram, 33, 94, 138, 196, 246, 279, 339, 384, 411, 415, 423, 425-428, 435, 437, 439, 440, 445, 449, 450, 452,

Marjoram *(continued)*, 467, 472, 473
Marrubenol, 29, 419
Marrubiin, 29, 419
Marrubium, 29, 91, 135, 193, 244, 277, 335, 381, 464, 467
Marrubium vulgare, 29, 91, 135, 193, 244, 277, 335, 381, 464, 467
Marsh parsley, 467
Marshmallow, 5, 70, 113, 170, 229, 259, 303, 365, 409, 415, 419, 421, 425, 467
Massoilactone, 24, 419, 445
Mat, absolute, 132, 445
Matricaria chamomilla, 29, 91, 135, 193, 244, 277, 335, 382, 459, 465
Matricaria recutita, 463
Matteucia struthiopteris, 91, 135, 244, 468
May lily, 467
Mayapple, 97, 141, 199, 248, 282, 343, 386, 408, 410, 414, 418, 420, 422, 423, 450, 467
Maybush, 467
Maypop, 467
Maysin, 51, 419
Meadow eyebright, 467
Meadow saffron, 14, 79, 122, 235, 266, 316, 372, 412, 414
Meadow sage, 467
Meadow sweet, 20, 84, 128, 185, 239, 271, 323, 376, 408, 415, 416, 424, 426, 427, 442, 443, 450, 451, 467
Mealberry, 467
Medicago sativa, 458, 466, 469
Medicinal tea tree, 30, 92, 136, 194, 244, 277, 336, 382, 412, 422, 426, 435, 439, 442, 451, 452, 467
Melaleuca, 30, 92, 136, 194, 244, 277, 336, 382, 412, 422, 426, 435, 439, 442, 451, 452, 467
Melaleuca alternifolia, 30, 92, 136, 194, 244, 277, 336, 382, 467
Melaleuca cajuputi, 30
Melilot, 30, 92, 136, 194, 244, 277, 336, 382, 412-414, 423, 425, 437, 438, 441, 445, 448, 467, 474
Melilotic acid, 136, 445
Melilotus arvensis, 467
Melilotus officinalis, 30, 92, 136, 244, 277, 336, 382, 465, 472, 474
Melissa, 30, 92, 136, 194, 244, 277, 336, 382, 456, 466, 467
Mentha spicata, 471
Mentha x piperita, 30, 92, 136, 194, 469
Menthol, 3, 30, 31, 115, 136, 419, 445
Menthone, 10, 30, 117, 136, 419, 443, 445
Menthy-2-octane, 131, 445
Mesaconitine, 3, 68, 419
Methanol, 113, 445
Methylalliin, 113, 446
Methyl acetate, 136, 446
Methyl anthranilate, 152, 446
Methyl benzoate, 118, 446
Methyl chavicol, 8, 32, 112, 115, 128, 140, 148, 161, 219, 296, 446
Methyl cinnamate, 446
Methyl ethers, 136, 446
Methyl eugenol, 112, 121, 144, 149, 446
Methyl heptenone, 446
Methyl ionone, 124, 446
Methyl isoeugenol, 446
Methyl jasmonate, 132, 446
Methyl ketones, 41, 419, 446
Methyl nonyl ketones, 145, 446
Methyl palmitate, 116, 446
Methyl phenyl esters, 446
Methyl salicylaldehyde, 121, 446
Methyl salicylate, 37, 118, 128, 129, 138, 142, 149, 150, 419, 446
Methyl sticin, 36, 419
Methyl xanthines, 10, 419
Mexican mint marigold, 46, 104, 149, 208, 254, 289, 356, 392, 415, 416, 421, 423, 426, 434, 440, 446, 467
Mexican tea, 467
Milfoil, 467

Milk thistle, 44, 61, 63, 64, 103, 148, 161, 164, 206, 253, 288, 296, 353, 391, 414, 425, 434, 440, 445, 448, 467
Milkwort, 467
Milsuba, 16, 467
Mistletoe (American), 35, 95, 140, 198, 247, 281, 341, 385, 410, 421, 428, 467
Mistletoe (European), 50, 107, 152, 411, 415, 418, 423
Mitsuba, 80, 123, 181, 236, 267, 318, 373, 431, 432, 438, 440, 441, 445, 446, 448, 467
Mitsubene, 123, 446
Moccasin flower, 467
Monarda odoratissima, 31, 92, 136, 194, 245, 278, 336, 382, 464
Monarda punctata, 464
Monkshood, 3, 63, 68, 112, 168, 221, 228, 258, 301, 364, 407, 409, 410, 417, 419, 420, 454, 467
Mono-acids, 140, 446
Monoacylglycerols, 152, 446
Monocarboxylic acid, 133, 446
Monoenes, 139, 446
Monogalactosylglycerol, 446
Monoterpene, 35, 51, 118, 122, 141, 147, 441, 446, 447
Monoterpene alcohols, 446
Monoterpene aldehydes, 118, 447
Monoterpene hydrocarbons, 35, 147, 447
Monoterpenoid, 33, 34, 56, 133, 151, 158, 164, 222, 447
Monoterpenols, 149, 447
Monotropein, 7, 419
Moose elm, 467
Mormon tea, 19, 83, 127, 184, 238, 269, 322, 375, 422, 467
Morphine, 34, 197, 419
Mortification root, 467
Mother's heart, 467
Motherwort, 27, 89, 133, 191, 242, 275, 332, 380, 411, 412, 418, 423-425, 445, 448, 451, 467
Mountain ash, 44, 103, 148, 207, 253, 288, 354, 392, 421, 426, 434, 441, 443, 445, 453, 467
Mountain cranberry, 106, 211, 255, 291, 359, 394, 408, 453, 467
Mountain fly honeysuckle, 467
Mountain holly, 467
Mountain mahogany, 467
Mountain mint, 10, 74, 117, 175, 232, 263, 309, 368, 415, 417, 419, 450, 467
Mountain tea, 457, 467
Mountain tobacco, 467
Mu dan pi, 34, 94, 196, 246, 339, 384, 407, 409, 410, 415, 419-421; 467
Mucilage, 4, 5, 9, 10, 13, 14, 19, 22, 24-27, 33, 36, 43, 45, 47-50, 115, 150, 419, 447
Mugwort, 8, 72, 116, 172, 230, 261, 306, 367, 412, 417, 423, 426-428, 437-439, 450-452, 461, 468
Mullein, 49, 106, 151, 211, 256, 291, 360, 394, 419, 424, 454, 468
Murolene, 153, 447
Mycene, 121, 140, 447
Myoinositol, 33, 420
Myosotis scorpioides, 31, 92, 136, 194, 245, 466
Myrcene, 13, 30, 115, 117, 121, 124, 126, 128, 129, 132, 139, 140, 145, 146, 420, 435
Myrica penxylvanica, 31, 92, 137, 195, 245, 278, 337, 383, 456
Myricylalchol, 420
Myristic, 6, 24, 29, 114, 130, 138, 148, 420, 447
Myristic acid, 24, 114, 420
Myristica fragrans, 31, 93, 137, 195, 245, 278, 337, 383, 403, 468
Myristicin, 31, 35, 137, 139, 147, 420, 447
Myristoleic acid, 114, 447
Myrobalan, 35, 96, 140, 198, 247, 281, 341, 385, 414, 421, 428, 437, 439, 444, 468

Myrosin, 47, 420
Myrrh, 15, 79, 122, 180, 235, 267, 316, 372, 427, 441, 442, 468
Myrrica cerifera, 31
Myrtenal, 127, 447
Myrtle, 31, 93, 137, 195, 245, 278, 337, 383, 415, 433, 435-437, 439, 444, 447, 468
Myrtocyan, 420
Myrtus communis, 31, 93, 137, 195, 245, 278, 337, 383, 468

n-Alkanes, 138, 447
n-Amyl alcohol, 130, 447
n-Butyldenephthalide, 6, 420
n-Butyric acid, 130, 447
n-Capric acid, 130, 447
n-Caprylic acid, 130, 447
n-Decane, 118, 447
n-Docosanol, 142, 447
n-Heptadecane, 153, 447
n-Hexanal, 142, 447
n-Methyl anabasine, 43, 420
n-Nonadecane, 153, 447
n-Tetracosanol, 39, 420, 447
n-Trans-coumaroyltyramine, 47, 420
n-Trans-feruloyltyramine, 47, 420
n-Undecylic acid, 116, 447
Naked ladies, 468
Napelline, 3, 420
Naphthalene, 149, 420, 447
Naphthalene glycosides, 420
Narcissine, 137, 447
Narcissus pseudonarcissus, 59, 93, 137, 245, 278, 383, 460, 466
Nasturtium, 31, 47, 93, 105, 137, 150, 195, 209, 245, 255, 278, 290, 337, 358, 383, 393, 415, 420, 421, 425, 428, 434, 468, 473
Nasturtium officinale, 31, 93, 137, 195, 245, 278, 337, 383, 468, 473
Neoherculin, 51, 420
Neoisopulegol, 127, 447
Neolignan ketone, 55, 420
Neoline, 3, 420
Neomenthol, 136, 447
Neoruscogenin, 41, 420
Nepeta cataria, 32, 93, 137, 195, 245, 279, 337, 383, 458
Neral, 124, 136, 144, 153, 447
Nerol, 115, 117, 121, 124, 130, 133, 139, 144, 150, 153, 447
Nerolidol, 137, 138, 156, 447
Neryl acetate, 130, 447
Nettle, 26, 48, 52, 89, 106, 133, 151, 154, 191, 210, 242, 255, 275, 291, 332, 359, 380, 394, 407, 409, 412, 415, 416, 422-426, 428, 433, 438, 441-443, 445, 448, 450, 451, 457, 461, 468, 472
New England pine, 468
Nicotine, 19, 192, 420
Nicotinic acid, 17, 420
Night willow herb, 468
Nitribine, 144, 447
Nobilin, 420
Nonadecane, 137, 153, 447, 448
Nonanal, 128, 448
Nor-alpha-trans-bergamotenone, 448
Nor-lapachol, 128, 448
Norboldine, 420
Nordihydroguaiaretic acid, 420
Norpinene, 144, 448
Northern pine, 468
Northern prickly ash, 468
Nostoclide, 34, 66, 420
Nupharine, 32, 420
Nutmeg, 31, 93, 137, 195, 245, 278, 337, 383, 397, 403, 420, 423, 433, 435, 436, 438, 440, 441, 444, 447-450, 452, 468
Nymphaea alba, 32, 93, 137, 195, 245, 279, 338, 383, 473
Nymphaeine, 32, 420

o-Cresol, 448

Oak (white), 468
Oat, 8, 54, 73, 116, 173, 231, 261, 293, 307, 367, 411, 422, 424, 428, 433, 435, 448, 468
Ocimene, 114, 115, 126, 128, 129, 132, 146, 433, 435, 438, 440, 448, 453
Ocimum basilicum, 32, 93, 137, 195, 245, 279, 338, 383, 456
Ocotea bullata, 32, 55, 93, 137, 195, 245, 279, 338, 383, 472
Octadecatetraenic acid, 420
Octadecatrienoic acid, 138, 420, 432, 438
Octanol, 118, 131, 153, 432, 448
Octyl alcohol, 131, 448
Octyl esters, 135, 448
Oenothera biennis, 32, 60, 94, 138, 158, 196, 217, 245, 279, 293, 338, 383, 462, 468
Old man, 468
Olea europaea, 32, 196, 279, 339, 383, 468
Oleanic acid, 13, 420
Oleasterol, 32, 420
Olefinic terpenes, 145, 448
Oleic acid, 12, 27, 31, 116, 123, 135, 138, 141, 145-147, 150, 151, 420, 448
Oleo-resin, 3, 420
Oleoropine, 32, 421
Oligomeric polyphenols, 421
Oligopeptides, 38, 55, 421
Olive, 18, 32, 83, 92, 94, 126, 138, 154, 160, 183, 196, 218, 238, 246, 269, 279, 321, 326, 339, 375, 383, 410, 411, 414, 417, 418, 420, 421, 432, 443, 446, 448, 452, 453, 468, 470
Onion, 4, 69, 113, 169, 228, 259, 295, 303, 364, 389, 408, 411, 425, 438-440, 446, 468
Oplopanax horridus, 18, 33, 65, 83, 94, 126, 138, 165, 183, 196, 238, 246, 269, 279, 321, 339, 374, 384, 461
Opuntia fragilis, 33, 94, 138, 196, 246, 279, 339, 384, 469
Oregano, 33, 94, 138, 196, 246, 279, 339, 384, 415, 423, 425-427, 435, 437, 439, 440, 449, 452, 468
Oregon grape, 9, 29, 73, 91, 116, 135, 173, 193, 231, 244, 262, 277, 308, 367, 440, 468
Oriental sweet gum, 28, 90, 134, 192, 243, 276, 333, 380, 418, 434, 438, 440, 449, 452, 453, 468
Origanum majorana, 33, 94, 138, 196, 246, 279, 339, 384, 472
Origanum vulgare, 33, 94, 138, 196, 246, 279, 339, 384, 467, 468, 473
Origanum vulgare subsp. *hirtum*, 33, 94, 138, 196, 246, 279, 339, 384, 467, 468
Oripavine, 34, 421
Orthocoumaric acid, 136, 448
Ostrich fern, 30, 91, 135, 194, 244, 277, 335, 382, 409-411, 414, 421, 422, 425, 427, 436, 443, 448, 449, 468
Oxalic acid, 47, 421
Oxalis acetosella, 471, 473, 474
Oxyberberine, 29, 421
Oxygenated monoterpenes, 147
Oxygenated sesquiterpenes, 147
Oxytocics, 31, 421

p-Anisic acid, 140, 448
p-Coumaric, 47, 421
p-Cresyl, 118, 448
p-Cymene, 121, 124, 128, 153, 448
p-Hydroxybenzaldehyde, 138, 448
p-Hydroxybenzoic, 30, 421
p-Hydroxycinnamic acid methyl ester, 126, 448
p-Mentha-1,3,8-triene, 448
Pacific myrtle, 468
Paclitaxel, 46, 421
Paeonia lactiflora, 33, 94, 138, 161, 196, 246, 280, 339, 384, 456, 473
Paeonia officinalis, 462
Paeonia suffruiticosa, 467, 473
Paeonine, 421

Paeonol, 33, 138, 421, 448
Paliloleic, 6, 114, 448
Palmarosa, 17, 237, 268, 319, 374, 412, 468
Palmitic acid, 3, 30, 49, 116, 123, 131, 135, 142, 145-147, 448
Palmitoleic, 130, 135, 138, 449
Panax ginseng, 34, 94, 139, 180, 196, 246, 280, 340, 384, 463
Panax notoginseng, 463
Panax quinquefolium, 61, 161, 463
Panaxosides, 421
Pansy, 50, 107, 152, 212, 256, 292, 361, 395, 419, 424, 426, 428, 431, 432, 449, 454, 468
Papain, 11, 421
Papaver bracteatum, 34, 95, 139, 246, 465
Papaver rhoens, 460, 469, 470
Papaya, 11, 75, 119, 176, 233, 264, 311, 369, 411, 412, 421, 434, 445, 448, 451, 468
Paraffin, 123, 144, 436, 449
Pariwinkle, 468
Parsley, 16, 24, 35, 80, 87, 95, 123, 131, 139, 181, 188, 197, 236, 241, 247, 267, 273, 281, 318, 328, 341, 373, 378, 385, 409, 413, 415, 416, 420-423, 425, 431, 432, 434, 436, 438, 440, 441, 443-449, 451, 452, 460, 465, 467, 468, 470
Passiflora incarnata, 34, 95, 139, 156, 197, 246, 280, 340, 384, 455, 467, 468
Passion flower, 34, 95, 139, 197, 246, 280, 340, 384, 408, 413, 414, 417, 419, 431, 433-435, 437, 441, 443, 445, 448, 449, 453, 468
Passion vine, 468
Patience dock, 468
Pectin, 5, 18, 21, 39-41, 46, 50, 421
Pedunculagin, 421
Pelargonic acid, 449
Pelargunium capitatum, 470, 473
Pelargunium graveolens, 466
Peltigera canina, 34, 66, 95, 139, 197, 247, 280, 340, 385, 461, 464
Pennyroyal mint, 468
Penta-(1,8z)-diene, 126, 449
Pentacyclic oxindole alkaloids, 48, 66, 421
Pentadeca-8-en-2-one, 126, 449
Pentagallotannin, 33, 421
Pentagalloyl glucoside, 421
Pentane, 421, 449
Pentoses, 421
Pepper (cayenne), 468
Pepper (chilli), 75, 118, 176, 232, 263, 311, 369, 411, 428, 436, 453, 468.
Pepper (English), 468
Pepper (sweet), 468
Peppermint, 30, 92, 136, 194, 245, 278, 294, 297, 336, 382, 404, 419, 432, 439, 443-445, 449, 452, 469
Peregrinine, 33, 421
Periwinkle, 12, 50, 76, 107, 110, 119, 152, 177, 212, 223, 233, 256, 264, 292, 312, 361, 369, 395, 407, 408, 413, 420, 421, 423, 424, 426, 428, 432, 451, 466, 469
Perlolyrin, 14, 421
Persicariol, 142, 449
Petroselaidic, 6, 114, 449
Petroselinum crispum, 16, 35, 95, 139, 197, 236, 247, 281, 318, 341, 373, 385, 468
Phataris canariensis, 35, 95, 140, 197, 247, 281, 341, 385, 458
Phelandrine, 8, 421
Phellodendron chinensis, 35, 95, 140, 247, 281, 385, 459, 464
Phenethyl, 113, 449
Phenolic acids, 37, 421
Phenols, 21, 31, 37, 43, 117, 118, 439, 449
Phenyl ethanol, 144, 449
Phenylacetic acid, 132, 449
Phenylacetic aldehyde, 449
Phenylethyl isothiocynate, 449
Phenylpropanes, 112, 449

Phenylpropionic acid, 144, 449
Phenylpropyl alcohol, 134, 449
Phenylpropyl cinnamate, 27, 28, 421
Phenylroparnoids, 18, 421
Philanthropos, 469
Phoradendron flavescens, 463, 467
Phoradendron leucarpum, 35, 95, 140, 152, 247, 256, 281, 341, 385, 467
Phoradendron serotinum, 35
Phoratoxin, 35, 421
Phosphatidylcholine, 152, 449
Phospholipids, 148, 421, 449
Phthalides, 27, 35, 114, 134, 408, 421
Phenylethyl alcohol, 139, 143, 449
Phyllandrene, 6, 14, 36, 114, 117-119, 121, 124, 126, 128, 136, 137, 139, 145, 147, 148, 433, 435, 439
Phyllanthus emblica, 35, 57, 63, 96, 140, 198, 247, 281, 341, 385, 461, 468
Physalin, 28, 35
Physalis alkekengi, 35, 96, 140, 198, 247, 281, 341, 385, 457, 459, 463, 472
Physcion, 41, 422
Phytoene, 16, 422
Phytofluene, 422
Phytol, 132, 139, 443, 449
Phytol ester, 449
Phytolacca americana, 35, 96, 140, 198, 247, 281, 342, 385, 469
Phytosterol, 25, 422
Picea ables, 471
Picea glauca, 471
Picea mariana, 35, 96, 140, 198, 247, 281, 342, 386, 471
Picrosalvin, 40, 422
Pilewort, 15, 39, 79, 99, 122, 143, 180, 201, 235, 250, 266, 284, 316, 347, 372, 388, 408, 424, 426, 433, 435, 437, 439-442, 448, 469
Pimpinella anisum, 36, 96, 140, 157, 198, 216, 248, 282, 342, 386, 455, 472
Pimpinellin, 131, 443, 449
Pinckly ash, 469
Pineapple, 5, 70, 113, 170, 229, 259, 304, 365, 407, 410, 428, 436, 439, 440, 443-445, 449, 450, 469
Pinecamphene, 25, 422
Pinenes, 8, 36, 422
Pinocarvone, 127, 151, 450
Pinus albicaulis, 473
Pinus contorta, 466
Pinus mugo var. Pumilio, 248
Pinus palustris, 471
Pinus strobus, 468, 473
Piper methysticum, 29, 36, 91, 97, 135, 141, 193, 199, 244, 248, 277, 282, 335, 343, 381, 386, 465
Piper nigrum, 36, 468
Piperidine, 28, 36, 141, 450
Piperine, 11, 36, 141, 422, 450
Piperitone, 136, 450
Plantago asiatica, 36, 97, 141, 199, 248, 282, 343, 386, 469
Plantago lanceolata, 36, 469
Plantago major, 469
Plantago psyllium, 469
Plantains, 97, 141, 248, 432, 434, 436, 437, 441-445, 450-452, 469
Planteose trisaccharides, 141, 450
Plastoquinones, 11, 422
Podophyllin, 450
Podophyllum peltatum, 37, 97, 141, 199, 248, 282, 343, 386, 461, 464, 466, 467, 473
Podophyllum resin, 37, 422
Poison flag, 469
Pokeweed, 35, 96, 140, 198, 247, 281, 342, 385, 411, 417, 439, 446, 451, 469
Polyacetylenes, 17, 18, 422
Polygala senega, 37, 60, 64, 97, 142, 199, 248, 282, 297, 343, 386, 467, 470, 471
Polygalitol, 37, 422
Polygonum bistorta, 37, 97, 142, 200, 248, 283, 344, 386, 457, 468, 471
Polygonum multiflorum, 97, 142, 283, 457,

Polygonum multiflorum (continued), 462, 463
Polyine, 112, 450
Polymeric, 43, 422
Polypeptides, 10, 422
Polyphenolic acid, 7
Polyphenols, 10, 20, 24, 52, 214, 421, 422
Polysaccharides, 4, 14, 15, 18, 20, 21, 28, 34, 48, 422
Polyynes, 33, 422
Poorman's treacle, 469
Poplar, 37, 200, 249, 283, 344, 387, 389, 414, 416, 456, 457, 469
Popotillo, 469
Poppy, 19, 34, 84, 95, 127, 139, 184, 197, 238, 246, 270, 280, 294, 322, 340, 375, 384, 411-414, 419, 421, 422, 427, 445, 448, 451, 458, 460, 462, 465, 469, 470
Populene, 142, 450
Populus balsamifera, 37, 97, 142, 200, 249, 283, 344, 387, 456, 469
Populus candicans, 469
Populus multiflorum, 456
Populus nigra, 456, 457
Porphyrins, 30, 422
Potassium, 422
Pregeijerene, 145, 450
Prickly pear, 33, 94, 138, 196, 217, 246, 279, 339, 384, 411, 419, 424, 469
Priest's crown, 469
Prim, 469
Primeverin, 142
Primrose, 32, 37, 60, 94, 98, 138, 142, 158, 163, 196, 200, 217, 221, 245, 249, 279, 283, 338, 344, 383, 387, 414, 416, 418, 420, 424, 426, 433, 435, 438, 442, 445, 454, 462, 469
Primula veris, 283, 460, 462, 465
Primula vulgaris, 37, 98, 142, 200, 249, 283, 344, 387, 469
Privet, 27, 90, 134, 192, 243, 275, 333, 380, 421, 426, 447, 469
Proanthocyanidins, 16, 23, 62, 163, 422
Proazulenes, 120, 135, 450
Progesteron, 422
Propanal, 126, 136, 431, 450
Protein, 10, 22, 35, 36, 44, 47, 50, 52, 54, 65, 95, 107, 154, 196, 422
Protoalnulin, 5, 422
Protocanemonin, 143, 450
Protocatechuic, 30, 422
Protopine, 19, 422
Prunasin, 38, 422
Prunella vulgaris, 38, 98, 142, 200, 249, 283, 344, 387, 455, 464
Prunin, 38, 422
Prunus affiricana, 455
Prunus dulcis, 455
Prunus mume, 54, 55, 166, 223, 455
Prunus serotina, 457, 473
Prunus virginiana, 459
Pseudoephedrine, 19, 413, 422
Pseudohypericin, 25, 422
Psoralen, 24, 423
Psyllic acid, 423
Psyllium, 36, 97, 141, 199, 248, 282, 343, 386, 414, 418, 420, 421, 441, 445, 448, 469
Pueraria lobata, 38, 98, 143, 161, 201, 219, 249, 283, 346, 387, 465
Puerarin, 38, 423
Pulegone, 10, 23, 31, 92, 117, 130, 136, 423, 450
Pumpkin, 16, 81, 123, 181, 236, 267, 318, 373, 413, 414, 422, 423, 445, 448, 469
Puncture vine, 47, 105, 150, 209, 254, 290, 357, 393, 420, 424-427, 440, 469
Purging buckthorn, 99, 144, 440, 469
Purine, 10, 423
Purple fire top, 469
Purple loosestrife, 29, 91, 135, 193, 244, 277, 334, 381, 424, 425, 427, 429, 445, 453, 454, 469
Purple medic, 469
Purpurea glycosides, 423

Pussy willow, 41, 101, 145, 203, 251, 286, 350, 389, 415, 416, 424, 425, 427, 453, 469
Putin, 423
Pygeum, 39, 98, 143, 201, 249, 284, 346, 387, 414, 420, 425-427, 469
Pygeum africanum, 39, 98, 143, 201, 249, 284, 346, 387, 469
Pyrethrins, 13, 423
Pyrethrum, 13, 39, 77, 98, 120, 143, 166, 178, 201, 234, 249, 265, 284, 298, 314, 346, 370, 387, 412, 423, 434, 435, 437, 440, 444, 452, 469
Pyrobetulin, 117, 450
Pyrocatechol, 117, 450
Pyrogallol, 27, 423
Pyrrolizidine, 43, 45, 47, 53, 57, 60, 117, 161, 218, 293, 423, 450
Pyrrolizidine alkaloids, 45, 57, 60, 161, 218, 450
Pyruvate, 5, 423
Pyruvic, 114, 450

Qing hao, 7, 72, 115, 230, 407, 409, 428, 432, 434, 435, 469
Quackgrass, 469
Quebrachitol, 116, 450
Queen Anne's lace, 469
Queen of the meadow, 469
Quercimeritrin, 135, 450
Quercitin, 15, 16, 22-25, 48, 142, 423, 443, 450
Quercus alba, 99, 143, 201, 249, 284, 346, 388, 468, 472
Quickbean, 469
Quinate, 5, 423
Quinic acid, 144, 423, 450

Radish, 39, 99, 143, 201, 250, 284, 347, 388, 409, 416, 428, 454, 469
Rainbow weed, 470
Ranunculus occidentalis, 39, 99, 143, 249, 284, 388, 458
Raphanus sativus, 39, 99, 143, 201, 250, 284, 347, 388, 469
Raspberry, 41, 100, 145, 202, 251, 285, 349, 389, 399, 408, 415, 421, 426, 428, 441, 443, 445, 448, 452, 464, 470
Rattle snake root, 470
Rebaudiosides, 45, 423
Red clover, 470
Red poppy, 470
Red puccoon, 470
Red-veined dock, 470
Redroot, 470
Resin, 3, 7-9, 13, 15, 25, 26, 30, 33, 37, 42, 46, 50, 51, 111, 122, 146, 168, 174, 180, 199, 262, 267, 420, 422, 423
Resmarinic acid, 33, 423
Rhamnus cathartica, 458, 462
Rhamnus frangula, 455, 457, 463
Rhamnus purshiana, 458, 459
Rhein anthrones, 39, 423
Rheum officinal, 470
Rheum palmatum, 39, 99, 144, 201, 250, 284, 348, 388, 459
Rheum tanguticum, 470
Rhodiola rosea, 39, 66, 99, 144, 202, 250, 284, 348, 388, 470
Rhodioloside, 43, 423
Rhubarb, 39, 99, 144, 201, 250, 284, 348, 388, 407, 409, 411, 412, 414, 415, 423, 426, 434, 435, 437, 441, 449, 459, 470
Rhubarb (Chinese), 39, 99, 144, 201, 250, 284, 348, 388, 407, 411, 412, 414, 415, 423, 426, 434, 435, 437, 441, 449, 459, 470
Rhus radicans, 40, 100, 144, 202, 250, 285, 348, 388, 469
Rhus toxicodendron, 469, 470
Ribes lacustre, 463
Ribes nigrum, 40, 100, 144, 202, 250, 285, 348, 388, 457
Robertium macrorrhizum, 40, 100, 144, 250, 285, 388, 463

Roman chamomile, 470
Root of the Holy Ghost, 470
Rosa canina, 40, 100, 144, 202, 250, 285, 349, 389, 461, 470
Rosa damascena, 460, 470
Rose (Damask), 470
Rose (Dog), 470
Rose geranium, 34, 95, 139, 197, 247, 280, 340, 385, 433, 435, 436, 438, 442-444, 448, 470, 473
Roselle, 470
Rosemary, 40, 100, 145, 164, 202, 205, 222, 250, 285, 349, 389, 409-413, 423, 435-437, 439, 441, 444, 447-453, 470
Roseroot, 39, 40, 43, 99, 100, 103, 144, 147, 202, 205, 250, 252, 284, 287, 348, 352, 388, 391, 414, 420, 423, 424, 435, 439, 452, 454, 470
Rosmarinic acid, 40, 45, 423
Rosmarinus officinalis, 40, 100, 145, 164, 202, 222, 250, 285, 349, 389, 468, 470
Rowan tree, 470
Royal jasmine, 25, 242, 274, 331, 379, 407, 417, 427, 470
Rubus chamaemorus, 40, 100, 145, 159, 202, 250, 285, 295, 349, 389, 459
Rubus fruiticosus, 457
Rubus idaeus, 41, 100, 145, 202, 251, 285, 349, 389, 470
Rue, 41, 101, 145, 203, 251, 286, 350, 389, 409, 412, 415, 419, 423, 431, 437, 440, 442, 444, 446, 447, 449, 450, 463, 470
Rumex acetosa, 463
Rumex acetosella, 41, 100, 145, 203, 251, 285, 350, 389, 471
Rumex obtusifolius, 457, 458, 470
Ruscogenin, 41, 423
Ruscus aculeatus, 41, 100, 145, 203, 251, 285, 350, 389, 458
Russian olive, 18, 83, 126, 183, 238, 269, 321, 375, 410, 414, 417, 453, 470
Ruta graveolens, 41, 62, 101, 145, 203, 220, 251, 286, 350, 389, 460, 463, 464, 470
Rutaverine, 41, 423
Rutin, 8, 16, 20, 25, 42, 135, 423, 450
Rutoside, 41, 423

Sabinene, 48, 112, 115, 121, 124, 125, 129, 132, 133, 137, 138, 140, 141, 144, 145, 151, 152, 412, 427, 439, 450
Safflower, 11, 56, 75, 119, 176, 216, 233, 264, 311, 369, 411, 418, 422, 433, 440, 445, 470
Saffron, 14, 16, 79, 80, 122, 123, 181, 235, 236, 266, 267, 316, 317, 372, 373, 410, 412-414, 422, 437, 450, 456, 470
Safranal, 123, 450
Safrole, 31, 32, 42, 57, 86, 118, 121, 130, 137, 147, 423, 450
Sage, 41, 60, 101, 146, 203, 219, 251, 286, 296, 350, 389, 410-412, 415, 419, 421, 422, 424, 426, 427, 433, 435-437, 444, 445, 447-449, 459, 463, 467, 470
Sagebrush, 8, 72, 116, 172, 230, 261, 306, 367, 415, 421, 431, 447, 448, 456, 470
Salicarin, 29, 424
Salicin, 37, 41, 128, 424, 443, 451
Salicortin, 41, 424
Salicylaldehyde, 121, 451
Salicylates, 13, 20, 424
Salicylic acid, 10, 25, 40, 128, 424, 451
Salix alba, 41, 101, 145, 203, 251, 286, 350, 389, 473, 474
Salix discolour, 469
Salvia officinalis, 41, 101, 146, 203, 251, 286, 350, 389, 463, 467, 470
Salvia sclarea, 459
Salvin, 41, 424
Sambucus nigra, 41, 101, 146, 203, 251,

Sambucus nigra (continued), 286, 351, 389, 458, 461, 462
Sambucus racemosa, 461
Sandalwood, 42, 102, 146, 204, 251, 286, 351, 390, 413, 424, 433, 435, 436, 438, 440, 448, 451-453, 470
Sanguinaria canadensis, 42, 102, 146, 251, 286, 390, 457, 465, 470, 473
Sanguinarine, 42, 65, 424
Sanicle, 455, 457, 470
Sanicle (American), 455
Sanicula marilandica, 42, 102, 146, 204, 286, 351, 390, 455, 457, 470
Santalol, 102, 114, 451
Santalone, 146, 451
Santalum album, 42, 102, 146, 204, 251, 286, 351, 390, 470
Santene, 145, 146, 451
Sapogenin, 42, 424
Saponaria officinalis, 42, 57, 102, 146, 204, 217, 251, 286, 351, 390, 471
Saponin, 3, 8, 19, 22, 29, 32, 41, 51, 148, 424
Sarsaparilla (American), 455
Saskatoon, 5, 62, 70, 113, 162, 170, 229, 259, 297, 304, 365, 403, 408, 412, 415, 421, 423, 441, 470
Sassafras, 42, 57, 102, 147, 204, 252, 287, 351, 390, 410-412, 417, 419, 420, 423, 434, 436, 445, 447, 449, 450, 452, 455, 459, 470, 471
Sassafras albidum, 42, 102, 147, 204, 252, 287, 351, 390, 455, 459, 470, 471
Satin flower, 470
Sativene, 119, 451
Saturated acids, 140, 151, 451
Satureja hortensis, 42, 102, 147, 205, 252, 287, 390, 470
Satureja montana, 474
Savory, 42, 102, 147, 205, 252, 287, 352, 390, 411, 413, 418, 427, 437, 439, 442, 444, 452, 470, 474
Saw palmetto, 44, 52, 103, 148, 154, 163, 205, 221, 252, 287, 353, 391, 424-426, 435, 436, 444, 445, 448, 451, 452, 470
Scented fern, 470
Schisandra, 43, 102, 147, 205, 252, 287, 352, 390, 470, 474
Schisandra chinesis, 43, 474
Sciadopitysin, 21, 424
Scoparoside, 17, 424
Scopoletin, 11, 38, 424
Scordinins, 4, 424
Scouring rush, 470
Scutellaria baicalensis, 43, 66, 102, 147, 205, 223, 252, 287, 352, 390, 456, 465, 471
Scutellaria galericulata, 471
Scutellaria lateriflora, 457, 464, 466, 473
Scutellarin, 43, 147, 424, 451
Sea buckthorn, 24, 59, 61, 87, 131, 155, 161, 164, 189, 214, 219, 241, 273, 295, 329, 378, 379, 400, 401, 411, 414, 425, 428, 432, 440, 445, 448, 449, 452, 470
Sea parsley, 470
Seaweed absolute, 129, 451
Sedacrine, 43, 424
Sedacryptine, 43, 424
Sedanolide, 114, 451
Sedanonic acid, 134, 451
Sedinine, 43, 424
Sedum acre, 43, 102, 147, 205, 252, 287, 352, 391, 471, 472
Sedum rosea subsp. *integrifolium*, 40, 250, 348, 388
Selenium, 86, 424
Self heal, 344
Selinene, 114, 133, 149, 153, 433, 435, 451
Sellnadiene, 133, 451
Sempervivum tectorum, 43, 52, 53, 103, 147, 205, 214, 252, 287, 352, 391, 464, 465
Seneca snakeroot, 37, 97, 142, 199, 248, 282, 343, 386, 419, 421, 422, 424,

Seneca snakeroot *(continued)*, 425, 437, 443, 446-448, 471
Senecio vulgaris, 53, 293, 463, 464
Senecionine, 43, 424
Seneciphyline, 43, 424
Senna (Alexandra), 471
Senna (American), 471
Senna (Tinnevelly), 471
Senna alexandrina, 43, 103, 147, 205, 252, 287, 353, 391, 471
Sennosides, 11, 424
Serenoa repens, 44, 103, 148, 205, 252, 287, 353, 391, 470
Serotonin, 48, 424
Sesame, 44, 61, 103, 148, 206, 220, 253, 288, 353, 391, 410, 413, 415, 416, 418, 420, 422, 427, 428, 445, 448, 449, 451, 453, 471
Sesamum indicum, 44, 54, 61, 103, 148, 206, 220, 253, 288, 353, 391, 471
Sesamum orientale, 471
Sesquiterpene alcohol, 451
Sesquiterpene hydrocarbon, 146, 451
Sesquiterpene lactones, 3, 7, 20, 59, 122, 132, 142, 160, 451
Sesquiterpenes, 12, 45, 66, 112, 114, 115, 117, 118, 121, 130, 134, 139, 143, 147, 149, 152, 153, 166, 424, 451
Sesquiterpenoids, 17, 120, 451
Sesquiterphenol, 140, 451
Shamrock, 471
Sheep sorrel, 471
Shen yao, 18, 183, 269, 320, 424, 471
Shepherdia canadensis, 44, 103, 148, 206, 253, 288, 353, 391, 458
Shi chang pu, 3, 68, 112, 168, 228, 258, 301, 364, 407, 409, 418-421, 424, 432, 434-436, 444, 446, 449, 451, 471
Shogaols, 51, 424
Shyobunone, 112, 451
Siberian ginseng, 18, 57, 83, 127, 183, 217, 238, 269, 321, 375, 413, 414, 417, 418, 421, 423, 424, 471
Silibinin, 44, 63, 425
Silicates, 425
Silicic acid, 9, 19, 47, 425
Silver pine, 471
Silybum marianum, 44, 61, 64, 103, 148, 161, 164, 206, 253, 288, 296, 353, 391, 466, 467
Silymarin, 206, 425
Simmondsia chinensis, 44, 103, 148, 206, 253, 288, 353, 391, 465
Sinapine, 9, 425
Sinigrin, 7, 425
Sitosterol, 11-13, 18, 19, 27, 28, 30, 33, 38, 39, 43, 44, 47, 48, 116, 131, 138, 141, 142, 148, 410, 425, 435, 451
Skullcap, 43, 102, 147, 205, 252, 287, 352, 390, 409, 411, 414, 424, 425, 431, 432, 448, 451, 456, 457, 471, 473
Slack birch, 471
Slippery elm, 48, 106, 150, 210, 255, 290, 358, 394, 410, 411, 420, 421, 426, 471
Small age, 471
Small houseleek, 205, 252, 352, 391, 414, 420, 424, 453, 471
Small-flowered willow herb, 19, 83, 127, 184, 238, 270, 322, 375, 414, 415, 425, 471
Small-leaved lime, 105, 411, 422, 426, 427, 441, 471
Smelling stick, 471
Snakeroot, 37, 42, 97, 102, 142, 146, 199, 204, 248, 282, 286, 343, 351, 386, 390, 408, 411, 415, 419, 421, 422, 424, 425, 437, 443, 446-448, 454, 457, 458, 471
Snakeweed, 471
Snapping hazel, 471
Snowball tree, 471
Snowflake, 471
Soapwort, 42, 102, 146, 204, 251, 286, 351, 390, 423-425, 453, 471
Solidago odora, 457, 472

Solidago virgaurea, 44, 103, 148, 206, 253, 288, 354, 391, 463
Sorbitol, 28, 425
Sorbus aucuparia, 44, 103, 148, 207, 253, 288, 354, 392, 467, 469, 470
Sorrel, 41, 100, 145, 203, 251, 285, 350, 389, 408, 421, 422, 441, 450, 463, 471, 473, 474
Southern pitch pine, 141, 199, 282, 342, 386, 408, 410, 415, 419, 423, 425, 428, 431, 435, 443, 444, 449, 471
Soybean, 22, 86, 129, 187, 240, 272, 326, 377, 409, 411, 414, 417, 418, 422, 428, 451, 471
Soybean oil, 129, 451
Sparteine, 17, 425
Spathulenol, 132, 451
Spearmint, 30, 297, 404, 419, 471
Speedwell, 49, 106, 151, 211, 256, 291, 360, 394, 407, 409, 415, 426, 428, 471
Sphondin, 24, 425
Spiked loosestrife, 471
Spikenard, 6, 71, 114, 171, 230, 260, 305, 366, 413, 426, 434, 471
Spiracoside, 128, 451
Spiraein, 128, 451
Spiroether, 120, 451
Spotted thistle, 471
Sprondrin, 131, 451
Spruce (black), 471
Spruce (Norway), 471
Spruce (white), 471
Squawroot, 471
Stachydrine, 29, 30, 45, 425
Stachyose, 425
Stachys officinalis, 45, 104, 148, 207, 253, 288, 354, 392, 456
Stag's horn, 472
Star flower, 472
Star thistle, 12, 76, 119, 177, 233, 264, 313, 370, 411, 415, 429, 437, 472
Star weed, 472
Stave oak, 472
Stearic acid, 27, 31, 51, 425, 451
Stearopten, 144, 451
Stellaria, 45, 104, 149, 207, 253, 288, 355, 392, 459, 470, 472
Stellaria media, 45, 104, 149, 207, 253, 288, 355, 392, 459, 470, 472
Sterins, 14, 425
Sterols, 3, 4, 8, 13, 33, 41, 43, 47, 154, 425
Sterylglucoside, 152, 432, 452
Stevia, 45, 104, 149, 207, 253, 288, 355, 392, 410, 411, 422, 423, 425, 426, 433, 435, 447, 472
Stevia rebaudiana, 45, 104, 149, 253, 288, 355, 392, 472
Stevioside, 45, 425
Stickwort, 472
Stigmasterol, 30, 44, 48, 141, 425, 452
Stinging nettle, 48, 52, 106, 151, 154, 210, 255, 291, 359, 394, 407, 409, 412, 415, 416, 422-426, 428, 433, 442, 445, 448, 450, 451, 472
Stinking weed, 472
Stinking willie, 472
Stinkwood, 32, 93, 137, 195, 245, 279, 338, 383, 417, 420, 433, 447, 450, 451, 472
Stone oak, 472
Stonecrop, 43, 102, 147, 205, 252, 287, 352, 391, 414, 420, 424, 454, 472
Storax, 27, 28, 90, 134, 192, 243, 276, 333, 380, 412, 418, 421, 427, 472
Strawberry, 21, 85, 128, 185, 239, 271, 324, 376, 394, 426, 428, 433, 436, 445, 446, 448, 472-474
Strawberry tomato, 472
Styrene, 134, 452
Suberins, 425
Sugar wrack, 472
Sulphoxide, 4, 425
Sulphur compounds, 425
Sunflower, 23, 86, 130, 156, 188, 240, 273, 328, 378, 398, 409, 418, 420, 421, 433, 452, 472, 473

Sunflower oil, 156, 398, 452
Suterberry, 472
Sweet annie, 472
Sweet balm, 472
Sweet bay, 472
Sweet birch, 472
Sweet chestnut, 472
Sweet clover, 30, 92, 136, 194, 244, 277, 336, 382, 412-414, 423, 425, 437, 438, 441, 445, 448, 472
Sweet cumin, 472
Sweet elm, 48, 106, 150, 210, 255, 290, 358, 394, 410, 411, 420, 421, 426, 472
Sweet fennel, 472
Sweet flag, 68, 168, 228, 258, 301, 364, 407, 409, 418-421, 424, 432, 434-436, 445, 446, 449, 451, 472
Sweet goldenrod, 472
Sweet grass, 24, 87, 131, 189, 273, 329, 472
Sweet herb of Paraguay, 45, 104, 149, 207, 253, 288, 355, 392, 410, 411, 422, 423, 425, 426, 433, 435, 447, 472
Sweet lucerne, 472
Sweet marjoram, 33, 94, 138, 196, 246, 279, 339, 384, 411, 415, 423, 426, 428, 437, 445, 450, 452, 472
Sweet weed, 472
Sweet wrack, 472
Sympyhtum officinale, 457, 464
Syringa vulgaris, 45, 54, 104, 149, 156, 208, 215, 253, 289, 355, 392, 466
Syzygium aromaticum, 20, 45, 84, 104, 110, 128, 149, 164, 184, 208, 239, 253, 270, 289, 322, 356, 392, 459

Tagetes lucida, 46, 104, 149, 155, 208, 254, 289, 356, 392, 467
Tanacetum arthenium, 462
Tanacetum cinerarifolium, 469
Tanacetum parthenium, 46, 104, 149, 208, 254, 289, 356, 392, 462
Tanacetum vulgare, 13, 77, 120, 179, 234, 265, 314, 370, 470, 472
Tangleweed, 472
Tangshenoside, 14
Tannins, 3-9, 11, 13-16, 18-21, 23-26, 28-33, 35-51, 58, 147, 218, 425
Tansy, 13, 46, 77, 104, 120, 149, 179, 208, 234, 254, 265, 289, 314, 356, 370, 393, 427, 435-437, 439, 444, 472
Taraxacin, 46
Taraxacum officinale, 46, 105, 149, 208, 254, 289, 357, 393, 461, 469, 473
Taraxasterol, 46, 426
Taraxerol, 5, 46, 47, 426
Tarragon, 8, 72, 115, 172, 230, 261, 306, 367, 412, 414, 415, 417, 421, 423, 426, 433-436, 440, 444-448, 450, 453, 463, 472
Tartaric acid, 15, 16, 51, 114
Tauremisin, 116, 452
Taxol, 46, 65, 223, 426
Taxus brevifolia, 65, 474
Taxus x media, 46, 105, 150, 208, 254, 289, 357, 393, 474
Teaberry, 21, 239, 325, 377, 419, 473
Telltime, 473
Teresantol, 146, 452
Terpene alcohols, 121, 452
Terpene-d-limoneene, 452
Terpenes, 111, 118, 120, 131, 134, 143-145, 152, 448, 452
Terpenic acid, 30, 426
Terpenic oxides, 452
Terpenoids, 23, 63, 164, 221, 426
Terpinene-4-ol, 128, 129, 452
Terpinenes, 452
Terpineol, 8, 16, 26, 27, 33, 114, 116, 118, 124, 127-130, 137-139, 153, 426, 433, 452
Terpins, 33, 426
Terpinolene, 153, 452
Terrestriamide, 47, 426
Tetain, 13, 426
Tetracosanol, 39, 116, 142, 420, 447, 452
Tetracyclic oxindole alkaloid, 426

Tetracyclic triterpenes, 140, 452
Tetrahydronaphthalene, 148, 427
Tetrahydrosesquiterpene hydrocarbon, 452
Tetrahydro-cannabinols, 10, 426
Tetramethozyally benzene, 139, 452
Tetterwort, 473
Thamnolic, 28, 34, 427
Thebaine, 34, 427
Theobromine, 14
Theophylline, 14
Throwwort, 473
Thujene, 48, 121, 128, 145, 427, 433, 452
Thujyl alcohol, 7, 452
Thyme (garden), 473
Thyme (lemon), 473
Thymolhydroquinone, 115, 452
Thymol, 31, 32, 42, 46, 115, 118, 136-138, 147, 150, 427, 452
Thymol hydroquinone dimethyl ether, 115, 452
Thymus citriodorus, 46, 105, 150, 209, 254, 289, 357, 393, 466, 473
Thymus serpyllum, 460, 474
Thymus vulgaris, 46, 105, 150, 209, 254, 289, 357, 473
Tiglic, 120, 452
Tiliadine, 47, 427
Tillia cordata, 47, 105, 150, 209, 254, 290, 357, 393, 466, 471
Tinnins, 6, 21, 32, 43, 427
Titerpenoids, 138, 452
Tocopherol, 22, 40, 60, 117, 119, 130, 131, 138, 145, 148, 427, 433, 452
Toothache tree, 51, 108, 153, 213, 257, 292, 362, 395, 420, 423, 426, 442, 446, 473
Torreyol, 138, 452
Toxicodendrol, 40, 427
trans-2-Hexanol, 122
trans-3,7-Dimethyll, 146, 453
trans-10-11-Dihydroztlantones, 119
trans-Anethole, 112, 139, 140, 453
trans-Asarone, 125, 453
trans-Beta-farnesene, 120, 453
trans-Ferutic acid esters, 453
trans-Ocimene, 126, 453
Tree peony, 34, 94, 138, 196, 246, 280, 339, 384, 409, 419, 473
Tri-acylglycerols, 453
Tri-consane, 453
Tri-cyclene, 453
Tri-cyclo-ekasantalal, 146, 453
Tri-terpenoid acid, 453
Triandrin, 427
Tribulus terrestris, 47, 59, 105, 150, 209, 254, 290, 357, 393, 469
Trifoil, 473
Trifolium pratense, 47, 105, 150, 209, 254, 290, 358, 393, 456, 459, 460, 470, 473
Triglycerols, 152
Trigonella foenum-graecum, 457, 463
Triterpene, 13, 22, 27, 28, 427
Triterpenoids, 9, 25, 427
Tropaeolum majus, 35, 47, 95, 105, 150, 209, 255, 290, 341, 358, 385, 393, 468
True wood sorrel, 473
Turmeric, 373, 423, 473
Turmeron, 124, 453
Tussilago farfara, 47, 105, 150, 210, 255, 290, 358, 393, 460

Ulmus rubra, 48, 106, 150, 210, 255, 290, 358, 394, 465, 467, 471, 472
Umbelliferone, 29, 427
Umbellularia california, 56, 61, 158, 161, 468
Umbellulone, 48, 151, 427, 453
Uncaria guianensis, 459
Uncaria tomentosa, 58, 66, 106, 109, 151, 210, 255, 290, 359, 394, 458
Undecane, 131, 453
Uric acid, 3, 427
Ursolic acid, 40, 427
Urtica dioica, 48, 106, 151, 210, 255, 291,

Urtica dioica (continued), 359, 394, 472
Urtica urens, 472
Usnic acid, 28, 34, 427
Uva-ursi, 7, 60, 71, 115, 171, 230, 260, 305, 366, 409, 419, 426, 441, 456, 473

Vaccinium macrocarpon, 48, 106, 151, 255, 291, 294, 394, 460
Vaccinium myrtilloides, 457
Vaccinium myrtillus, 55, 156, 216
Vaccinium oreophyilum, 49, 106, 151, 211, 255, 291, 360, 394, 456
Vaccinum vitis-idaea, 48, 106, 151, 211, 255, 291, 359, 394, 460, 462, 466
Valepotriates, 21, 49, 151, 427, 453
Valerian, 49, 106, 151, 211, 255, 291, 360, 394, 408, 417, 427, 435, 436, 443, 453, 455, 473
Valerian (American), 455
Valeriana officinalis, 49, 106, 151, 211, 255, 291, 360, 394, 473
Valerianic acid, 5, 131, 453
Valeric acid, 6, 118, 133, 453
Valtrate, 49, 427
Vanillin, 5, 20, 47, 126, 134, 152, 427, 453
Verbascum thapsus, 49, 106, 151, 211, 256, 291, 360, 394, 458, 462, 468
Verbenalin, 15, 427
Verbenol, 117, 453
Veronica officinalis, 49, 106, 151, 211, 256, 291, 360, 394, 464, 466, 471
Vetivene, 49, 152, 427, 453
Vetiver, 49, 107, 116, 152, 154, 211, 256, 291, 360, 395, 409, 421, 427, 433, 435, 443, 444, 451-454, 473
Vetiver oil, 154, 453
Vetiveria zizaniodes, 107, 152, 211
Vetiverol, 152, 453
Viburnum opulus, 49, 107, 152, 211, 256, 291, 361, 395, 460, 463, 464, 471
Vinblastine, 50, 428
Vinca Minor, 12, 50, 107, 119, 152, 177, 212, 233, 256, 264, 292, 312, 361, 369, 395, 466, 468, 469
Vincristine, 50, 428
Viola tricolor, 50, 107, 152, 212, 256, 292, 361, 395, 464, 465, 468
Violin, 50, 428
Virginia dogwood, 473
Virginia skullcap, 43, 102, 147, 205, 252, 287, 352, 390, 409, 411, 414, 424, 425, 431, 432, 448, 451, 473
Viscotoxin, 50, 428
Viscum album, 50, 107, 152, 212, 256, 292, 361, 395, 467
Vitamin A, 5, 13, 39, 41, 428
Vitamin B, 8, 17, 38, 46, 428
Vitamin B complex, 8, 17, 428
Vitamin B1, 11, 20, 40, 428
Vitamin B12, 6, 428
Vitamin B2, 428
Vitamin C, 15, 16, 20, 21, 24, 33, 35, 39-41, 43-45, 48, 49, 62, 428
Vitamin D, 428
Vitamin E, 11, 56, 216, 428
Vitamin K, 194, 429
Vitamin P, 429
Vitex agnus-castus, 50, 65, 107, 152, 165, 212, 256, 292, 361, 395, 459
Vitex negundo, 459
Vitis labrusca, 51, 107, 152, 212, 256, 292, 362, 395, 463
Vitis vinifera, 463

Walnuts, 274, 331, 379, 473
Water lily, 32, 93, 137, 195, 279, 338, 383, 420, 427, 435, 441-443, 448, 453, 473
Watercress, 31, 93, 137, 195, 245, 278, 337, 383, 411, 416, 417, 419, 428, 440, 473
White ash, 21, 85, 129, 186, 239, 271, 325, 376, 412, 414, 427, 454, 473
White peony, 33, 94, 138, 196, 246, 280, 339, 384, 407, 409, 410, 415, 419-

White Peony *(continued)*, 419-421, 434, 436, 438, 444-449, 451, 473
White pine, 36, 96, 141, 199, 248, 282, 342, 386, 408, 410, 415, 419, 423, 425, 428, 431, 435, 443, 444, 449, 473
White water lily, 32, 93, 137, 195, 279, 338, 383, 420, 427, 435, 441-443, 448, 453, 473
White willow, 41, 101, 145, 203, 251, 286, 350, 389, 415, 416, 424, 425, 427, 454, 473
Wild celery, 71, 473
Wild chamomile, 473
Wild cherry, 38, 98, 200, 249, 283, 345, 387, 409, 422, 434, 443, 473
Wild fennel, 473
Wild ginger, 8, 72, 116, 173, 231, 261, 307, 367, 409, 410, 418, 422, 426, 434, 436, 439, 441, 442, 444, 446, 449, 452, 473
Wild indigo, 8, 73, 116, 173, 231, 261, 307, 367, 408, 412, 414, 417, 422, 473
Wild lemon, 473
Wild licorice, 473
Wild marjoram, 473
Wild rose geranium, 34, 95, 139, 197, 247, 280, 340, 385, 433, 435, 436, 438, 442-444, 448, 473
Wild senna, 473
Wild strawberry, 473
Wild sunflower, 473
Wild thyme, 474
Willow (white), 474
Willow herb, 19, 83, 127, 184, 238, 270, 322, 375, 414, 415, 425, 468, 471, 474
Winter bloom, 474
Winter green, 474
Winter savory, 474
Witch hazel, 23, 86, 130, 187, 240, 272, 327, 377, 411, 414, 416, 417, 422, 423, 426, 453, 474
Witch's bells, 474
Withania somnifera, 51, 63, 108, 152, 155, 212, 256, 292, 297, 362, 395, 404, 405, 456, 465
Wolfberry, 28, 91, 135, 193, 243, 276, 334, 381, 410-412, 421, 423, 425, 428, 445, 459, 474
Wolfberry (Chinese), 474
Wood sorrel, 473, 474
Woodland strawberry, 474
Wormseed, 13, 77, 120, 178, 234, 265, 313, 370, 409, 415, 420, 424, 434, 442, 446, 474
Wormwood, 7, 72, 115, 172, 230, 261, 306, 367, 407, 409, 426, 428, 432-435, 449, 452, 474
Wormwood (Chinese), 474
Wu wei zi, 43, 102, 147, 205, 252, 352, 390, 418, 425, 428, 453, 474

Xanthone, 12, 429
Xylenol, 117, 454

Yam (Chinese), 474
Yam (Mexican), 474
Yangonin, 36, 429, 454
Yarrow, 3, 68, 112, 168, 228, 258, 301, 364, 407, 411, 419, 424-427, 432, 434, 436, 437, 442, 444, 445, 449-451, 474
Yellow ginseng, 474
Yellow indian shoe, 474
Yellow melilot, 474
Yellow puccoon, 474
Yellow rochet, 474
Yew, 46, 65, 105, 150, 208, 223, 254, 289, 357, 393, 421, 423, 426, 431, 432, 474
Ylang-ylang, 10, 74, 118, 175, 232, 263, 310, 368, 415, 424, 434, 437, 439, 441, 442, 445, 446, 448-450, 452, 453, 474
Yu chin, 474

Yucca, 51, 108, 153, 165, 212, 256, 292, 362, 395, 416, 424, 431-434, 436, 438, 442, 444, 445, 447, 448, 450, 452, 474
Yucca aloifolia, 51, 108, 153, 212, 256, 292, 362, 395, 474

Zanthoxylum americanum, 51, 108, 153, 257, 292, 395, 455, 468, 469, 472, 473
Zea mays, 51, 108, 153, 213, 257, 292, 363, 396, 460
Zingiber officinale, 51, 53, 62, 108, 153, 155, 163, 213, 221, 257, 292, 363, 396, 463
Zingiberene, 153, 454
Zizanoic acid, 152, 454